D0948750

STATISTICAL MECHANICS FOR CHEMISTS

STATISTICAL MECHANICS FOR CHEMISTS

JERRY GOODISMAN
Department of Chemistry
Syracuse University
Syracuse, NY

A Wiley-Interscience Publication
JOHN WILEY & SONS, INC.
New York / Chichester / Weinheim / Brisbane / Singapore / Toronto

This text is printed on acid-free paper.

Library of Congress Cataloging in Publication Data:

Goodisman, Jerry.
Statistical mechanics for chemists / Jerry Goodisman.
p. cm.
"A Wiley-Interscience publication."
ISBN 0-471-16812-2 (cloth : alk. paper)
1. Statistical mechanics. 2. Chemistry, Physical and theoretical.
I. Title.
QD455.3.M3G66 1997
541'.0153013--dc20

0-471-16812-2
Printed in the United States of America

10 9 8 7 6 5 4 3 2 1

CONTENTS

PREFACE

The discipline of statistical mechanics makes the connection between thermodynamic and other macroscopic properties (properties of collections of molecules) and properties of individual molecules. Since modern chemistry is involved as much (or probably more) with matter on the molecular level as with macroscopic behavior, all chemists need to understand some statistical mechanics. Physical chemists need to know more statistical mechanics than others, and some theoretical chemists make it their main area of research; there is still work to be done in the field. This book, an introductory graduate-level text, is intended for all chemists, rather than for those who will specialize in statistical mechanics. Designed for a one- or two-semester graduate course in statistical mechanics, it introduces the important concepts and equations and gives applications to a variety of areas of interest to chemists. The material cannot all be covered in one semester, and probably not in two. After going through most of Chapters 1 to 4, the student or the professor should choose some or all of each of Chapters 5 to 8 to study from its beginning.

Because of the important role it plays in modern chemistry, some study of statistical mechanics enters the elementary courses in the typical undergraduate chemistry program. In the first-year chemistry course, the Maxwell–Boltzmann velocity distribution for molecules in a gas may be discussed and perhaps the Boltzmann distribution for molecules in a gravitational field. Some more concepts of statistical mechanics (partition functions, calculation of thermodynamic properties) are found in the undergraduate physical chemistry course. Thus some of what is discussed in Chapters 1, 2, and 3 of this book should be familiar, and hence not too intimidating, to the average reader. On the other hand, the field of mathematics does not seem to lose its power to intimidate; even familiarity with it never leads to contempt.

Physical chemistry, and statistical mechanics in particular, involves a fair amount of mathematics. For those planning to specialize in statistical mechanics, this is not a problem (it may even have helped attract such students to the subject), but it *is* a barrier for the average first-year graduate student. There is no way to avoid mathematics in presenting this subject, but an attempt has been made to take the reader through mathematical manipulations step by step and to explain what is being done. Although the "step-by-step" presentation increases the number of formulas and equations, it was preferred to the "it-can-be-shown" approach. At the end of each chapter will be found a number of problems, some with solutions. Doing these will likewise give the reader some practice in mathematics.

This book was written because, although there are good statistical mechanics textbooks and monographs available, the author of the present text found none of them satisfactory for the course at Syracuse. A one-semester course in statistical mechanics has been offered for many years to first- or second-year graduate students in chemistry, chemical engineering, and related fields. Some existing texts were written for specialists in the field, or for those wanting to become specialists. Others, for physicists or mathematicians, are too theoretical and mathematical; some emphasize basic problems in statistical mechanics at the expense of applications of interest to chemists. Finally, some, written for chemists, are too old and miss topics of contemporary interest. None of the problems with these books has prevented the author from drawing heavily on them. These sources are listed in the Bibliography. For each, the sections are given of the present book for which the earlier source can be particularly helpful.

Chapters 1 to 3 contain general and introductory material. Section 1.2 is meant to be a summary of important formulas that are referred to later, not a short course in thermodynamics. The grand canonical ensemble, Section 2.4, is not always discussed in statistical mechanics courses for chemists, but it has important applications (Sections 4.4 and 4.5, Chapters 5 and 7). For chemists, the most important applications of statistical mechanics are to calculation of thermodynamic functions of gases and chemical equilibrium constants from molecular properties (some textbooks are concerned almost completely with this). Thus Chapter 4, or at least the first few sections, should not be omitted.

Sections 5.1 and 5.2, which introduce Fermi–Dirac and Bose–Einstein statistics, also should be studied. How deeply one wants to go into the applications, to electrons in metals and semiconductors or to photons, is a matter of personal choice. Chapter 6 should be dealt with like Chapter 5. Sections 1 to 3, which are general and introductory, should be covered, whereas how much one wants to go into the applications to imperfect gases or to dipole moments in electric or magnetic fields is for the individual to decide. Should fluids (Chapter 7) be discussed? The author thinks so: we spend so much course time on gases, and most chemistry occurs in liquids.

The most important concepts are found in Sections 1, 4, and 5. Of course, everyone knows something about Debye–Hückel theory (Section 6); it is interesting to see how it arises from correlation functions.

Finally, there is Chapter 8 on time dependence, which the author included in spite of some advice against it. Since it is a hard subject, it was suggested that it be left for a later course, but that course may never arrive. The author believes that some understanding of the Liouville equation and time-correlation functions is good for a chemist's soul (and understanding), and has sometimes inflicted Sections 1 to 3 on his students. Others may wish to try it, too.

There is certainly a lot of material here, but it is important. Chemists should be conversant with the main ideas of statistical mechanics and know how to apply statistical mechanics to a variety of problems. Gaining the necessary understanding requires work. The hope is that this textbook can make understanding a little easier.

Syracuse University JERRY GOODISMAN

CHAPTER 1

INTRODUCTION

1.1. STATISTICAL MECHANICS AND DISTRIBUTIONS

Statistical mechanics shows how the properties of macroscopic systems are related to the properties of the molecules of which they are composed. Of particular interest are thermodynamic properties, such as energy, entropy, and temperature. Molecular properties, in principle, are expectation values of operators; a molecule's state is defined by quantum numbers, which specify the energy of the state and the wave function used to calculate the expectation values. In contrast, the state of a macroscopic system is defined by giving the values of thermodynamic and other properties. It is remarkable that, to describe a macroscopic system, we generally need to give values for only a few properties, because relations between properties are given by the laws of thermodynamics.

In thermodynamics, one starts from a few basic assumptions and definitions, including the laws of thermodynamics, and derives relations between properties mathematically. Thus thermodynamics shows how the value of one property can be calculated from the values of others. It does not tell us how to get a value for any particular property of any particular system. The laws of thermodynamics make no mention of molecules and would be valid even if molecules did not exist.

As already noted, the way we describe a molecular system is quite different from the way we describe a macroscopic system. For a molecular system, we have to set up and solve the Schrödinger equation, using what we know about the forces between the nuclei and electrons of which molecules are composed. Solution of the Schrödinger eigenvalue equation, $H\Psi_j = E_j\Psi_j$, gives the eigenfunctions Ψ_j and the eigenvalues E_j. (Here, j represents the set of values of all the quantum numbers, which identify the state.) The eigenvalues are the possible energies for the molecular system; from the corresponding wave functions Ψ_j, one calculates values for other properties by evaluating expectation values of the relevant operators.

Energy plays a fundamental role in a quantum mechanical description and also in thermodynamics. Indeed, the first and second laws of thermodynamics are concerned with defining energy and energy transfer. Because the concept of energy is meaningful both to molecules and to macroscopic systems, energy will play a fundamental role in our development of statistical mechanics. It seems obvious how the energy of a macroscopic system relates to the energies of the molecules in it: the energy of the system is the sum of the energies of the molecules. This idea also will be used in our development.

In contrast with energy, other thermodynamic properties are not appropriate for the description of molecules. We will see that neither the temperature nor the entropy of a molecule is a meaningful concept. On the other hand, quantum numbers are not particularly useful in describing a macroscopic system. In principle, one *could* consider a system containing 10^{20} (0.17 millimoles) atoms as one giant molecule, but, with 10^{20} nuclei, there would be about 3×10^{20} quantum numbers to specify without even considering the electrons. It would be impossible to list them even if one could ascertain what they are. Furthermore, writing them down would be less than useless, because of a second problem: molecules in a macroscopic system are always changing their states. Giving all 3×10^{20} quantum numbers would specify the state the system used to be in at some previous instant.

Measurement of any property of a macroscopic system takes some time, anyway. During the time that the measuring instrument is connected to the system, the molecules change their states. Thus the state (meaning the set of quantum numbers) of the macroscopic system is always changing. If the system is being held at constant temperature by being immersed in a temperature bath, its energy is not fixed because it continually transfers heat to and from its surroundings. Even if the system is insulated from its surroundings, so that its energy is constant, its state can change. For a large system, the degree of degeneracy is large, so there are many states with the same energy which are accessible. (More correctly, one should note that the energy can only be fixed within some uncertainty, say ΔE, and the states with energy between E and $E + \Delta E$ are accessible.)

Looking at the problems this way suggests something about their solution. The operative word here is "averaging." The values obtained for properties of a macroscopic system are actually averages over the time of the measurement. The large size of the system also can be dealt with by invoking averaging, in this case over molecules. Even at one instant, only the average energy or the average speed of a molecule is of interest. The total energy is the number of molecules times the average energy of one molecule.

J. W. Gibbs suggested that the averaging over molecules could be equivalent to the averaging over time. Imagine a system containing a large number of identical molecules, and suppose we follow one molecule over time, keeping track of its change of state. Although there were no motion pictures in Gibbs' day, we can imagine we make a movie or a video of the molecule, on which we can see what state the molecule is in at each instant. Then we turn the movie into a series of still pictures and put them on the wall: Exhibit A. Imagine also that we take one big still picture of the whole assembly of molecules and cut it up into pieces to make a collection of pictures of individual molecules at the same instant: Exhibit B. The equivalence of molecule-averaging and time-averaging means that Exhibit A and Exhibit B look the same. This idea is the basis of the notion of the "ensemble," which will be instrumental in the development of the laws of statistical mechanics in Chapter 2.

Clearly, we are not interested in which molecule is in which state. To carry out the averaging that will enable us to get the properties of the system, we need to know only how many molecules are in each state. Let the number of molecules in the system be N, and let N_j be the number of molecules in state j, which has energy E_j. Then the total energy of the system is $\sum_j N_j E_j$. The probability that a molecule is in state j is

$$p_j = \frac{N_j}{N} \tag{1.1}$$

so the total energy is $N\sum_j p_j E_j$ and the average energy of one molecule is $\sum_j p_j E_j$. Suppose that a molecule in the state j had a value P_j for some property P (P_j would be the expectation value of the quantum mechanical operator corresponding to P). Then the average value of the property is $\sum_j p_j P_j$.

If we are considering one molecule over time, p_j is the fraction of its life that this molecule spends in state j. If we are considering the whole system, p_j is the fraction of the molecules in state j, as in (1.1). Gibbs' hypothesis is that the two definitions of p_j are equivalent. Under what circumstances this is true is still a subject of research in statistical mechanics. We will just assume it is true for all the systems of interest to us.

The set of $\{p_j\}$ (we use the curly bracket to indicate a collection of things) is called a distribution. In the present case, it is the distribution of molecules over states. If the sum, $\sum_j p_j$, is unity, we say the distribution is normalized. Much of the work of Chapter 2 involves finding the distribution for certain kinds of systems composed of molecules. For a system that is totally isolated, so that its energy is fixed, the distribution will be called the "microcanonical distribution." For a closed system held at constant temperature, the relevant distribution is the "canonical distribution." For an open system, which can exchange matter with the surroundings, we will consider the "grand canonical distribution." There are other distributions, but they are less important than these three.

The distributions referred to so far have been discrete. Because j takes on discrete values (j refers to a set of quantum numbers), the probabilities could be listed in a table, giving p_j for each j. It may be necessary to consider a variable, such as the momentum of a particle, which can take on any value in a range. Then we have a continuous distribution. A continuous distribution is given as a function of the variable instead of a table of values.

An example is the Maxwell–Boltzmann speed distribution, specifying the probability of finding a molecule with speed between u and $u + du$ in a gas of molecules at temperature T:

$$p(u)\,du = \frac{dN}{N} = 4\pi\left(\frac{m}{2\pi kT}\right)^{3/2} e^{-mu^2/2kT}u^2\,du \tag{1.2}$$

In (1.2), dN is the number of molecules with speed between u and $u + du$, and N is the total number of molecules. Thus $p(u)\,du$ is the probability that any molecule will have its speed between u and $u + du$; $p(u)$ is a probability density. This continuous probability distribution is normalized according to

$$\int_0^{\infty} p(u)\,du = 1 \tag{1.3}$$

since the possible values of speed range from 0 to ∞. To find the average value of a function of u, $f(u)$, one multiplies $p(u)$ by $f(u)$ and integrates. For the kinetic energy $f(u) = \frac{1}{2}mu^2$, so the average kinetic energy is

$$\langle T \rangle = \int_0^{\infty} p(u)\tfrac{1}{2}mu^2\,du$$

For a continuous distribution, the sum over probabilities (like $\Sigma_j\, p_j P_j$ for the average value of the property P) is replaced by an integral over a probability density.

An important example of a discrete probability distribution is the result of a series of tosses of a coin, each of which can give heads (H) or tails (T) with equal probability. Suppose the coin is tossed N times, so there are 2^N possible outcomes, each of which is a series of Hs and Ts. The probability that there will be N heads is $(\frac{1}{2})^N$ because the probability of H on each toss is $\frac{1}{2}$ and, since the tosses are independent, the probabilities are multiplied together. The probability of getting N_H heads (and $N - N_H$ tails) is $(\frac{1}{2})^{N_H}(\frac{1}{2})^{N-N_H} = (\frac{1}{2})^N$, multiplied by the binomial coefficient,

$$C(N, N_H) = \frac{N!}{N_H!(N - N_H)!} \tag{1.4}$$

$C(N, N_H)$ is the number of ways of choosing which of the N tosses gives heads, or the number of arrangements of N_H heads and $(N - N_H)$ tails. It is equal to 1 when $N_H = N$ or $N_H = 0$ and has its maximum value for $N_H = \frac{1}{2}N$. The sum of the probabilities is unity because it is simply a binomial expansion:

$$\sum_{k=1}^{N} C(N,k)\left(\tfrac{1}{2}\right)^N = \sum_{k=1}^{N} \frac{N!}{k!(N-k)!}\left(\tfrac{1}{2}\right)^k\left(\tfrac{1}{2}\right)^{N-k} = \left(\tfrac{1}{2} + \tfrac{1}{2}\right)^N$$

The maximum value, $C(N, \frac{1}{2}N)$, becomes larger for larger values of N.

Furthermore, $C(N, j)$ is more sharply peaked as a function of j for larger N. This is shown in Figure 1.1, which gives $C(N, j)$ for $N = 10$, 40, and 160 (the Stirling approximation, discussed below, was used to evaluate the larger factorials). We will be interested in probability distributions for molecules over states. If there are N molecules and each can be in either state 1 or state 2, the probability that M of the molecules are in state 1 is $C(N, M)(\frac{1}{2})^N$. It is thus most likely that we will find half the molecules in state 1 and half in state 2. Because N is extremely large (10^{20}, perhaps), it is extremely unlikely that many more molecules will be found in one state than the other. Thus, the most probable distribution, corresponding to $M = \frac{1}{2}N$, is overwhelmingly more probable than distributions which differ substantially from it.

Something else happens when N is very large. Except when M is close to 0 or N, the difference between $C(N, M)$ and $C(N, M - 1)$ is very small compared to $C(N, M)$. The probabilities, $C(N, M)(\frac{1}{2})^N$, start to look like a continuous distribution; N and M can be considered as continuous variables. Since we will have to deal with very large numbers in discussing macroscopic systems, it will be very useful to approximate the factorial of a large number by a continuous analytic function. The approximation we use, referred to as the Stirling approximation, is

$$\ln(n!) \cong n \ln n - n \tag{1.5}$$

The derivation of (1.5) is given as Problem 1.2.

As can be seen in Figure 1.2, or by some experiments on a hand calculator, the Stirling approximation gets better as n gets larger. (Naturally, we are unable to do the calculation for *really* large values of n.) It may also be seen in the figure that (1.5) is too low for all n. The difference can be mostly made up by adding $\ln(n^{1/2})$ and a constant to (1.5). This improved Stirling approximation is also derived in Problem 1.2.

If (1.5) is used in $C(N, N_H)$, equation (1.4), we obtain

$$\ln[C(N, N_H)] = [N \ln N - N] - [N_H \ln N_H - N_H] - [(N - N_H)\ln(N - N_H) - (N - N_H)]$$

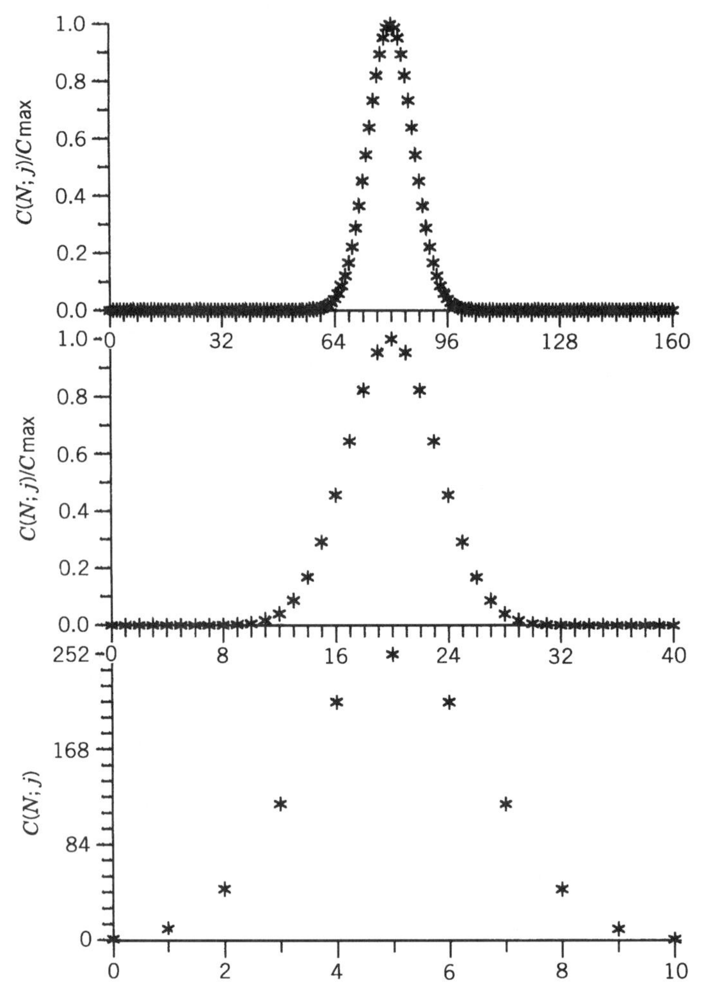

Figure 1.1. Binomial coefficients $C(N, j)$ plotted vs. j for $N = 10$, 40, and 160. Increased N makes $C(N, j)$ more sharply peaked.

where N_H can now be considered as a continuous variable. To find the maximum value of $C(N, N_H)$ one can differentiate $\ln[C(N, N_H)]$ with respect to N_H to obtain

$$-[\ln N_H + 1 - 1] - [-\ln(N - N_H) - 1 + 1] = 0$$

which yields $N - N_H = N_H$ or $N_H = N/2$.

The Stirling approximation will be almost universally used for $\ln(n!)$ in the derivations of Chapter 2. There we will derive the probability distributions for

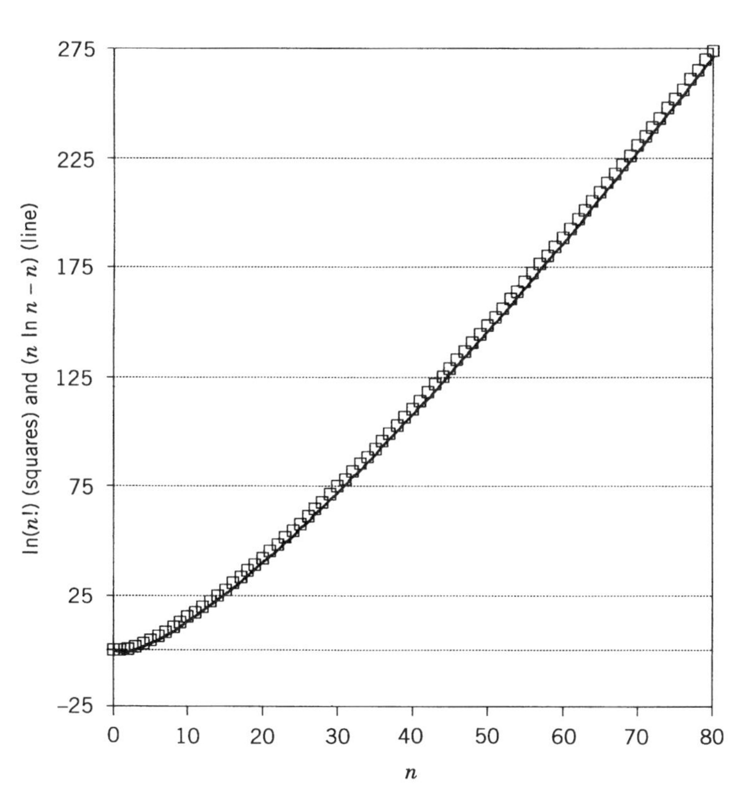

Figure 1.2. Logarithm of $n!$ (squares) compared with the Stirling approximation, $n \ln n - n$ (line). The error, $\ln(n!) - (n \ln n - n)$, is fairly constant, so the relative error decreases with n.

the microcanonical and canonical ensembles. The latter, corresponding to a system at constant temperature, is more interesting, and also more complicated, than the former. A fair amount of discussion will be required to complete its definition and to make the connection between the distribution and thermodynamic variables. However, once the connection is made, we will have formulas of general applicability to a great variety of macroscopic systems. We will also consider the ensemble for describing an open system, and make the connection to thermodynamics. In Chapter 3, we will see how the formalism simplifies for systems which are composed of many equivalent and independent parts, such as gases. Applications are given in Chapters 3 and 4, with Chapter 4 devoted to gases of atoms and molecules. We will show how the thermodynamic properties of gases of atoms and molecules can be calculated from their molecular properties.

Although we assume our systems are described by quantum mechanics, the symmetry with respect to exchange of identical particles will be taken into account only approximately. This is done correctly in Chapter 5, where the grand canonical ensemble turns out to be useful in developing quantum statistics. Applications are to gases of electrons and photons. It also will be important to consider the classical mechanical limit, for situations when the state of the system is appropriately defined in classical mechanical terms. This is done in Chapter 6, with some applications. The main application is to imperfect gases and liquids (Chapter 7). Finally, we give an introduction to the vast subject of time dependence in Chapter 8.

1.2. REVIEW OF THERMODYNAMICS

Before getting to the ensembles, we review some important concepts and formulas of thermodynamics. These will be referred to throughout the book. An understanding of thermodynamics will be important in making the connection between statistical mechanics and thermodynamics, discussed in Chapter 2. On the other hand, the results of statistical mechanics are used to give meaning to thermodynamic concepts and to explain thermodynamic variables in molecular terms.

Perhaps the most important idea in thermodynamics is that of a state function. A state function or state variable is a property of a system that depends on the state of the system and not on its history. The state of a system is defined by giving values for a small number of thermodynamic variables; values for all others are then fixed. If the system undergoes a process in which some of the thermodynamic variables change their values, its state is changed, but, if the system returns to its original state, all the thermodynamic variables recover their original values.

The first law of thermodynamics states that the energy is a state function, and that the energy of a system can be changed by the input of work or heat from the surroundings. For an infinitesimal change, one writes

$$dE = đq + đw \tag{1.6}$$

Here, dE is the differential of the energy, but $đq$ and $đw$ are not the differentials of anything: they represent small amounts of heat or work. According to (1.6), heat and work are equivalent in that they are both ways to change the energy of a system.

To say that dE is the differential of the energy, which is a state function, is to say that, when a system, originally in state a, undergoes a process which leaves it in state b, the change in energy is $\int dE = E_b - E_a$. The situation is different for $đq$ and $đw$; to know the heat absorbed by, or the work done on, the system, one must know not only the states a and b, but the path by which the system was taken from a to b. As an example, consider a mole of ideal gas at pressure P_a, volume V_a, and temperature T, which ends up at pressure $P_b = 2P_a$, volume $V_b = V_a/2$, and the same temperature, so that the overall energy change is 0. One could: (1) double the temperature at constant volume so that the pressure became P_b and then cool the gas at constant pressure P_b until the volume decreased by half, or (2) compress the gas at constant temperature until the pressure was doubled and the volume halved. The work done and the heat absorbed would be different for paths (1) and (2).

Work is defined in the surroundings, in mechanical terms. The work done on the system is equal to the product of the force exerted on the system (by devices located in the surroundings) and the distance this force moves. When the only kind of work that can be done is pressure–volume work, $đw = -P_{sur}\, dV$, with P_{sur} the pressure applied to the system by the surroundings and V the volume of the system. If the process is carried out carefully enough so that the system and the surroundings are (almost) in equilibrium at each stage of the process, P_{sur} is equal to P, the pressure of the system. The process is then said to be reversible, because an infinitesimal change in P would cause the process to occur in reverse, and one could return system and surroundings to their original states. For an irreversible process, the system is in equilibrium only at the beginning and the end.

There are other kinds of work besides mechanical work: electrostatic work, surface work, and so on. In every case, $đw$ is given by the product of an intensive variable, measured in the surroundings, and the change in an extensive variable of the system. For electrostatic work, the intensive variable is the electrical potential and the extensive variable the electric charge; for surface work, the intensive variable is the surface free energy and the extensive variable the surface area.

It is well known that the state of a system, and its energy, can be changed without work being done. The change in energy in this case defines the heat transferred between system and surroundings. Heat can be transferred when system and surroundings are at different temperatures, provided that the surface separating system and surroundings is not completely insulating. Heat always flows from higher temperature to lower.

When heat is transferred to the system, its energy and its temperature usually increase. If no work is being done, $dE = đq$. The ratio of the (infinitesimal) heat to the change in temperature is defined to be the heat capacity. If only P–V work is possible and no work is done, the volume is constant, so the heat capacity is the heat capacity at constant volume:

$$C_V = \frac{đq}{dT} = \left(\frac{\partial E}{\partial T}\right)_V \tag{1.7}$$

The transfer of heat from the surroundings must be accompanied by a decrease in the energy of the surroundings, but, if the surroundings are much larger than the system, the decrease in the surroundings' temperature will be negligible. In this situation, we refer to the surroundings as a constant-temperature bath. If the system is much larger than its surroundings, heat transfer may result in a change in temperature of the surroundings with negligible change in the system; the surroundings are then acting as a thermometer.

The second law of thermodynamics defines another state function in addition to the energy E. This is the entropy, denoted by S. The entropy has a close connection with heat, since, for an infinitesimal reversible change in a system,

$$dS = \frac{đq}{T} \tag{1.8}$$

where T is the temperature. At constant volume, $đq = C_V \, dT$ so

$$\left(\frac{\partial S}{\partial T}\right)_V = \frac{C_V}{T} \tag{1.9}$$

However, S may change without heat being transferred.

The meaning of entropy is rather more difficult to understand than that of energy, but it is often associated with disorder or uncertainty on the molecular scale. For example, S for a gas increases with volume at constant temperature, because an increase in volume means there are more places each molecule of the gas can be found. In thermodynamics, of course, such an "explanation" is off limits, and one can say only that $(\partial S/\partial V)_T$ is positive for a gas.

Entropy being a state function, its value depends only on the state of the system. Thus the difference in S between state A and state B depends only on the states, and not on the process used to go from A to B:

$$\Delta S_{A \to B} = S_B - S_A$$

Equation (1.8) may be used to calculate $\Delta S_{A \to B}$, however, only if the heat is measured while the system is taken from A to B by a reversible process. For an irreversible process, $dS \geq đq/T$. (In fact, S and the second law are important because they distinguish between reversible and irreversible processes.) If the system passes from A to B reversibly, and then back to A by an irreversible process, $\Delta S_{\text{total}} = 0$, but the total amount of heat transferred to the system during this cyclic process may be negative (heat is rejected to the surroundings). This confirms that heat is not a state function and also makes the connection between heat and irreversibility, by way of entropy.

A process, or a change of state of a system, is either reversible, spontaneous, or unnatural (meaning that it will never occur in nature). The reverse process to an unnatural process is spontaneous. A reversible process, in which the system remains (almost) at equilibrium, can go forward or backward, but infinitely slowly. For an isolated system, spontaneous processes are characterized by an increase in S. (One statement of the second law is that the entropy of the universe, considered as a system with no surroundings, always increases.) Thus, for an isolated system, one can decide whether a process is reversible, spontaneous, or unnatural from the behavior of S.

For a reversible process, $dS = đq/T$ according to (1.8) and $đw = -P\,dV$. Then (1.6) can be written as

$$dE = T\,dS - P\,dV \tag{1.10}$$

Equation (1.10), which involves only state functions, is sometimes called the fundamental equation for E. It is a combination of the first and second laws. By giving the change in E due to a change in S and a change in V, it shows that the "natural variables" for E are S and V, with $(\partial E/\partial S)_V = T$ and $(\partial E/\partial V)_S = -P$.

If (1.10) is rearranged to

$$dS = \frac{1}{T}\,dE + \frac{P}{T}\,dV \tag{1.11}$$

it shows the the "natural variables" for S are E and V. In particular, $(\partial S/\partial E)_V = 1/T$ and $(\partial S/\partial V)_E = P/T$. Both of these quantities are always positive. Note that (1.10) and (1.11) refer to a change from one equilibrium state to another, and thus to reversible processes. Assuming only pressure–volume work is possible, constant V means no work is done and constant E means an isolated system.

According to (1.10), E is constant if S and V are constant. In life, controlling entropy is not as common, because not as convenient, as controlling temperature. Also, it is often easier to control pressure than volume. Thus one would like to define new thermodynamic functions for which T and P are natural variables. This is done by the Legendre transformations, in which independent variables are replaced by derivatives with respect to these variables. The enthalpy, defined by $H = E + PV$, is of interest because of the fundamental equation for H:

$$dH = dE + P\,dV + V\,dP = T\,dS + V\,dP.$$

The natural variables for H are S and P. For a system at constant P, $dH = T\,dS$ and, if only P–V work is possible, the heat is

$$đq_p = dE + P\,dV = dH$$

The ratio of the heat to the temperature change is

$$C_P = \frac{đq_p}{dT} = \left(\frac{\partial H}{\partial T}\right)_P \tag{1.12}$$

which is the heat capacity at constant pressure.

The free energies are of greater interest because they do not have S as a natural variable. The Helmholtz free energy, obtained by a Legendre transformation of E to replace S by T as independent variable, is defined as

$$A = E - TS \tag{1.13}$$

Taking the differential of (1.13) and combining it with (1.10), one obtains the fundamental equation for A:

$$dA = T\,dS - P\,dV - T\,dS - S\,dT = -P\,dV - S\,dT \tag{1.14}$$

This shows that the natural variables for A are V and T, with

$$\left(\frac{\partial A}{\partial V}\right)_T = -P \tag{1.15}$$

and

$$\left(\frac{\partial A}{\partial T}\right)_V = -S \tag{1.16}$$

The Helmholtz free energy is useful for discussing systems at constant T and V and will play a fundamental role in our discussion of the canonical ensemble. The change in A determines whether a process is reversible, spontaneous, or unnatural in a system at constant T and V.

Of even greater practical use in thermodynamics is the Gibbs free energy,

$$G = H - TS = A + PV$$

The differential,

$$dG = dA + P\,dV + V\,dP = -S\,dT + V\,dP \tag{1.17}$$

shows that the natural variables are T and P. Thus $(\partial G/\partial T)_P = -S$ and $(\partial G/\partial P)_T = V$. The change in G determines whether a process is reversible, spontaneous, or unnatural in a system at constant T and P, in which only pressure–volume work is possible. This is demonstrated by considering a system at constant T and P, together with its surroundings, as an isolated supersystem, and calculating the total entropy change for the supersystem.

The surroundings are considered to act as a constant-temperature bath and a constant-pressure environment. Suppose heat $đq$ is transferred from the surroundings to the system and work $đw$ is done by the surroundings on the system. The entropy of the system increases by dS and its energy increases by $dE = đq + đw = đq - P\,dV$. Because the temperature bath is always at equilibrium at temperature T, the loss of heat by the surroundings is a reversible process and the change in the entropy of the surroundings is

$(-đq)/T$. Then the supersystem's entropy changes by

$$\Delta S_{\text{super}} = dS - \frac{đq}{T} = dS - \frac{dE + P\,dV}{T} = \frac{T\,dS - dE - P\,dV}{T}$$

with all thermodynamic variables being those of the system. The change is spontaneous, unnatural, or reversible according to whether ΔS_{super} is positive, negative, or zero. For a system held at constant T and P, $dG = dE + P\,dV - T\,dS$, so that ΔS_{super} is just $-dG/T$. Since T is positive, we conclude that the change is spontaneous, unnatural, or reversible according to whether dG is negative, positive, or zero. The change in Gibbs free energy is thus the criterion of reversibility for systems at constant temperature and pressure, with a reversible process characterized by $dG_{T,P} = 0$.

From the fundamental equations for E, H, A, and G, a host of relations between the thermodynamic variables and their derivatives can be obtained with no further assumptions about nature. These relate apparently different measurable properties, making it unnecessary to measure everything. For example, we have derived equations (1.15) and (1.16) from the fact that A is a well-defined function of state. Since the mixed second derivative of a function is independent of the order of differentiation,

$$\left(\frac{\partial^2 A}{\partial T\,\partial V}\right) = \left(\frac{\partial^2 A}{\partial V\,\partial T}\right)$$

or

$$-(\partial P/\partial T)_V = -(\partial S/\partial V)_T$$

One important relation which can be established in this way is that between C_P and C_V.

So far, we have considered closed systems, for which no matter can be exchanged between system and surroundings. To extend the discussion to open systems, one has only to note that matter coming into the system carries energy. Then the fundamental equation for E, equation (1.10), must be amended to read

$$dE = T\,dS - P\,dV + \sum_i \mu_i\,dn_i \tag{1.18}$$

where n_i is the number of moles of species i in the system and

$$\mu_i = \left(\frac{\partial E}{\partial n_i}\right)_{S,V,n_j(j\neq i)}$$

is the chemical potential of species i. It gives the change in E due to the addition of dn_i moles of i to the system. By subtracting $d(TS)$ from (1.18), one gets

$$dA = -S\,dT - P\,dV + \sum_i \mu_i\,dn_i$$

and, by adding $d(PV)$,

$$dG = -S\,dT + V\,dP + \sum_i \mu_i\,dn_i \tag{1.19}$$

which implies

$$\mu_i = \left(\frac{\partial G}{\partial n_i}\right)_{T,P,n_j(j\neq i)} \tag{1.20}$$

This is another definition of the chemical potential.

The chemical potential is an intensive variable, like the temperature or the pressure. It depends on the composition of a system, and on T and P. Suppose one has a very small system and increases its size, at constant T and P, by adding matter of various species, keeping the same proportions of different species so composition remains constant. Then all chemical potentials are constant, and (1.19) may be integrated to give

$$G = \sum_i \mu_i n_i \tag{1.21}$$

The differential of this is

$$dG = \sum_i [\mu_i\,dn_i + n_i\,d\mu_i]$$

and, comparing with (1.19), we find

$$-S\,dT + V\,dP = \sum_i n_i\,d\mu_i$$

the Gibbs–Duhem relation. It shows the chemical potentials are not independent of each other.

The free energies have mole numbers among their natural variables. These can be replaced by chemical potentials by a Legendre transformation. Thus, a new thermodynamic function may be defined by

$$J = A - \sum_i n_i\,\mu_i$$

so that

$$dJ = -S\,dT - P\,dV - \sum_i n_i\,d\mu_i \tag{1.22}$$

J has T, V, and $\{\mu_i\}$ as natural variables, it will be used in Section 2.5.

It is useful to consider chemical reactions in a closed system as if the system were open. The conversion of one chemical species into another is thought of as removal of the first species and addition of the second. A chemical reaction is conventionally written as

$$\sum_i \nu_i A_i = 0$$

where the $\{A_i\}$ are the chemical species and the ν_i are the stoichiometric coefficients; ν_i is positive for a product and negative for a reactant. The stoichiometric coefficients give the relative numbers of moles of the different species in the reaction. This may be expressed by defining the degree of advancement ξ such that

$$d\xi = \frac{dn_j}{\nu_i} \tag{1.23}$$

for all i. If the reaction occurs at constant T and P, changing the number of moles of i (reactant or product) by dn_i,

$$dG = \sum_i \mu_i\,dn_i = \sum_i \mu_i \nu_i\,d\xi. \tag{1.24}$$

The reaction will be a spontaneous process if dG is negative.

Whether the reaction goes forward or backwards spontaneously depends on the sign of $A_f \equiv \sum_i \mu_i \nu_i$. For negative A_f, dG is negative when $d\xi$ is positive, that is, the reaction is spontaneous in the forward direction, as written. For positive A_f, the reaction is spontaneous in the reverse direction, since then $d\xi < 0$ makes $dG < 0$. If $A_f = 0$, the reaction is at equilibrium, and can occur reversibly. The quantity A_f is sometimes referred to as the affinity, but more often as the (Gibbs) free energy of reaction:

$$\Delta_r G = \sum_i \mu_i \nu_i$$

The stoichiometric coefficients are fixed, but the chemical potentials depend on temperature, pressure, and composition.

Each chemical potential μ_i may be written as a standard chemical potential μ_i^0 plus a term defining the activity a_i:

$$\mu_i = \mu_i^0 + RT \ln a_i, \tag{1.25}$$

The activity is unity when the substance is in its standard state (defined differently for solids, liquids, gases, or solutes, but usually referring to a pressure of 1 atm). Inserting (1.25) in the free energy of reaction yields

$$\begin{aligned} \Delta_r G &= \sum_i \mu_i^0 \nu_i + RT \sum_i \nu_i \ln a_i \equiv \Delta_r G^0 + RT \sum_i \ln a_i^{\nu_i} \\ &= \Delta_r G^0 + RT \ln\left[\prod_i a_i^{\nu_i}\right] \end{aligned} \tag{1.26}$$

Here, $\Delta_r G^0$ is the standard free energy of reaction.

If $\Delta_r G = 0$, the reaction is at equilibrium. According to (1.26), one will then have

$$\ln\left[\prod_i a_i^{\nu_i}\right] = -\frac{(\Delta_r G^0)}{RT} \tag{1.27}$$

The quantity $\exp(-\Delta_r G^0/RT)$ is called the equilibrium constant. Equation (1.27) gives the relation between activities, and hence concentrations or pressures, of reactants and products that must hold at equilibrium.

PROBLEMS

1.1. Show that the Maxwell–Boltzmann distribution, equation (1.2), is normalized, that is, that it obeys equation (1.3). What is the most probable speed for a molecule in a gas at temperature T? What is the average speed? What is the root-mean-square speed (the square root of the average value of u^2)?

1.2. To derive the Stirling approximation for $\ln(n!)$, we seek a function $f(x)$ to approximate $\ln(x!)$. For any well-behaved function $f(x)$, power series expansions about $x + \frac{1}{2}$ yield:

$$\begin{aligned} f(x+1) - f(x) = &\left[f(x+\tfrac{1}{2}) + \tfrac{1}{2}f'(x+\tfrac{1}{2}) + \tfrac{1}{8}f''(x+\tfrac{1}{2}) + \cdots\right] \\ &- \left[f(x+\tfrac{1}{2}) - \tfrac{1}{2}f'(x+\tfrac{1}{2}) + \tfrac{1}{8}f''(x+\tfrac{1}{2}) + \cdots\right] \cong f'(x+\tfrac{1}{2}) \end{aligned}$$

From the definition of the factorial function, $(n+1)! = (n+1)n!$, so $\ln[(n+1)!] - \ln(n!) = \ln(n+1)$. If $f(x)$ is to approximate $\ln(x!)$, one must have $\ln(x+1) = f'(x+\frac{1}{2})$ and

$$f(n) - f(1) = \int_1^n f'(y)\,dy = \int_1^n \ln(y+\tfrac{1}{2})\,dy$$

Here $f(1) = \ln(1) = 0$. Carry out the integration to obtain the improved Stirling approximation:

$$\ln(n!) = n \ln n - n + \ln n^{1/2} + c$$

and find the value of the constant c. (You will have to use

$$\ln(n+\tfrac{1}{2}) = \ln(n) + \ln\left(1 + \frac{1}{2n}\right) \cong \ln(n) + \frac{1}{2n}$$

which is valid for large n.) (*Answer:* $f(n) = \int_{3/2}^{n+1/2} \ln u\,du = (n+\frac{1}{2})\ln(n+\frac{1}{2}) - (n+\frac{1}{2}) - \frac{3}{2}\ln(\frac{3}{2}) + \frac{3}{2} \cong (n+\frac{1}{2})(\ln n) + (n+\frac{1}{2})(2n)^{-1} - n + 1 - \frac{3}{2}\ln(\frac{3}{2}) = n\ln n - n + \frac{1}{2}\ln n + [\frac{3}{2} - \frac{3}{2}\ln\frac{3}{2}]$.)

1.3. Evaluate $\ln(n!)$ for $n = 10$, 20, 40, and 100, and calculate $n \ln n - n$ for the same values. What is the relative error in each case? Evaluate $\ln(n!)$ for these same n-values using the result of Problem 1.2 and obtain the relative errors.

CHAPTER 2

ENSEMBLES

2.1. MICROCANONICAL AND CANONICAL ENSEMBLES

We want to describe a macroscopic system, either at constant temperature or at constant energy, and relate its properties, especially thermodynamic properties, to a description of the system at the molecular level. At the molecular level, we consider quantum states, specified by sets of quantum numbers. The problem, as discussed in Section 1.1, is that the macroscopic system is always changing its state, so that its properties will have to be calculated as average values. The average value of a property P, which would have the value P_j if the system were in the state j, is

$$\langle P \rangle = \sum_j p_j P_j$$

Here, p_j is the probability of finding the system in the jth quantum state with energy E_j. It is equal to the fraction of time the system spends in state j.

To find the probability distribution, we consider an ensemble, a collection of N systems identical to the system of interest. This means that each system in the ensemble has the same volume and the same numbers of the same kinds of particles as the system of interest. If the system has its energy fixed, each system in the ensemble has the same energy as well (because of degeneracy, this does not mean all the systems in the ensemble are in the same state). The ensemble is to be constructed to represent the system, in the sense that the fraction of systems in the ensemble in state j at one instant is equal to p_j for the system. Thus, if N_j systems in the ensemble are in state j,

$$p_j = \frac{N_j}{N} \tag{2.1}$$

and

$$\sum_j p_j = 1 \tag{2.2}$$

as should be true for probabilities.

The ensemble for representing a closed system at constant energy and volume is called the microcanonical ensemble. The ensemble for a closed system at constant temperature and volume is called the canonical ensemble. Other ensembles exist for describing closed systems under other conditions, but they turn out to be less useful. For open systems, an ensemble called the grand canonical ensemble can be constructed. We discuss the grand canonical ensemble in Section 2.5.

The number of systems in any ensemble, N, is supposed to be very large. This means that the probability distribution for the ensemble will not differ much from the most likely probability distribution, so it is the most likely probability distribution that we seek. Like the possible results of a series of coin tosses (see Section 1.1), each possible probability distribution has a statistical weight W, representing the number of ways of choosing which members of the ensembles are in each accessible state j. The most likely probability distribution is the one with the largest value of W.

The microcanonical ensemble consists of N copies of the system, each isolated from the others. The canonical ensemble, on the other hand, is supposed to represent a system held at constant temperature, so that each member of the ensemble exchanges energy with its surroundings. This is equivalent to putting all the members of the ensemble into thermal contact, so energy can be exchanged between them. Then the ensemble as a whole serves as a temperature bath for each system. Because the ensemble is so large (N should approach infinity), the total energy of the ensemble may be considered as fixed.

The problem for the microcanonical ensemble is then to find the most likely probability distribution (set of N_j) for a collection of N isolated systems with the same energy and volume. For the canonical ensemble, the problem is to find the most likely probability distribution for N systems in thermal contact, with the total energy of the ensemble fixed. In either case, one can assume that the probability p_j depends only on E_j, the energy of state j.

The problem for the microcanonical ensemble is easily solved. Suppose the energy is E and there are g states with this energy (g is the degeneracy of the energy level). More correctly, because the energy cannot be exactly known, suppose the energy is known to be E with an uncertainty of ΔE and there are g states with energy between E and $E + \Delta E$. If the probability depends only on energy, all p_j must be equal. Since they must add to unity, we must have

$$p_j = g^{-1}, \qquad j = 1 \cdots g \tag{2.3}$$

One can also derive (2.3) by noting that the statistical weight of a distribution (set of $\{p_j\}$) is

$$W = \frac{N!}{N_1!\, N_2! \cdots} \tag{2.4}$$

and minimizing W with respect to the $\{N_j\}$, subject to the constraint

$$\sum_{j=1}^{g} N_j = N \tag{2.5}$$

Because of the constraint, the $\{N_j\}$ are not independent, so, before differentiating W with respect to each, one of the N_j must be expressed in terms of the others. This is done using (2.5) (see Problem 2.1).

2.2. CANONICAL ENSEMBLE: DERIVATION

For the canonical ensemble, things are somewhat more complicated. As for the microcanonical ensemble, the number of ways of having N_i of the N systems in the ensemble in state i is given by equation (2.4). W is to be maximized with respect to the occupation numbers N_i with *two* constraints. The total number of systems is fixed at N,

$$\sum_j N_j = N \tag{2.6}$$

and the total energy of the ensemble is fixed at E^T,

$$\sum_j N_j E_j = E^T. \tag{2.7}$$

It is convenient to maximize, instead of W,

$$\begin{aligned} \ln W &= \ln(N!) - \sum_{i=1}^{\infty} \ln(N_i!) = N \ln N - N - \sum_{i=1}^{\infty} [N_i \ln N_i - N_i] \\ &= N \ln N - \sum_i (N_i \ln N_i) \end{aligned} \tag{2.8}$$

because we can use the Stirling approximation for factorials of large numbers,

$$\ln(n!) \cong n \ln n - n$$

(see Section 1.1 and Problem 1.2).

One cannot simply differentiate $\ln W$ with respect to each N_i and set equal to zero becuase the N_i are not independent variables. They are connected by the constraints of equations (2.6) and (2.7). To write $\ln W$ in terms of independent variables, we rewrite the constraint equations as

$$N = N_1 + N_2 + \sum_{i=3}^{\infty} N_i \quad \text{and} \quad E^T = N_1 E_1 + N_2 E_2 + \sum_{i=3}^{\infty} N_i E_i$$

and consider these as simultaneous linear equations for N_1 and N_2. Solving, we obtain

$$N_1 = \frac{E^T - NE_2 + \sum_{i=3}^{\infty} N_i(E_2 - E_i)}{E_1 - E_2}$$

and (2.9)

$$N_2 = \frac{NE_1 - E^T - \sum_{i=3}^{\infty} N_i(E_1 - E_i)}{E_1 - E_2}$$

Then we write (2.8) as

$$\ln W = N \ln N - N_1 \ln N_1 - N_2 \ln N_2 - \sum_{i=3}^{\infty} N_i \ln N_i$$

and substitute N_1 and N_2 from (2.9). Then $\ln W$ is in terms of the other $\{N_i\}$, which are independent parameters. This permits us to put

$$\begin{aligned}
\frac{\partial \ln W}{\partial N_i}(i > 2) &= -\left(\frac{N_1}{N_1} + \ln N_1\right)\left(\frac{\partial N_1}{\partial N_i}\right) \\
&\quad -\left(\frac{N_2}{N_2} + \ln N_2\right)\left(\frac{\partial N_2}{\partial N_i}\right) - \left(\frac{N_i}{N_i} + \ln N_i\right) \\
&= -\left(\frac{N_1}{N_1} + \ln N_1\right)\left(\frac{E_2 - E_i}{E_1 - E_2}\right) - \left(\frac{N_2}{N_2} + \ln N_2\right)\left(\frac{E_i - E_1}{E_1 - E_2}\right) \\
&\quad -\left(\frac{N_i}{N_i} + \ln N_i\right) = 0
\end{aligned}$$

This equation is easily solved for $\ln N_i$ to give, for $i > 2$,

$$\ln N_i = \frac{E_1 \ln N_2 - E_2 \ln N_1}{E_1 - E_2} - E_i \frac{\ln N_2 - \ln N_1}{E_1 - E_2} \equiv \alpha - \beta E_i \quad (2.10)$$

The first fraction, which is independent of i, has been denoted by α and the second, also independent of i, by β. It is easy to see that (2.10) is satisfied for $i = 1$ and $i = 2$ as well. The problem now is to determine the parameters α and β.

Before doing this, we shall obtain the result (2.10) by the method of Lagrange multipliers. We want to maximize $\ln W$, that is,

$$\delta(\ln W) = 0$$

where $\delta(X)$ means the variation in X due to the variation of the $\{N_j\}$. The two constraints, (2.6) and (2.7), may be written as $\delta(N) = 0$ and $\delta(E^T) = 0$. Let α and β be two constants; for any values of α and β,

$$\delta(\ln W) + \alpha\delta\left(\sum_i N_i\right) - \beta\delta\left(\sum_i N_i E_i\right) = 0$$

Using expression (2.8) for $\ln W$, this becomes

$$-\sum_i (1 + \ln N_i)\,\delta N_i + \alpha\left(\sum_i \delta N_i\right) - \beta\left(\sum_i E_i\,\delta N_i\right) = 0 \qquad (2.11)$$

and $\Sigma_i(1)\delta N_i = 0$. The Lagrange multipliers α and β, so far arbitrary, are now supposed to be chosen to make the coefficients of δN_1 and δN_2 (or any two δN_i) vanish. Then (2.11) will involve only the other δN_i, which are independent, so one can set the coefficient of each equal to zero. Therefore

$$-\ln N_i + \alpha - \beta E_i = 0$$

for *all* i, which is identical to (2.10).

The canonical distribution, just derived, may be written as

$$N_i = e^{\alpha - \beta E_i} \qquad (2.12)$$

To determine the value of α, we note that

$$N = \sum_{i=1}^{\infty} N_i = \sum_{i=1}^{\infty} e^{\alpha} e^{-\beta E_i}$$

so $e^{\alpha} = N/Q$ where the partition function Q is defined by

$$Q = \sum_{i=1}^{\infty} e^{-\beta E_i} \tag{2.13}$$

Now

$$N_i = \frac{N e^{-\beta E_i}}{Q} \tag{2.14}$$

and $p_i = N_i/N$.

To get some idea about the meaning of β, we consider a composite ensemble, which contains two kinds of systems, A and B. The A and B type systems differ by virtue of their compositions or their volumes or both. Let there be N_A systems of type A and N_B systems of type B, all in thermal contact, and let the number of A systems in state $i(A)$ be N_i^A and the number of B systems in state $j(B)$ be N_j^B. The energy of an A system in state $i(A)$ is E_i^A and the energy of a B system in state $j(B)$ is E_j^B. We seek the most likely probability distribution, or set of $\{N_i^A\}$ and $\{N_i^B\}$.

The statistical weight for a distribution of the composite ensemble is

$$W = \left(\frac{N_A!}{(N_1^A)!\,(N_2^A)!\cdots} \right) \left(\frac{N_B!}{(N_1^B)!\,(N_2^B)!\cdots} \right) \tag{2.15}$$

We seek the values of $\{N_i^A\}$ and $\{N_i^B\}$ which maximize $\ln W$, with three constraints:

$$\sum_{i=1}^{\infty} N_i^A = N_A, \qquad \sum_{i=1}^{\infty} N_i^B = N_B \quad \text{and} \quad \sum_{i=1}^{\infty} \left(N_i^A E_i^A + N_i^B E_i^B \right) = E^T \tag{2.16}$$

The numbers of A and B systems are separately conserved, because an A system cannot be converted into a B system, but it is only the total energy that is constant, because A and B systems exchange energy. We maximize $\ln W$ using the technique of Lagrange multipliers. If the Lagrange multipliers corresponding to the three constraints of (2.16) are called α, γ, and $-\beta$, respectively, the maximization condition is

$$\delta\left(\ln W + \alpha N^A + \gamma N^B - \beta E^T\right) = 0$$

$$= \sum_i^{(A)} \left[-\left(1 + \ln N_i^A\right) + \alpha - \beta E_i^A \right] \delta N_i^A$$

$$+ \sum_i^{(B)} \left[-\left(1 + \ln N_i^A\right) + \gamma - \beta E_i^B \right] \delta N_i^B$$

The Lagrange multipliers may be chosen to make the coefficients of any three of $\{\delta N_i^A\}$ and $\{\delta N_i^B\}$ vanish. Then the coefficients of all the others, which may be considered as independent variations, must vanish.

The result is that

$$N_i^A = e^{\alpha - \beta E_i^A} \quad \text{and} \quad N_i^B = e^{\gamma - \beta E_i^B} \tag{2.17}$$

where $e^\alpha = N_A/Q_A$ and $e^\gamma = N_B/Q_B$. Q_A is the partition function for an A system and Q_B the partition function for a B system. What is striking about (2.17) is that the two distributions, for the two kinds of system, share the same value of β. The only thermodynamic property they share, however, is the temperature. The conclusion is that β is related to the shared temperature. The relation will be made more specific below.

2.3. THERMODYNAMIC PROPERTIES

Returning to the ensemble containing only one kind of system, we suppose that the occupation probability of the state j is given by $p_j = N_j/N = e^{-\beta E_j}/Q$, as in (2.14), with Q defined by (2.13). We also suppose that the only parameter one can vary is the volume V, so that the energies E_j depend on V only. First, the average energy is to be calculated, and then the average pressure.

The average energy is $\Sigma_j \, p_j E_j$ or

$$E = \frac{1}{Q} \sum_i E_i \, e^{-\beta E_i} = \frac{-1}{Q} \sum_i \left(\frac{\partial e^{-\beta E_i}}{\partial \beta} \right)_V = -\left(\frac{\partial \ln Q}{\partial \beta} \right)_V \tag{2.18}$$

where in the last member the differentiation of Q with respect to β is carried out term by term. To get the average pressure, we calculate $\Sigma_j p_j P_j$ where P_j, the pressure in the state j, can only mean the negative derivative of E_j with respect to V. Then the average pressure is

$$P = \frac{1}{Q} \sum_i P_i \, e^{-\beta E_i} = \frac{1}{Q} \sum_i \left(\frac{-\partial E_i}{\partial V} \right)_\beta e^{-\beta E_i} = \frac{1}{\beta} \left(\frac{\partial \ln Q}{\partial V} \right)_\beta \tag{2.19}$$

The enthalpy can now be constructed as $E + PV$ using (2.18) and (2.19). The calculation of the entropy remains to be elucidated.

Writing the energy as $E = \Sigma_j \, p_j E_j$, we have for the change in the energy

$$dE = \sum_j (p_j \, dE_j + E_j \, dp_j) = \sum_j \left(p_j \frac{dE_j}{dV} \, dV + \frac{\ln Q + \ln p_j}{-\beta} \, dp_j \right) \tag{2.20}$$

where the fact that E_j depends only on V is used in the first group of terms and

$$\ln p_j = -\beta E_j - \ln Q$$

is used to substitute for E_j in the second. The first sum on the right-hand side of (2.20) is $-P\,dV$. Since $\Sigma_j\, dp_j$ vanishes, the terms multiplying $(-\ln Q/\beta)$ sum to zero. Therefore

$$dE = -P\,dV - \frac{1}{\beta}\sum_j \ln p_j\, dp_j \tag{2.21}$$

Now

$$\sum_j (\ln p_j\, dp_j) = \sum_j d(p_j \ln p_j) - \sum_j dp_j$$

and the last sum vanishes, so we have

$$dE = -P\,dV - \frac{1}{\beta}\sum_j d(p_j \ln p_j) \tag{2.22}$$

This last expression should be compared with the fundamental equation for energy (1.10), $dE = T\,dS - P\,dV$. The second term in (2.22) must be equal to $T\,dS$. This strongly suggests that β^{-1} is kT, where k is a so far unknown constant, and $\Sigma_j(p_j \ln p_j)$ is $-S/k$. The negative sign has been associated with the sum because the entropy S must be positive or zero and every term in the sum is either negative or zero. The value of the constant k will be determined below. With

$$S = -k\sum_j (p_j \ln p_j) \tag{2.23}$$

we have completed the determination of the thermodynamic properties for the canonical ensemble.

The expressions for the thermodynamic properties can now be rewritten in terms of k and T. Since β is $1/kT$, a derivative with respect to β is $(-kT^2)$ multiplied by a derivative with respect to T, and the condition of constant β may be replaced by constant T. Thus the energy (2.18) becomes

$$E = kT^2\left(\frac{\partial \ln Q}{\partial T}\right)_V \tag{2.24}$$

and the pressure (2.19) is

$$P = kT\left(\frac{\partial \ln Q}{\partial V}\right)_T \tag{2.25}$$

The entropy of (2.23) becomes

$$S = -k \sum_i p_i \ln p_i = -k \sum_i p_i\left(\frac{-E_i}{kT} - \ln Q\right) = \frac{E}{T} + k \ln Q \tag{2.26}$$

This makes the Helmholtz free energy

$$A = E - TS = -kT \ln Q \tag{2.27}$$

The simplicity of this relation reflects the fact that A is the natural variable [see (1.14)] for describing systems at constant T and V.

The effect of a change in the zero of energy on the thermodynamic properties should be pointed out. Suppose a constant C is added to all the $\{E_j\}$, corresponding to a decrease in the zero of energy by C. Then the partition function Q changes to

$$Q' = \sum_j e^{-\beta(E_j+C)} = e^{-\beta C} \sum_j e^{-\beta E_j} = e^{-\beta C} Q$$

so that $\ln Q' = \ln Q - C/kT$. The energy, given by (2.24), changes by $kT^2(C/kT^2) = C$, as could be expected. The pressure (2.25) does not change. The change in the entropy is $C/T + k(-C/kT) = 0$. The Helmholtz free energy is increased by C because the energy is increased by C.

The thermodynamic energy, given by (2.18) or (2.24), is the *average* energy for a system at temperature T. The system is always changing its state, so the energy fluctuates. To investigate the importance of these fluctuations we differentiate (2.18) a second time with respect to β.

$$\left(\frac{\partial^2 \ln Q}{\partial \beta^2}\right)_V = \frac{\partial}{\partial \beta}\left(\frac{-\Sigma_i E_i e^{-\beta E_i}}{Q}\right) = \left(\frac{\Sigma_i E_i^2 e^{-\beta E_i}}{Q}\right) - \left(\frac{\Sigma_i E_i e^{-\beta E_i}}{Q}\right)^2 \tag{2.28}$$

Since $e^{-\beta E_i}/Q$ is the probability of finding the system in state i, the first term is the average value of the squared energy, which we denote by $\langle E^2 \rangle$. The second term is the square of the average energy, (2.18). Then (2.28) is

$\langle E^2 \rangle - E^2$, the mean-square fluctuation of the energy of a system in the canonical ensemble from its mean value:

$$\langle E^2 \rangle - E^2 = \sum_i \frac{(E_i - E)^2 e^{-E_i/kT}}{Q} \tag{2.29}$$

On the other hand, we obtained it as

$$\left(\frac{\partial^2 \ln Q}{\partial \beta^2} \right)_V = -\left(\frac{\partial E}{\partial \beta} \right)_V = kT^2 \left(\frac{\partial E}{\partial T} \right)_V \tag{2.30}$$

and $(\partial E/\partial T)_V$ is the heat capacity at constant volume C_V. Thus (2.30) relates a measurable macroscopic quantity, C_V, to a statistical mechanical quantity, $\langle E^2 \rangle - E^2$.

For small variations about the mean, the probability of finding the system with an energy F may be assumed to be given by the normalized Gaussian distribution

$$P(F) = \frac{1}{\sqrt{2\pi}\,\sigma_E} \exp\left[\frac{-(F - E)^2}{2\sigma_E^2} \right] \tag{2.31}$$

where $\sigma_E^2 = \langle E^2 \rangle - E^2$. (According to the central limit theorem of statistics, the probability distribution for a variable which is the mean of many random variables is Gaussian.) In this distribution, E is the mean and σ_E the width of the distribution.

E is an extensive parameter (its value is proportional to the size of the system), as is C_V. Then (2.28) to (2.30) show that the mean-square fluctuation is also proportional to the size of the system. This means that for a system at constant temperature, represented by the canonical ensemble, the *relative* root-mean-square fluctuation

$$\frac{\sigma_E}{E} = \frac{\sqrt{\langle E^2 \rangle - E^2}}{E}$$

is smaller, the larger the system. For example, the energy of N ideal gas atoms is $3NkT/2$ and $C_V = 3Nk/2$, so $\langle E^2 \rangle - E^2 = 3Nk^2T^2/2$. Then

$$\frac{\sigma_E}{E} = \sqrt{\frac{2}{3N}}$$

which approaches zero for large N.

For a macroscopic system, the relative fluctuation in the energy is so small that one can say the energy is fixed. The difference between what one gets from the microcanonical and canonical ensembles is unimportant for most practical purposes. The exception is when C_V becomes very large, which means the energy can change at constant temperature. This occurs, for example, when there is a phase equilibrium.

Now consider the chemical potential. For a system containing n moles of substance, the chemical potential is

$$\mu = \left(\frac{\partial A}{\partial n}\right)_{V,T} = \frac{-1}{\beta}\left(\frac{\partial \ln Q}{\partial n}\right)_{V,T}$$

More generally, if there are different chemical substances present, the chemical potential of substance j is

$$\mu_j = \left(\frac{\partial A}{\partial n_j}\right)_{V,T,n_i(i \neq j)} = \frac{-1}{\beta}\left(\frac{\partial \ln Q}{\partial n_j}\right)_{V,T,n_i(i \neq j)} \tag{2.32}$$

The derivatives with respect to the number of moles of substance j are taken holding the numbers of moles of all other substances constant. Although we have not emphasized it in our notation, the partition function depends on the numbers of moles, or numbers of molecules, of all species. The chemical potentials play an important role in open systems, which we now discuss.

2.4. GRAND CANONICAL ENSEMBLE

An open system is one which can exchange matter with its surroundings, so that the numbers of molecules of different species fluctuate. The ensemble which represents an open system at thermal equilibrium is called the grand canonical ensemble. It consists of a large number of systems of the same volume in thermal contact and with walls permeable to molecules of all species, so molecules as well as heat can be interchanged between them. For each system in the ensemble, the other systems form a "molecule bath," analogous to temperature bath they provide in the canonical ensemble. The total energy of the ensemble is fixed, as are the total numbers of molecules of all species in the ensemble. A representation of the grand canonical ensemble is shown in Figure 2.1. The cells separated by dashed lines are the systems in the ensemble. There are two species of molecules, represented by the circles and the triangles.

We seek the probability that a system in the ensemble has N_a molecules of species a, N_b molecules of species $b, \ldots,$ N_k molecules of species $k, \ldots,$ and is in the jth energy state possible for these molecule numbers. In quantum mechanical terms, the energy states correspond to eigenstates of the Hamiltonian, and the Hamiltonian depends on the numbers of molecules. Suppose there are N systems in the ensemble ($N \to \infty$) and let $N(N_a N_b, \ldots, N_k, \ldots, j)$,

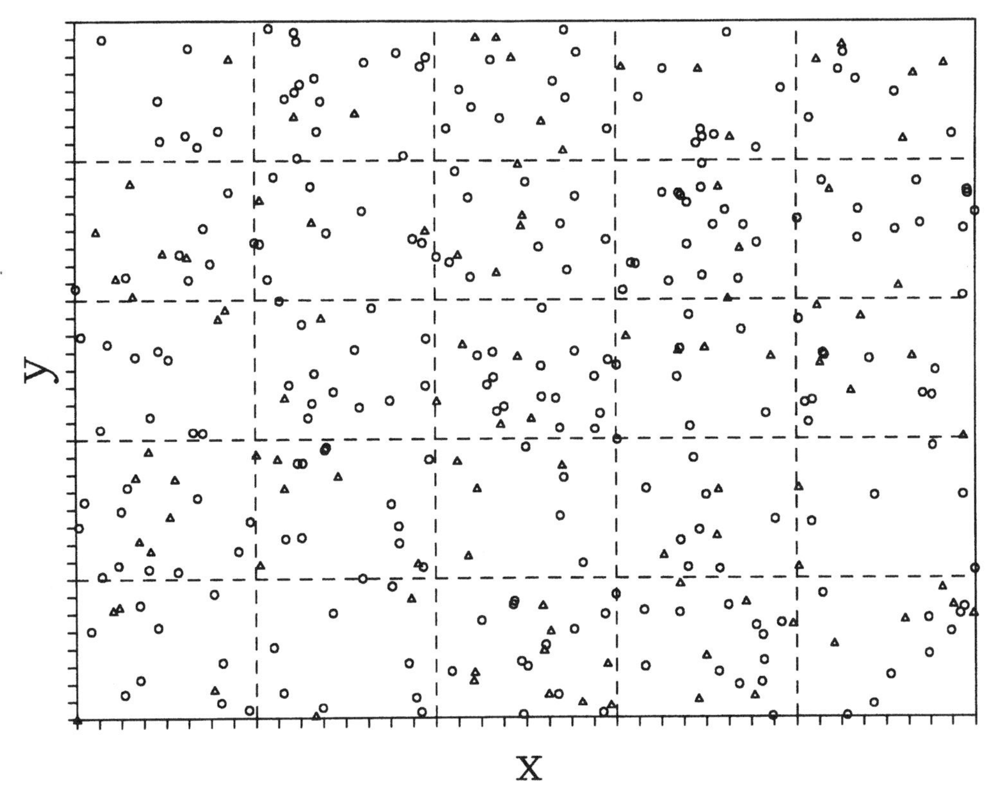

Figure 2.1. Representation of the grand ensemble, representing an open system. Each cell is a system, containing some numbers of molecules of two species (circles and triangles) which can pass from system to system.

or $N(\{N_k\}; j)$ for short, be the number of systems with N_k molecules of species k for each k, and in the jth energy state. Then the probability that a system has these molecule numbers and quantum numbers is

$$p(\{N_k\}; j) = \frac{N(\{N_k\}; j)}{N} \tag{2.33}$$

The distribution or set of probabilities we seek is the most likely one, the one with the highest statistical weight. The statistical weight, W, is the number of ways of distributing the N systems so that $N(\{N_k\}; j)$ systems have N_k molecules of species k and are in the jth energy state.

W is equal to $N!$ divided by the product of factorials of the $N(\{N_k\}; j)$ [compare (2.4), the expression used in deriving the canonical ensemble]. We want to maximize

$$\ln W = \ln(N!) - \sum_{N_a} \sum_{N_b} \sum_{N_c} \cdots \sum_{N_k} \cdots \sum_{j} \ln[N(\{N_k\};j)!]$$

with respect to all the $N(\{N_k\};j)$, with these occupation numbers satisfying certain constraints. First, the total number of the systems in the ensemble is fixed:

$$\sum_{N_a N_b \cdots j} N(\{N_k\};j) = N \tag{2.34}$$

Second, the total number of molecules of each species in the ensemble is fixed:

$$\sum_{N_a N_b \cdots j} N(\{N_k\};j) N_m = \mathcal{N}_m \tag{2.35}$$

where m is one of the species k. Finally, the total energy of the ensemble is fixed:

$$\sum_{N_a N_b \cdots j} N(\{N_k\};j) E_j(\{N_k\}) = E^T \tag{2.36}$$

The energies depend on V as well as on the $\{N_k\}$, but the V dependence is suppressed to simplify the notation.

Using the method of Lagrange multipliers, we consider

$$\delta \ln W + \alpha\, \delta N + \sum_{m} \gamma_m\, \delta \mathcal{N}_m - \beta\, \delta E^T = 0 \tag{2.37}$$

For each factorial, the Stirling approximation, $\ln n! \cong n \ln n - n$, is used; $\delta(\ln n!)$ then becomes $\ln n\, \delta n$. Then (2.37) becomes

$$-\sum_{N_a N_b \cdots j} \ln[N(\{N_k\};j)]\, \delta N(\{N_k\};j) + \alpha \sum_{N_a N_b \cdots j} \delta N(\{N_k\};j)$$
$$+ \sum_{m} \gamma_m \sum_{N_a N_b \cdots j} \delta N(\{N_k\};j) N_m - \beta \sum_{N_a N_b \cdots j} \delta N(\{N_k\};j) E_j(\{N_k\}) = 0 \tag{2.38}$$

If there are s species, there are $s + 2$ constraints. The $s + 2$ Lagrange multipliers may be chosen to annihilate the coefficients of $s + 2$ of the $\delta N(\{N_k\};j)$ in (2.38). The remaining $N(\{N_k\};j)$ being independent, the coefficients of the $\delta N(\{N_k\}; j)$ must also vanish. Therefore we may write

$$-\ln[N(\{N_k\};j)] + \alpha + \sum_m \gamma_m N_m - \beta E_j(\{N_k\}) = 0$$

and

$$N(\{N_k\};j) = e^{\alpha} \exp\left[\sum_m \gamma_m N_m - \beta E_j(\{N_k\})\right]$$

for all k.

The value of the parameter α is chosen so that (2.34) is satisfied. Then

$$N(\{N_k\};j) = \frac{N \exp[\Sigma_m \gamma_m N_m - \beta E_j(\{N_k\})]}{\Sigma_{N_a N_b} \exp(\Sigma_m \gamma_m N_m) Q(N_a N_b, \ldots, V, \beta)} \tag{2.39}$$

where

$$Q(N_a, N_b, \ldots, V, \beta) = \sum_j e^{-\beta E_j(\{N_k\})} \tag{2.40}$$

It may be verified that summing (2.39) over all species numbers and all energy states gives N. The expression $Q(N_a, N_b, \ldots, V, \beta)$ is just the canonical partition function for a system having N_a molecules of species a, N_b molecules of species b, and so on. The parameter β is equal to $1/kT$, just as for the canonical ensemble. The meaning of the parameters $\{\gamma_m\}$ is still to be elucidated, but they relate to equilibrium with respect to interchange of matter between the systems. Since the equality of chemical potentials characterizes equilibrium with respect to interchange of matter, one can anticipate that γ_k involves the chemical potential of the species k.

The denominator of (2.39), a weighted sum of partition functions for different numbers of molecules, is called the grand partition function and denoted by Ξ (capital Greek xi).

$$\Xi = \sum_{N_a N_b \cdots} \exp\left(\sum_m \gamma_m N_m\right) Q(N_a N_b, \ldots, V, \beta) \tag{2.41}$$

According to (2.39), the probability that a system in the ensemble has N_k molecules of species k for each k is

$$p(\{N_k\};j) = \frac{\Sigma_j \exp[\Sigma_m \gamma_m N_m - \beta E_j(\{N_k\})]}{\Xi}$$

$$= \frac{\exp[\Sigma_m \gamma_m N_m] Q(N_a N_b, \ldots, V, \beta)}{\Xi}$$

where we have summed over the states of a system with the given set of molecule numbers. The average number of molecules of species a is therefore

$$\langle N^a \rangle = \sum_{N_a N_b \cdots} p(\{N_k\}; j) N_a = \left(\frac{\partial \ln \Xi}{\partial \gamma_a} \right)_{V, \beta, \gamma_j (j \neq a)} \tag{2.42}$$

The average energy $\langle E \rangle$ (normally denoted just by E) is

$$\frac{\sum_{N_a N_b} \ldots \sum_j \exp\left[\sum_m \gamma_m N_m - \beta E_j(\{N_k\}) \right] E_j(\{N_k\})}{\Xi} = -\left(\frac{\partial \ln \Xi}{\partial \beta} \right)_{V, \{\gamma_k\}} \tag{2.43}$$

and the average pressure is

$$P = \frac{\sum_{N_a N_b} \ldots \sum_j \exp\left[\sum_m \gamma_m N_m - \beta E_j(\{N_k\}) \right] \left(-\partial E_j(\{N_k\}) / \partial V \right)}{\Xi}$$

$$= \beta^{-1} \left(\frac{\partial \ln \Xi}{\partial V} \right)_{\beta, \{\gamma_k\}} \tag{2.44}$$

Equations (2.42) to (2.44) are useful in establishing the meaning of the grand partition function in thermodynamic terms.

The grand partition function Ξ is a function of the parameters β, $\{\gamma_k\}$, and V:

$$d \ln \Xi = \left(\frac{\partial \ln \Xi}{\partial \beta} \right)_{V, \{\gamma_k\}} d\beta + \sum_a \left(\frac{\partial \ln \Xi}{\partial \gamma_a} \right)_{V, \beta, \gamma_k (j \neq a)} d\gamma_a$$

$$+ \left(\frac{\partial \ln \Xi}{\partial V} \right)_{\beta, \{\gamma_k\}} dV$$

$$= -\langle E \rangle \, d\beta + \sum_a \langle N^a \rangle \, d\gamma_a + \beta P \, dV \tag{2.45}$$

and we expect γ_a to be related to the chemical potential of species a, μ_a. In Section 1.2, we defined the thermodynamic function J as $A - \sum_m \mu_m n_m$, so that J has the chemical potentials $\{\mu_m\}$, along with T and V, as natural variables. According to (1.22),

$$dJ = -S \, dT - P \, dV - \sum_b n_b \, d\mu_b$$

Using $S = k\beta(E - A)$ and $dT = -d\beta/(k\beta^2)$, this is rewritten

$$\beta\, dJ = (E - A)\, d\beta - \beta P\, dV - \beta \sum_b n_b\, d\mu_b$$

Then

$$\beta\, dJ + J\, d\beta = E\, d\beta - \beta P\, dV - \sum_b n_b(\beta\, d\mu_b + \mu_b\, d\beta) \tag{2.46}$$

This is identical to (2.45) if $\ln \Xi = -\beta J$ and $N^b \gamma_b = n_b \beta \mu_b$. The ratio of the number of molecules of a to the number of moles of a, N^a/n_a, is Avogadro's number.

The logarithm of the grand partition function Ξ is now identified with $(-A + \sum_j n_j \mu_j)/kT$. According to (1.21), $\sum_j n_j \mu_j$ is equal to G, so

$$\ln \Xi = \frac{PV}{kT} \tag{2.47}$$

This means that in addition to (2.44) one has the simpler relation,

$$P = \frac{kT \ln \Xi}{V}$$

We have also found that the Lagrange multiplier γ_b is $\beta\mu_b/\mathcal{N}_A$ ($\mathcal{N}_A$ = Avogadro's number) for each species b, with the conventional definition of the chemical potential as a partial derivative with respect to number of moles (and not molecules). Sometimes the expression $e^{\beta\mu_a/\mathcal{N}_A}$ is referred to as the fugacity of substance a, and some authors refer to $\Lambda^{-3} e^{\beta\mu_a/\mathcal{N}_A}$ as the activity of substance a. Here Λ is the de Broglie wave length, $\Lambda = (h^2\beta/2\pi m)^{1/2}$; its significance will be made clear in Section 3.2.

The partition function itself, using the fact that $\ln Q(\{N_b\}), V, T)$ is $-\beta A(\{N_b\}, V, T)$ where $A(\{N_b\}, V, T)$ is the free energy for a system of N_b particles of species b, volume V, and temperature T, may be rewritten as

$$\Xi = \sum_{N_a N_b \cdots} \exp\left(\sum_m \gamma_m N_m\right) \exp(-\beta A(N_a N_b, \ldots, V, \beta)) \tag{2.48}$$

The average value of the Helmholtz free energy, which is identified with the thermodynamic free energy, is

$$A = \Xi^{-1} \sum_{N_a N_b \cdots} A(N_a N_b, \ldots, V, \beta) \times \exp\left(\sum_m \gamma_m N_m\right) \exp(-\beta A(N_a N_b, \ldots, V, \beta)) \tag{2.49}$$

The average entropy is, similarly,

$$S = \Xi^{-1} \sum_{N_a N_b \cdots} S(N_a N_b, \ldots, V, \beta)$$

$$\times \exp\left(\sum_m \gamma_m N_m\right) \exp(-\beta A(N_a N_b, \ldots, V, \beta))$$

where $S(N_a N_b, \ldots, V, \beta) = -(\partial A(N_a N_b, \ldots, V, T)/\partial T)_{V,\{N_m\}}$. One is here averaging over all possible particle numbers.

Note that the probability of finding N_a particles of species a is

$$P(N_a) = \Xi^{-1} \sum_{N_b \cdots} \exp\left(\sum_m \gamma_m N_m\right) \exp(-\beta A(N_a N_b, \ldots, V, \beta)) \quad (2.50)$$

The most probable value of N_a may be obtained by differentiating $P(N_a)$ with respect to N_a and setting the derivative equal to zero. This result is

$$\gamma_a \sum_{N_b \cdots} \exp\left(\sum_m \gamma_m N_m\right) \exp(-\beta A(N_a N_b, \ldots, V, \beta))$$

$$= \beta \sum_{N_b \cdots} \exp\left(\sum_m \gamma_m N_m\right) \exp(-\beta A(N_a N_b, \ldots, V, \beta)) \left(\frac{\partial A}{\partial N_a}\right)_{N_b \cdots TV}$$

Since $(\partial A/\partial N_a)_{N_b \cdots TV} = \mu_a(N_a N_b, \ldots, V, \beta)/\mathscr{N}_A$, this shows that, for the most probable value of N_a, the chemical potential is equal to the average chemical potential. This is related to the fact that there is only a small probability of finding a value of N_a differing much from the most probable value.

To investigate the fluctuations in N_a, we consider a one-component system, for which the algebra and notation are easier to follow. The grand partition function is

$$\Xi = \sum_N e^{\gamma N} Q(N, V, T) \quad (2.51)$$

with N the number of molecules and $\langle N \rangle = -(\partial \ln \Xi/\partial \gamma)_{T,V}$. The second derivative is

$$\left(\frac{\partial^2 \ln \Xi}{\partial \gamma^2}\right)_{T,V} = \frac{\Sigma_N N^2 e^{\gamma N} Q(N, V, T)}{\Xi} - \frac{\left(\Sigma_N e^{\gamma N} Q(N, V, T)\right)^2}{\Xi^2}$$

$$= \langle N^2 \rangle - \langle N \rangle^2$$

Like the corresponding mean-square fluctuation of the energy, $\sigma_N^2 = \langle N^2\rangle - \langle N\rangle^2$ is positive and proportional to the size of the system (note that $\ln \Xi$ is extensive and γ, being a chemical potential divided by the temperature, is intensive). Then σ_N/N decreases as the size of the system increases, and the fluctuations in particle numbers, like the fluctuations in energy, become quite unimportant for a macroscopic system.

To interpret σ_N in terms of thermodynamic variables, we note that, since $\gamma = \beta\mu/\mathcal{N}_A$,

$$\left(\frac{\partial^2 \ln \Xi}{\partial \gamma^2}\right)_{T,V} = \frac{\mathcal{N}_A}{\beta}\left(\frac{\partial N}{\partial \mu}\right)_{T,V} = \frac{\mathcal{N}_A}{\beta}\left(\frac{\partial N}{\partial P}\right)_{T,V}\left(\frac{\partial P}{\partial \mu}\right)_{T,V}$$

From the Gibbs–Duhem relation, $S\,dT - V\,dP + (N/\mathcal{N}_A)\,d\mu = 0$, $(\partial P/\partial\mu)_T$ is $(N/V\mathcal{N}_A)$. Then

$$\left(\frac{\partial^2 \ln \Xi}{\partial \gamma^2}\right)_{T,V} = \frac{V}{\beta}\left(\frac{\partial [N/V]}{\partial P}\right)_{T,V}\left(\frac{N}{V}\right) \tag{2.52}$$

where N/V is the number density ρ. Now note that

$$\left(\frac{\partial \rho}{\partial P}\right)_{T,V} = N\left(\frac{\partial V^{-1}}{\partial P}\right)_{T,N} = \frac{-N}{V^2}\left(\frac{\partial V}{\partial P}\right)_{T,N}$$

so, finally,

$$\langle N^2\rangle - \langle N\rangle^2 = \left(\frac{\partial^2 \ln \Xi}{\partial \gamma^2}\right)_{T,V} = \frac{-N^2}{\beta V^2}\left(\frac{\partial V}{\partial P}\right)_{T,N} = \frac{N^2}{\beta V}\kappa_T \tag{2.53}$$

where κ_T is the isothermal compressibility. Equation (2.53) may be written

$$\sigma_N^2 = N\rho\kappa_T$$

to emphasize that σ_N/N decreases with the size of the system.

For any reasonable macroscopic system, σ_N/N will be negligibly small, so the number of molecules in the volume V is effectively fixed, even if the system is open (example: the number of molecules of air in 1 cm^3 of the volume of this room). Thus an open macroscopic system is equivalent to a closed one in the sense that the canonical and grand canonical distributions give the same values for thermodynamic properties. The exception to this is when density fluctuations are very important, such as near a critical point. In such cases, the compressibility becomes very large, and may even become infinite.

PROBLEMS

2.1. The microcanonical ensemble represents a system at fixed energy. Suppose there are g states with this energy, with N_j ($j = 1 \cdots g$) being the number of systems in the ensemble in state j. The statistical weight for the distribution of N systems (the number of systems in the ensemble) into g states is

$$W = \frac{N!}{N_1!\, N_2!\, N_3! \cdots}$$

and the occupation numbers must obey the constraint

$$\sum_i N_i = N$$

Use the constraint to express N_1 in terms of the other N_j, substitute into W, and, by differentiation, find the values of N_j ($j > 1$) which maximize $\ln W$. (The N_j for $j > 1$ are independent variables.)

2.2. Use Lagrange multipliers to find the probability distribution for the microcanonical ensemble. This means setting

$$\delta(\ln W + \varepsilon N) = 0$$

where ε is the Lagrange multiplier and setting the coefficient of each δN_j equal to 0. Find the value of ε which allows the constraint to be satisfied.

2.3. Using the expressions (2.23) to (2.27) and $\ln p_j = -\beta E_j - \ln Q$, show the thermodynamic relation

$$\left(\frac{\partial A}{\partial T}\right)_V = -S$$

is satisfied.

2.4. Consider a system which can be in one of four states, a nondegenerate ground state and three degenerate states at an energy 10^{-15} ergs above the ground state. If the system is held at constant temperature, what is the probability of finding the system in its ground state when $kT = 10^{-16}$ ergs, 5×10^{-16} ergs, 10^{-15} ergs, or 10^{-14} ergs? What is the average energy in each case?

2.5. Suppose a system has two quantum states, with energies $+\varepsilon$ and $-\varepsilon$. When in contact with a heat reservoir, the average energy of a system is $-\varepsilon/2$. Calculate the probability of finding the system in the lower-energy state, the partition function, and the value of β.

2.6. If the energy levels for a system are equally spaced, say $E_j = j\Theta$ with Θ a constant, the canonical partition function is an infinite geometric sum

$$Q = \sum_{j=1}^{\infty} e^{-\beta j \Theta} = \frac{e^{-\beta\Theta}}{1 - e^{-\beta\Theta}}$$

Calculate the average energy as a function of β and Θ, and show that, at a given temperature (given value of β), a larger energy spacing (larger Θ) gives rise to a higher average energy.

2.7. According to (2.23), the entropy is given by a sum over states: $S = -k \sum_j p_j \ln p_j$. Suppose the number of states is s. What is the lowest possible value of entropy? What is the highest possible value of the entropy? What are the probabilities p_j corresponding to the two cases?

2.8. Show that the average energy of a system at constant temperature, calculated by differentiating $\ln Q$, increases with temperature at fixed volume.

2.9. Consider the normalized Gaussian distribution for the energy (2.31). Show that the width at half-height is σ_E. What is the probability of finding an energy differing from the average value by more than σ_E? (You will have to estimate the value of an integral involving e^{-x^2}.)

2.10. For a monatomic ideal gas the energy is $3NkT/2$ and the heat capacity is $3Nk/2$, where N is the number of molecules and k the Boltzmann constant $k_B = 1.38 \times 10^{-16}$ ergs/K. Calculate σ_E for a system containing 0.001 moles of ideal gas at $T = 300$ K. Calculate the relative energy fluctuation, σ_E/E. Calculate E and σ_E/E for the same system at $T = 10$ K.

2.11. The Gibbs–Duhem relation, derived in Section 1.2, is

$$\sum_b n_b \, d\mu_b - V dP + S \, dT = 0$$

For the grand ensemble, representing a system at constant T and V, μ_a is $\gamma_a \mathcal{N}_A/\beta$ and $P = \ln \Xi/\beta V$. Use these and the definitions of the thermodynamic functions to prove the Gibbs–Duhem relation for T constant.

2.12. Suppose the partition function $Q(N, V, T)$ in (2.41) is $(\Lambda^3 V)^N/N!$, where $\Lambda^2 = (\beta h^2/2\pi m)$ (as we will show in the next chapter, this is the case of an ideal gas). Calculate the average energy and the average density in terms of β, γ, and V.

2.13. The expressions for the (average) Helmholtz free energy A and entropy S of an open system were given above, as well as the simpler expression for the average energy,

$$E = -\left(\frac{\partial \ln \Xi}{\partial \beta}\right)_{\{\gamma b\}, V}$$

Show that $A = E - TS$.

2.14. Defining the fugacity z_b as $\exp(\gamma_b) = \exp(\beta\mu_b)$, show that the free energy is

$$G = \sum_b N_b kT \ln z_b$$

2.15. Show that the number of particles of species a may be written in terms of the fugacity (Problem 2.14) as

$$N_a = \left(\frac{\partial \ln \Xi}{\partial \ln z_a}\right)$$

(*Hint:* (2.45) gives N_a as a derivative of $\ln \Xi$.)

2.16. The mean-square fluctuation of particle number from the mean is $\sigma_N^2 = N\rho\kappa_T$ where κ_T is the isothermal compressibility, $\kappa_T = -V^{-1}(\partial V/\partial P)_T$. Calculate σ_N for 0.001 moles of an ideal gas in a volume of 100 cm^3 at a temperature of 100 K and compare the result with the value of N. Repeat the calculation for 10^{-10} moles of gas at a temperature of 10 K.

CHAPTER 3

INDEPENDENT PARTICLES

3.1. INDEPENDENCE AND INDISTINGUISHABILITY

We consider a macroscopic system with fixed temperature, volume, and number of particles, meaning that it is described by the canonical distribution and canonical partition function Q. In addition, we suppose that the system is composed of independent parts, with no interactions, so the energy is exactly the sum of the energies of the parts. Then a state of the macroscopic system is specified by giving the states (sets of quantum numbers) for the independent subsystems. The most important example of this situation is a dilute gas, with the independent subsystems being the molecules.

Let the subsystems be numbered from 1 to N, and write $\varepsilon_n^{(a)}$ for the energy of the nth quantum state of subsystem a. Then the jth energy state of the system has energy

$$\varepsilon_j = \varepsilon_m^{(1)} + \varepsilon_n^{(2)} + \varepsilon_p^{(3)} + \cdots \tag{3.1}$$

and the system partition function is

$$Q = \sum_j e^{-\beta\varepsilon_j} = \sum_m \sum_n \sum_p \cdots e^{-\beta\varepsilon_m^{(1)}} e^{-\beta\varepsilon_n^{(2)}} e^{-\beta\varepsilon_p^{(3)}} \cdots \tag{3.2}$$

The sum over states j is replaced by sums over all the quantum numbers of the subsystems. Since there are no interactions between the subsystems, these sums are independent, and the exponential is a product of exponentials.

Therefore

$$Q = \left[\sum_m e^{-\beta\varepsilon_m^{(1)}}\right]\left[\sum_n e^{-\beta\varepsilon_n^{(2)}}\right]\left[\sum_p e^{-\beta\varepsilon_p^{(3)}}\right]\cdots$$

which is a product of partition functions for the individual subsystems; ln Q, and hence the thermodynamic functions, will be a sum of contributions of the subsystems.

If all the subsystems are identical, like the molecules in a gas, each has the same quantum states and energies available. Then

$$Q = \left[\sum_m e^{-\beta\varepsilon_m^{(1)}}\right]^N = q^N \tag{3.3}$$

where q is the partition function for a subsystem, or the molecular partition function. Equation (3.3), however, is correct only if the subsystems or molecules are distinguishable, so that one can differentiate between two situations such as: (1) having one molecule in state j and a second molecule in state k, and (2) having the first molecule in state k and the second in state j. The molecules would be distinguishable if they were labeled, or if they were localized in space; molecules in a gas are not distinguishable because they are all the same and continually changing places.

If the molecules are indistinguishable, the partition function is not (3.3). In replacing sums of products in (3.2) by a product of sums over subsystem states, we effectively included both the situations referred to in the previous paragraph. For indistinguishable molecules, the situations are one and the same: there is one molecule in state j and another in state k, and one cannot tell which is which. Therefore, (3.3) is too big because too many states were included in the sum. A reasonable approximation is to divide it by $N!$, the number of ways of putting each of N distinguishable molecules into a state.

The number of ways of assigning N distinguishable molecules to states is $N!$ if no more than one molecule is in any state, and less than $N!$ if more than one molecule is in some states. Almost always, the number of available states will be large compared to the number of molecules, so there will rarely be more than one molecule in a state, and dividing (3.3) by $N!$ is an excellent approximation. Then

$$Q = \frac{q^N}{N!} \tag{3.4}$$

for a system composed of N independent and indistinguishable subsystems or molecules. The correct way to take into account the indistinguishability of the molecules will be considered when we derive the Bose–Einstein and Fermi–Dirac statistics in Chapter 5, since indistinguishability of particles constitutes a restriction on their wave functions. We will then be able to treat situations in which many available states are occupied by more than one molecule and also to justify the division of Q by $N!$.

According to (3.4), the indistinguishability decreases $\ln Q$ by $\ln(N!)$. This means it will not affect properties like energy and pressure, which are derivatives of $\ln Q$. It will affect the free energies through the entropy, which is associated with the distribution of particles over states. The statistics generated from (3.4) is sometimes called "corrected Boltzmann statistics." It will be applied to molecules of an ideal gas, to molecular vibrations, and to spins, after showing how they are considered as independent systems.

3.2. IDEAL GAS

We will first apply corrected Boltzmann statistics to an ideal gas of N structureless noninteracting particles. The molecular partition function q will be calculated using the quantum mechanical formula for the energy levels of a particle in a box. From q, the pressure and energy will be calculated. They will depend on the constant k. By comparing the calculated pressure and energy to the ideal-gas values, we will determine the value of k.

For a particle of mass m confined to a one-dimensional box of length L, the possible energies are $h^2n^2/8mL^2$. Here, the quantum number n takes on positive integer values. For a three-dimensional box of volume $V = L^3$, the possible energy levels are indexed by three quantum numbers, all positive integers:

$$\varepsilon_{ijk} = \frac{h^2}{8mL^2}(i^2 + j^2 + k^2) \tag{3.5}$$

The molecular partition function,

$$q = \sum_{ijk} e^{-\beta\varepsilon_{ijk}}$$

breaks down into a product of three sums because the quantum numbers are independent.

$$q = \sum_{i=1}^{\infty} e^{-\beta h^2 i^2/8mL^2} \sum_{j=1}^{\infty} e^{-\beta h^2 j^2/8mL^2} \sum_{k=1}^{\infty} e^{-\beta h^2 k^2/8mL^2}$$

The three sums are identical. To evaluate one, we approximate it as an integral:

$$\sum_{k=1}^{\infty} e^{-\beta h^2 k^2/8mL^2} \cong \int_0^{\infty} e^{-\beta h^2 k^2/8mL^2}\, dk = \sqrt{\frac{2\pi mL^2}{\beta h^2}} \tag{3.6}$$

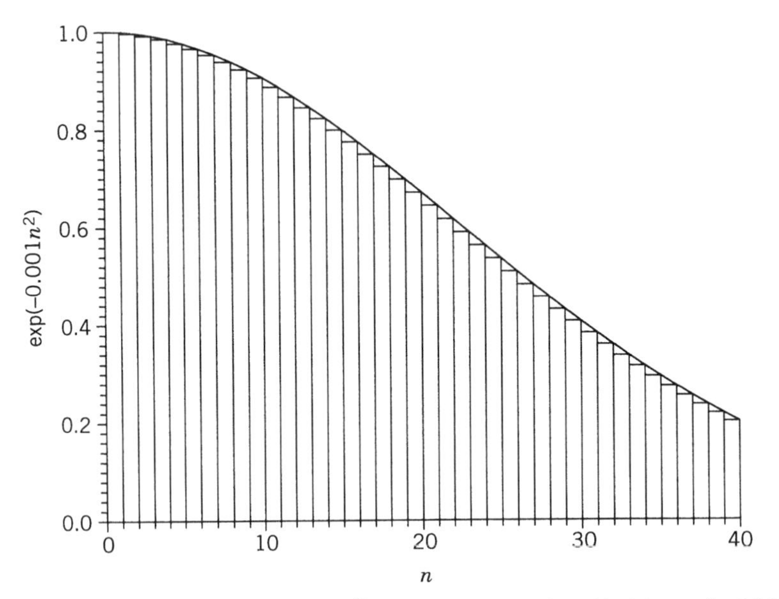

Figure 3.1. Plot of $f(n) = \exp(-0.001n^2)$ vs. n and rectangles of height $\exp(-0.001n^2)$ for integral n. The sum of the areas of the rectangles approximates the area under the curve better, the less $f(n)$ changes when n is increased by 1.

The curve in Figure 3.1 is the function $e^{-0.001n^2}$; the value of the integral is the area under the curve. The sum of the areas of the rectangles, each of width unity, is the value of the sum. The approximation of the sum by the integral will be better, the less the function changes with k is increased by 1. This means it will work better for small values of $\beta h^2/8mL^2$, and this quantity will be very, very small for all cases of practical interest (see Problems 3.2 and 3.3). The quantity

$$\Lambda = \sqrt{\frac{h^2\beta}{2\pi m}}$$

has the dimensions of length, and is referred to as the de Broglie wavelength. For the sum to be well approximated by the integral, the de Broglie wavelength should be small compared to L, the dimension of the system. When Λ is not small compared to L quantum mechanical effects are important and it is not a good idea to treat the system using classical mechanics. Also, (3.6) will not be a good approximation.

Using (3.6), the molecular partition function for an ideal-gas molecule becomes

$$q = \left(\frac{2\pi m L^2}{\beta h^2} \right)^{3/2}$$

and the partition function for a gas of N molecules, using (3.4), is

$$Q = \frac{1}{N!} \left(\frac{2\pi m}{\beta h^2} \right)^{3N/2} V^N \tag{3.7}$$

since $V = L^3$. The pressure is therefore

$$P = \beta^{-1} \left(\frac{\partial \ln Q}{\partial V} \right)_\beta = \frac{N}{V\beta}$$

or NkT/V. Since the pressure of an ideal gas is actually Nk_BT/V with k_B the Boltzmann constant, we can identify our constant k with k_B. One can also calculate the energy

$$E = -\left(\frac{\partial \ln Q}{\partial \beta} \right)_V = \frac{3N}{2\beta} = \frac{3NkT}{2} \tag{3.8}$$

and identify it with the ideal-gas energy of $\frac{3}{2}Nk_BT$. Again one sees k is k_B.

One can also describe the ideal gas in terms of the grand partition function Ξ. For a system of one component,

$$\Xi = \sum_{N=0}^{\infty} e^{\gamma N} Q_N = \sum_{N=0}^{\infty} e^{\gamma N} \frac{qN}{N!}$$

The sum will be recognized as the power-series expansion of $\exp(qe^\gamma)$ so that $\ln \Xi = qe^\gamma$. For an ideal gas, q is $(2\pi m/h^2\beta)^{3/2}V$ and, according to (2.47), $PV = kT \ln \Xi$. Thus

$$P = kT \left(\frac{2\pi m}{h^2 \beta} \right)^{3/2} e^\gamma \equiv Be^\gamma$$

Taking logarithms of both sides and remembering that $\gamma = \mu\beta/\mathcal{N}_A$, we have $\ln P$ equal to $\ln B + \mu\beta/\mathcal{N}_A$ or

$$\mu = \mathcal{N}_A kT(-\ln B + \ln P)$$

We know that for an ideal gas $\mu = \mu^\circ + RT \ln P$ with μ° independent of P, so that $\mathcal{N}_A kT$ must be equal to RT and k is $R/\mathcal{N}_A$, which is the Boltzmann constant.

3.3. VIBRATIONS IN A CRYSTAL

As another example of independent degrees of freedom, we consider the vibrational contribution to the thermodynamic properties of a crystalline solid. The solid is assumed to consist of N identical atoms, with the position of atom j represented by $\mathbf{r}_j$, interacting according to some potential energy function $u(\mathbf{r}_1, \mathbf{r}_2, \ldots, \mathbf{r}_N)$. The potential energy is a minimum when the atoms are arranged in a regular array, each atom then being at its equilibrium position. Because the atoms are localized in space, they must be considered as distinguishable.

For small displacements from the equilibrium positions, the potential energy may be expanded in a power series in the displacements, truncated after the quadratic terms (the linear terms vanish because we are expanding about a minimum):

$$u(\mathbf{r}_1, \mathbf{r}_2, \ldots, \mathbf{r}_N) = u_0 + \sum_{i=1}^{N} \sum_{j=1}^{N} \left[a_{ij} x_i x_j + b_{ij} y_i y_j + \cdots \right]$$

Here x_i is the x displacement of atom i from its equilibrium position and u_0 is the value of u when all atoms are in their equilibrium positions. The quantum mechanical operator for the vibrational kinetic energy of the atoms is a sum of $3N$ terms, each being a second derivative with respect to a coordinate, divided by a mass. It is always possible to choose $3N$ linear combinations of the $3N$ coordinates $\{x_i, y_i\, z_i\}$, say,

$$\xi_k = \sum_j \left(c_j^k x_j + d_j^k y_j + e_j^k z_j \right)$$

where the c_j^k, d_j^k, and e_j^k are constants, which simultaneously diagonalize the potential energy and the kinetic energy operators. This means that the Hamiltonian becomes

$$H = u_0 + \sum_{k=1}^{3N} B_k \xi_k^2 - \sum_{k=1}^{3N} C_k \frac{\partial^2}{\partial \xi_i^2} \tag{3.9}$$

where the negative sign has been used because the C_k will then be positive quantities. (Actually, there are only $3N - 6$ terms in the sums, since 6 of the $\{\xi_k\}$ correspond to rotation and translation of the crystal, but, since N is a very large number, $3N - 6$ and $3N$ are hardly different.)

Since (3.9) consists of a constant term and a sum of Hamiltonians, one for each vibrational coordinate, the $\{\xi_k\}$ are independent degrees of freedom. The eigenfunctions of H will be products and the eigenvalues of energy will be sums over the $3N$ degrees of freedom. The vibrational partition function will be

$$Q = e^{-u_0/kT} \prod_k q^{(k)} \tag{3.10}$$

where $q^{(k)}$ is the partition function for the kth vibration. Furthermore, the Hamiltonian for each vibrational coordinate is a harmonic oscillar Hamiltonian:

$$h^{(k)} = B_k \xi_k^2 - C_k \frac{\partial^2}{\partial \xi_k^2}$$

The eigenvalues for a harmonic oscillator are given by $(n + \frac{1}{2})h\nu_0$ where ν_0 is the frequency and the quantum number n takes on nonnegative integral values.

The question is: what are the vibrational coordinates and the corresponding frequencies for our crystal? Two approximate answers to this question, the Einstein and Debye models, will be considered below. First, however, we consider the thermodynamic properties of a one-dimensional harmonic oscillator with fundamental frequency ν_0.

The partition function is easily evaluated because it is an infinite geometric series, which can be summed:

$$q^{\text{vib}} = \sum_{n=0}^{\infty} e^{-(n+\frac{1}{2})h\nu_0/kT} = e^{-h\nu_0/2kT}[1 - e^{-h\nu_0/kT}]^{-1} \tag{3.11}$$

The average energy of this harmonic oscillator is then

$$E^{\text{vib}} = kT^2 \left(\frac{\partial \ln q^{\text{vib}}}{\partial T} \right)_V = \frac{h\nu_0}{2} + \frac{h\nu_0 \, e^{-h\nu_0/kT}}{1 - e^{-h\nu_0/kT}} \tag{3.12}$$

(Keeping V constant corresponds to keeping the vibrational frequency constant.) The first term, temperature independent, is the zero-point energy, which is the lowest possible energy the harmonic oscillator can have. As $T \to 0$, the second term approaches zero.

The (constant volume) heat capacity is obtained by differentiating E^{vib} with respect to T:

$$C^{\text{vib}} = h\nu_0 \frac{\partial}{\partial T} \left(\frac{1}{e^{h\nu_0/kT} - 1} \right) = k \left(\frac{h\nu_0}{kT} \right)^2 \frac{e^{h\nu_0/kT}}{(e^{h\nu_0/kT} - 1)^2} \tag{3.13}$$

The heat capacity, like the second term in the energy, approaches zero as T approaches 0. To see this, note that as $T \to 0$ the last factor becomes $\exp(-h\nu_0/kT)$, which approaches 0 more strongly than $(h\nu_0/kT)^2$ approaches infinity. We will also be interested in the behavior of the heat capacity at high temperatures. If $h\nu_0/kT$ is small enough to expand the exponentials in power series,

$$C^{\text{vib}} \to k\left(\frac{h\nu_0}{kT}\right)^2 \frac{1+\cdots}{(h\nu_0/kT+\cdots)^2} = k$$

independent of the vibrational frequency.

The simplest model for the vibrations of a crystal is the Einstein model, which pictures each atom as vibrating in a cage formed by its neighbors. Since the three spatial directions are equivalent, there are three identical (but distinguishable) one-dimensional vibrators per atom. Then (3.10) becomes

$$Q = e^{-u_0/kT}\left[q^{\text{vib}}\right]^{3N} \tag{3.14}$$

where q^{vib} is given by (3.11). The Helmholtz free energy is

$$A = -kT \ln Q = u_0 + \frac{3Nh\nu_0}{2} + 3NkT \ln(1 - e^{-h\nu_0/kT})$$

The energy is $u_0 + 3N$ times E^{vib} of (3.12). Since N is large, the contribution of the zero-point energy, $3Nh\nu_0/2$, can be appreciable.

The heat capacity in the Einstein model is

$$C^{\text{vib}} = 3Nk\left(\frac{h\nu_0}{kT}\right)^2 \frac{e^{h\nu_0/kT}}{(e^{h\nu_0/kT}-1)^2} \tag{3.15}$$

where we have used (3.13) for the heat capacity of a single oscillator. Note that the important parameter is $h\nu_0/kT$; $h\nu_0/k$, which has the dimensions of temperature, is often called the Einstein temperature. In Figure 3.2, we have plotted the measured molar heat capacities for carbon (diamond) which Einstein considered in deriving his model. The broken curve is the molar heat capacity calculated from (3.15), using 1320 K for the Einstein temperature. It is easy to see from (3.15) that $C^{\text{vib}} \to 0$ as $T \to 0$, and that $C^{\text{vib}} \to 3Nk$ as $T \to \infty$. Thus the high-temperature limit of the molar heat capacity for any crystal is $3R$ (the gas constant R is the product of the Boltzmann constant k and Avogadro's number).

That C^{vib} approached zero as $T \to 0$ was well known in the nineteenth century (when used to calculate entropies at low T, it was important to the derivation of the third law of thermodynamics), but the explanation of this

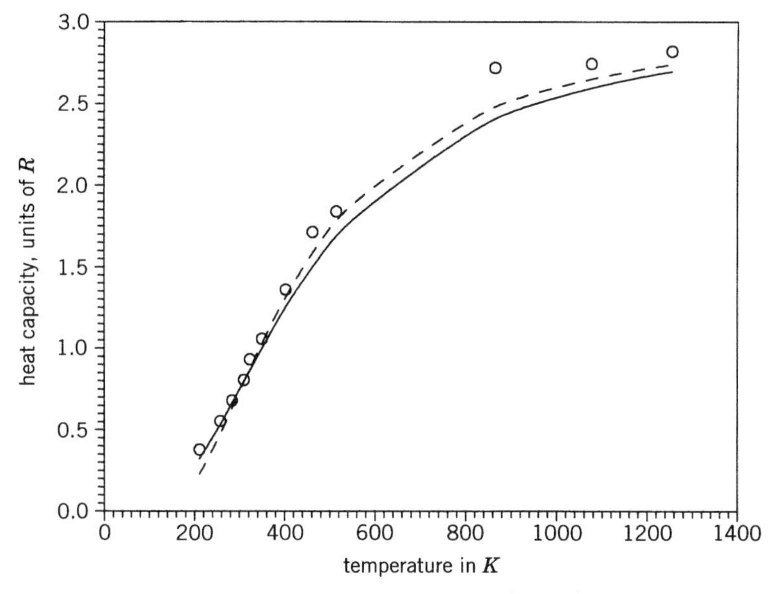

Figure 3.2. Heat capacity of diamond vs. temperature (circles), compared with predictions of Einstein model (dashed curve) and Debye model (solid curve). The Einstein temperature Θ_E is 1320 K, chosen to give the best fit to experimental data. The Debye temperature Θ_D is 1860 K, also chosen to give the best fit.

fact was a mystery. We now know it is a consequence of quantum mechanics applied to the harmonic oscillator; because the energy difference between the first excited state and the ground state is nonzero, the probability of occupying any state other than the ground state vanishes for low temperatures. The high-temperature limit of C^{vib} was also well known as the law of Dulong and Petit, who found that heat capacities of many solid elements at room temperature were about 6 cal/K. For some elements, room temperature heat capacities were significantly lower (the value for diamond was about 1.65 cal/K); the explanation of this, too, had to await quantum mechanics. Thus the Einstein model was a great advance in understanding crystals, as well as evidence of the validity of quantum mechanics. Its inadequacy was already recognized by Einstein; vibrational heat capacities of crystals are observed to become proportional to T^3 at low temperatures, which is not what the Einstein model predicts. The Debye model, an improvement on the Einstein model, corrects this error.

It is obviously an oversimplification to assume that all the vibrators have the same frequency. Also, the vibrators are not the individual atoms; instead, each vibrational coordinate is a linear combination of atomic displacements. We seek the frequency corresponding to each such linear combination.

Because the number of vibrators ($3N$) is so large, it is not convenient to list all the frequencies. Instead, we define $g(\nu)$ such that $g(\nu)\,d\nu$ is the number of vibrators with frequencies between ν and $\nu + d\nu$. The total number of vibrators is

$$\int_0^\infty g(\nu)\,d\nu = 3N \tag{3.16}$$

Instead of (3.10),

$$\ln Q = \frac{-u_0}{kT} + \sum_k \ln q^{(k)} \to \frac{-u_0}{kT} + \int_0^\infty g(\nu)\,d\nu \ln q(\nu)$$

where $q(\nu)$ is the partition function for a vibrator of frequency ν [see equation (3.11)]. Debye suggested how $g(\nu)$ might be estimated.

The vibrations are standing waves, the amplitude of the wave at the position of an atom giving the displacement of that atom. Consider first a one-dimensional crystal with n atoms, equally spaced along the x axis. The standing waves are of the form

$$W(x,t) = A\sin(2\pi k_x x)\sin(2\pi\nu t)$$

where the wavelength is $\lambda = 1/k_x$ and the frequency is ν. The product of wavelength and frequency, ν/k_x, is the wave velocity c. Let the interatomic spacing be a. To have standing waves, na must be an integral multiple of $\lambda/2$. This means $k_x na$ must be an integer or a half-integer, $k_x = n_x/(2na)$ with n_x a positive integer. In three dimensions, the wave amplitude is

$$W(\mathbf{r},t) = A\sin(2\pi\mathbf{k}\cdot\mathbf{r})\sin(2\pi\nu t)$$

where $\mathbf{k} = \{k_x, k_y, k_z\}$ with $k_x = n_x/(2na)$, $k_y = n_y/(2na)$, and $k_z = n_z/(2na)$. The wave velocity c is ν/k, where $k^2 = k_x^2 + k_y^2 + k_z^2$. Thus the possible frequencies are given by

$$\nu = c\left[k_x^2 + k_y^2 + k_z^2\right]^{1/2} = \frac{c}{2na}\left[n_x^2 + n_y^2 + n_z^2\right]^{1/2} \tag{3.17}$$

We need to calculate the number of vibrations with frequency between ν and $\nu + d\nu$.

The number of vibrations with frequency *less than* ν, according to (3.17), is the number of sets of three positive integers having the sum of the squares less than $(2\nu na/c)^2$. Imagine a three-dimensional grid with unit spacing, so that every grid point is associated with a set of three integers and with a unit cube. The volume of a sphere of radius r is numerically equal to the number of grid points within it, and each point is associated with three integers such that $n_x^2 + n_y^2 + n_z^2$ is less than r^2. We are interested in positive integers only so we consider an octant of a sphere, with volume $\pi r^3/6$. The number of sets of three positive integers having the sum of their squares less than $(2\nu na/c)^2$ is therefore

$$\frac{\pi}{6}\left(\frac{2\nu na}{c}\right)^3 = \frac{4\pi\nu^3 V}{3c^3}$$

where $V = (na)^3$ is the volume of the crystal. The number of vibrations with frequency between ν and $\nu + d\nu$ is obtained by differentiation:

$$g(\nu)\,d\nu = \frac{4\pi\nu^2 V}{c^3}\,d\nu$$

This result must now be used in the Debye model for vibrations of a crystal.

Note first that there are two kinds of standing waves, longitudinal and transverse, depending on whether the displacements of the atoms are parallel or perpendicular to **k**. There are two transverse directions, and longitudinal and transverse waves have different velocities. Therefore the above formula for $g(\nu)$ should be multiplied by 3, and some average velocity c_{av} should be substituted for c. More important is the fact that the formula cannot be correct at large frequencies. For large frequencies and small wavelengths, $W(\mathbf{r}, t)$ would have oscillations *between* the atomic positions, which are physically meaningless. There must be a maximum frequency. In fact, we know that the total number of vibrations is $3N$. Debye took this into consideration in a simple way, by writing

$$g(\nu) = \frac{12\pi\nu^2 V}{c_{\text{av}}^3}\,d\nu, \qquad \nu < \nu_{\max}$$
$$g(\nu) = 0, \qquad \nu < \nu_{\max} \tag{3.18}$$

The maximum frequency is obtained using (3.16):

$$\int_0^{\nu_{\max}} \frac{12\pi\nu^2 V}{c_{\text{av}}^3}\,d\nu = 3N$$

This gives $\nu_{\max}^3 = (3Nc_{\text{av}}^3/4\pi V)$, so that

$$g(\nu) = \frac{9N\nu^2}{\nu_{\max}^3}, \qquad \nu < \nu_{\max} \tag{3.19}$$

This may now be used in

$$\ln Q^{\text{vib}} = \frac{-u_0}{kT} + \int_0^\infty g(\nu)\, d\nu \ln q(\nu)$$

to calculate heat capacities.

The energy and the heat capacity are obtained by differentiating $\ln Q^{\text{vib}}$ with respect to T. Since only $q(\nu)$ depends on T, we may use (3.13) to obtain

$$C^{\text{vib}} = \int_0^{\nu_{\max}} \frac{9Nk\nu^2}{\nu_{\max}^3} \left(\frac{h\nu}{kT}\right)^2 \frac{e^{h\nu/kT}}{\left(e^{h\nu/kT} - 1\right)^2}\, d\nu \tag{3.20}$$

As $T \to \infty$, $e^{h\nu/kT} \to 1 + h\nu/kT$ and this becomes

$$C^{\text{vib}} = \int_0^{\nu_{\max}} \frac{9Nk\nu^2}{\nu_{\max}^3} \left(\frac{h\nu}{kT}\right)^2 \frac{1}{(h\nu/kT)^2}\, d\nu = 3Nk$$

which is the law of Dulong and Petit again. As $T \to 0$, $e^{h\nu/kT} \to \infty$ and the last fraction in (3.20) becomes $e^{-h\nu/kT}$. Since

$$\lim_{h\nu/kT \to 0} \left\{ \left(\frac{h\nu}{kT}\right)^2 e^{-h\nu/kT} \right\} = 0$$

$C^{\text{vib}} \to 0$ as $T \to 0$ (the contribution of each vibrator approaches 0).

It is usual to define the Debye temperature as

$$\Theta_{\text{D}} = \frac{h\nu_{\max}}{k} \tag{3.21}$$

In Figure 3.2, the solid curve is the calculated molar heat capacity for diamond, using $\Theta_{\text{D}} = 1860$ K. The Debye model is seen to give a better fit to the data at lower temperatures than the Einstein model. At high temperatures, the Debye model is more in error because the high-frequency vibrations, which become more important at high temperatures, are not treated correctly.

The heat capacity expression (3.20) is commonly written, putting x for $h\nu/kT$, as

$$C^{\text{vib}} = \int_0^{\Theta_{\text{D}}/T} 9Nk \left(\frac{kTx}{h}\right)^2 \frac{x^2}{\nu_{\max}^3} \frac{e^x}{(e^x - 1)^2} \frac{kT}{h}\, dx$$

$$= \frac{9Nk}{(\Theta_{\text{D}}/T)^3} \int_0^{\Theta_{\text{D}}/T} \frac{x^4 e^x}{(e^x - 1)^2}\, dx = 3Nk\text{D}(\Theta_{\text{D}}/T) \tag{3.22}$$

where D is the Debye function,

$$\mathrm{D}(u) = \frac{3}{u^3}\int_0^u \frac{x^4 e^x}{(e^x - 1)^2}\,dx \tag{3.23}$$

For high temperatures (small u) the exponentials may be expanded in power series in x so that

$$\begin{aligned}\mathrm{D}(u) &= \frac{3}{u^3}\int_0^u \frac{x^4\left(1 + x + \frac{1}{2}x^2\right)}{\left(x + \frac{1}{2}x^2 + \cdots\right)^2}\,dx \\ &= \frac{3}{u^3}\int_0^u x^2\left(1 + x + \tfrac{1}{2}x^2 \cdots\right)\left(1 - x + \frac{5x^2}{12} + \cdots\right)dx\end{aligned}$$

Integrating term by term then gives $\mathrm{D}(u) = 1 + (u^2/20) + \cdots$. Using this in (3.22) with $u = \Theta_\mathrm{D}/T$ confirms the law of Dulong and Petit for high temperatures. To consider the low-temperature (large-u) behavior, we write:

$$\begin{aligned}\mathrm{D}(u) &= \frac{3}{u^3}\left(\int_0^\infty \frac{x^4 e^x}{(e^x - 1)^2}\,dx - \int_u^\infty \frac{x^4 e^x}{(e^x - 1)^2}\,dx\right) \\ &= \frac{3}{u^3}\left(\frac{4\pi^4}{15} - \frac{u^4}{e^u - 1} - \int_u^\infty \frac{4x^3}{e^x - 1}\,dx\right) \\ &\cong \frac{4\pi^4}{5u^3} - \frac{3u}{e^u - 1} - 12e^{-u}\end{aligned}$$

Integration by parts has been used; the value of the first integral is found in tables of definite integrals. As $T \to 0$ so $u \to \infty$, the last two terms become negligible, and

$$C^{\mathrm{vib}} \to 3Nk\,\frac{4\pi^4 T^3}{5\Theta_\mathrm{D}^3} \tag{3.24}$$

so that the heat capacity is proportional to T^3 at low T.

It should be noted that in the Debye (or Einstein) model the heat capacity depends only on T/Θ_D [or T/Θ_E for the Einstein model, where $\Theta_\mathrm{E} = h\nu_0/k$; see (3.15)], which we may call the reduced temperature. Thus, if experimentally measured heat capacities for different substances are plotted against reduced temperature, they should all fall on a common curve. This is an example of a law of corresponding states: different substances having the same reduced temperatures are said to be in corresponding states.

Values for the Debye temperature Θ_D are chosen by fitting (3.22) to experimental heat-capacity data. If the data are for low temperatures, (3.24) may be used to find Θ_D, but values found in this way differ from what one obtains using (3.22) over a wide temperature range. This reminds us that the Debye model is only an approximation; $g(\nu)$ is correctly given by (3.18) only for low frequencies. Since Θ_D relates to ν_{max}, it is larger when interatomic interactions are stronger, and hence increases with the melting point of the substance. The value of 1860 K for *C* (diamond) is the highest found for any substance. The values for *Tl* and Pb, 95 K and 89 K respectively, are among the lowest. It should also be noted that, at low temperatures, the measured heat capacity has a contribution from the conduction electrons which is proportional to T (discussed in Section 5.3). Thus, at very low temperatures, it exceeds the contribution of the crystal vibrations.

For a real crystal the spectrum of vibrational frequencies, $g(\nu)$, is quite complex. For low frequencies (large wavelengths), $g(\nu)$ is proportional to ν^3, independent of the nature of the crystal, There is a cut-off or maximum frequency, but considerably higher than ν_{max}, and $g(\nu)$ drops gradually, not abruptly, to zero at the cut-off. There are several peaks in $g(\nu)$, corresponding to different modes of vibration. These details depend on the crystal structure and the interatomic bonding, and cannot be predicted by any simple model.

3.4. ANGULAR MOMENTUM

As another example of independent systems, we consider a collection of particles having angular momenta. In classical mechanics, the angular momentum **L** of a particle about the origin is a three-component vector, with magnitude equal to the product of **p**, its momentum, and **r**, its distance from the origin. It is written as a vector product: $\mathbf{L} = \mathbf{r} \times \mathbf{p}$ or, in terms of components,

$$L_x = yp_z - zp_y, \qquad L_y = zp_x - xp_z, \qquad L_z = xp_y - yp_x \tag{3.25}$$

The square of the magnitude of the angular momentum is $L^2 = L_x^2 + L_y^2 + L_z^2$. In quantum mechanics, the angular momentum operators are constructed from (3.25) by substituting position and momentum operators for the components of position and momentum. The operators for the three components of angular momentum do not commute, but instead

$$[L_x, L_y] = L_xL_y - L_yL_x = i\hbar L_z$$

and similarly for the other components, where $\hbar$ is Planck's constant divided by 2π. Then the definition of angular momentum in quantum mechanics is generalized to include any set of three operators obeying these commutation relations. The next three paragraphs review quantum mechanical angular momentum.

Because the operators do not commute, there can be no set of functions which are simultaneously eigenfunctions of all three operators. One can have functions which are eigenfunctions of L^2 and L_z (the operator for L^2 is the sum of the squares of the operators for L_x, L_y, and L_z). From the commutation properties of the operators, it can be shown that the eigenvalues of L^2 are of the form $\hbar^2 \ell(\ell + 1)$, where ℓ is a nonnegative integer or half-integer, and the eigenvalues of L_z are of the form $m\hbar$, where m ranges from $-\ell$ to ℓ in integer steps ($2\ell + 1$ values). Half-integral angular momenta do not exist in classical mechanics, and are explained as intrinsic angular momenta or spins (integral values of ℓ may arise from spin or orbital angular momenta). For example, an electron has a spin angular momentum of $\frac{1}{2}$ (as does a proton) so the possible m-values are $-\frac{1}{2}$ and $\frac{1}{2}$.

A magnetic moment is associated with the angular momentum of a particle if the particle is charged, since the movement of a charged particle constitutes a current and Faraday's law relates a circulating current to a magnetic moment. The magnetic moment is parallel to the angular momentum: $\boldsymbol{\mu} = q\mathbf{L}/2m$, where q and m are the charge and mass of the particle. The magnitude of the magnetic moment of an atomic electron with a nonzero orbital angular momentum l is $(e/2m)\hbar\sqrt{[l(l+1)]}$ since the magnitude of the angular momentum is the square root of the magnitude of L^2 and the eigenvalue of L^2 is $\hbar^2 l(l+1)$. The quantity $e\hbar/2m$ is called the Bohr magneton,

$$\mu_B = \frac{e\hbar}{2m} = 9.274 \times 10^{-24} \frac{J}{T} \tag{3.26}$$

where T = tesla. The magnetic moment associated with the electron spin is about twice what one expects from an orbital angular momentum of $\frac{1}{2}$: $\mu_S = g\mu_B \mathbf{S}$ where $\mathbf{S}$ is the spin angular momentum, and g is the free-electron g-factor, equal to 2.0023.

When there are several interacting angular momenta (spin and/or orbital), they couple together by vector addition to form angular momentum. The coupling of two angular momenta l_1 and l_2 yields several values of total angular momentum: $|l_1 - l_2|, |l_1 - l_2| + 1, \ldots, |l_1 + l_2|$. It may be verified that the total number of states is $(2l_1 + 1)(2l_2 + 1)$. For a light atom, the angular momenta of the electron spins couple together to form a total spin angular momentum $\mathbf{S}$, with an associated magnetic moment $g\mu_B\mathbf{S}$, and the orbital angular momenta of the electrons couple together to form a total orbital angular momentum $\mathbf{L}$, with an associated magnetic moment $\mu_B\mathbf{L}$. Then $\mathbf{L}$ and $\mathbf{S}$ couple together to form a total angular momentum $\mathbf{J}$. The magnetic moment associated with **J** is given by $\mu_J = g_L u_B \mathbf{J}$ where the Landé g-factor is

$$g_L = 1 + \frac{J(J+1) + S(S+1) - L(L+1)}{2J(J+1)} \tag{3.27}$$

For heavy atoms, the situation is somewhat different: the spin and orbital angular momenta for each electron couple first to form a total angular momentum **j** for that electron, and the one-electron **j**-values couple to give **J** for the atom.

We consider first a collection of identical rigid rotators, with no external field. The effect of electric and magnetic fields will be considered later. In general, a rotator is characterized by three moments of inertia. If it is a diatomic or linear molecule, the moment of inertia about the molecular axis is zero and the other two moments of inertia are equal. The rotational energy states are characterized by two angular momentum quantum numbers, l and m, where l takes on nonnegative integral values and m has any integral value between $-l$ and l. The rotational energy levels are

$$E(l,m) = \frac{\hbar^2}{2I} l(l+1) = hcBl(l+1)$$

where I is the moment of inertia and B is called the rotational constant. The units of B are cm^{-1} (wave numbers).

Since $E(l,m)$ is independent of m, the degeneracy of the lth level is $2l+1$ and the rotational partition function is

$$q_r = \sum_{l=1}^{\infty} (2l+1) e^{-hcBl(l+1)/kT} \tag{3.28}$$

(For homonuclear diatomics and symmetrical linear molecules, q_r must be divided by 2, as discussed below.) The probability of finding a rotor with energy $hcBl(l+1)$ is

$$P(l) = (2l+1) e^{-hcBl(l+1)/kT} / q_r.$$

Because the first factor increases with l and the second decreases, there is a most probable value of l, which depends on the temperature. To find it, we treat l as a continuous variable and differentiate:

$$\frac{dP(l)}{dl} = \left[2 - \frac{hc}{kT} B(2l+1)^2\right] e^{-hcBl(l+1)/kT} / q_r = 0$$

giving $l = (kT/2Bhc)^{1/2} - \frac{1}{2}$. The most likely value of l is the integer closest to this value. The maximum in $P(l)$ shows up in the intensities of lines in the rotational and rotational–vibrational spectra.

Only molecules which have permanent dipole moments can exhibit pure rotational spectra (this excludes homonuclear diatomics). In the rotational absorption or emission spectrum, the most intense lines are those for which l changes by unity. The transition frequencies (usually in the microwave or far infrared region) are equal to

$$h^{-1}[hcB(l+2)(l+1) - hcBl(l+1)] = 2Bc(l+1), \qquad l = 0, 1, 2, \ldots$$

The lines are therefore equally spaced. The intensity of a line is proportional to the population of the initial state $P(l)$ and to the Einstein coefficient, which involves the squares of the transition dipole moments averaged over initial, and summed over final m-values, and the transition frequency. The result is that the intensity of an absorption line at frequency $2Bc(l+1)$ is proportional to

$$[2Bc(l+1)]^4(2l+2)\, e^{-hcbl(l+1)/kT}$$

For an emission line in which the rotational quantum number changes from l to $l-1$ the intensity is proportional to

$$[2Bcl]^4(2l)\, e^{-hcBl(l+1)/kT}$$

The most intense line in either case is not the one starting from the most populated state (see Problem 3.20).

For rotational transitions accompanying a vibrational transition, the line intensities more closely reflect the population of the initial rotational states. This is because the transition frequency, which is the vibrational frequency plus or minus $2Bc(l+1)$, enters the above formulas in the place of $2Bc(l+1)$, and, since the vibrational frequency is much larger than $2Bc$, the transition frequency is almost independent of l.

To evaluate the partition function (3.28) in general, one must calculate individual terms until they become small. However, if hcB/kT is small, successive terms in q differ by very little and [see (3.6)] the sum may be approximated by an integral:

$$q_r \cong \int_0^{\infty} (2l+1)\, e^{-hcBl(l+1)/kT} dl = \frac{kT}{hcB} \tag{3.29}$$

The the average rotational energy is $kT^2\ (\partial \ln q/\partial T) = kT$, independent of B.

This high-temperature limit is valid when the quantity hcB/k, called the rotational temperature and denoted by θ_r, is small compared to T. For most molecules, θ_r is small; it is 2.86 K for N_2 and 0.116 K for Br_2. It is larger for lighter molecules, having the values 85.4 K for H_2, 64.0 K for HD, 42.7 K for D_2, and 30.3 K for HF. For these molecules, replacing the sum in q_r by an integral is a poor approximation for T much below room temperature, and the terms in the sum must be evaluated directly. When θ_r/T is large, however, evaluating the sum is not difficult because it converges rapidly.

For symmetrical linear molecules such as homonuclear diatomics, there is an additional factor in q_r. All rotational states are either symmetric or antisymmetric with respect to an inversion in the center, which corresponds to interchanging identical nuclei. Depending on whether the nuclei are fermions or bosons (see Section 5.1), the overall wave function of the molecule must either change sign or remain unchanged on interchange of identical nuclei, so either odd values of l or even values of l must be excluded from the sum. If the temperature is high enough for (3.29) to hold, this may be taken into account by dividing q_r by 2, the symmetry number.

We now consider the effect of an electric field on a rotating linear molecule. If the molecule is nonpolar, having no dipole moment, there is no effect. Suppose the molecule has a dipole moment μ, which must lie along the molecule axis. With no external field, the rotation averages the dipole moment along any space-fixed direction to zero. For example, the component of the dipole moment along the polar axis is

$$\frac{\int \psi_{lm}^* \cos\theta\, \psi_{lm}\, d\tau}{\int \psi_{lm}^* \psi_{lm}\, d\tau}$$

where ψ_{lm} is the eigenfunction for the rotational state characterized by the quantum numbers l and m. The integral in the numerator vanishes for all rotational eigenstates. If there is an external field which can define a preferred direction in space, it orients the rotating molecules and induces a dipole moment along the field direction. The effect on the rotational energy levels may be computed by perturbation theory.

The perturbation is the interaction energy between a dipole $\boldsymbol{\mu}$ and a field $\mathbf{E}$, $-\boldsymbol{\mu} \cdot \mathbf{E}$, or, if the field axis is the polar axis, $-\mu E \cos\theta$. Starting with the eigenfunction ψ_{JM} (J and M are now used for the rotational quantum numbers), the first-order energy is just $-\mu E$ multiplied by the expectation value of $\cos\theta$ which, as noted above, is 0. The second-order energy is

$$\frac{\mu^2 E^2}{2hBcJ(J+1)}\left[\frac{J(J+1) - 3M^2}{(2J-1)(2J+3)}\right]$$

for $J \neq 0$ and $-\mu^2 E^2/6hBc$ for $J = 0$. The M-degeneracy is partly lifted, but $\pm M$ states still have the same energies in an electric field. Since the energy is $-\boldsymbol{\mu} \cdot \mathbf{E}$, the dipole moment in the field direction may be calculated as the negative derivative of the energy with respect to E. This induced dipole moment (the dipole moment in the field direction is zero in the absence of the field) is proportional to E. It is positive for $J = 0$; for $J > 0$ it may be positive or negative, depending on the value of M.

For a collection of molecules at thermal equilibrium, the average induced dipole moment is a sum,

$$\mu_{av} = \sum_{J=1}^{\infty} \sum_{M=-J}^{J} P(J) \frac{-\mu^2 E}{hBcJ(J+1)} \left[\frac{J(J+1) - 3M^2}{(2J-1)(2J+3)} \right] \tag{3.30}$$

where $P(J)$, the occupation probability of the state (J, M), is proportional to $\exp[-hcBJ(J+1)/kT]$. Multiplying μ_{av} by the number of molecules per unit volume gives the dipole moment per unit volume, or electrical polarization, in the field direction. Dipole moments are generally given in Debyes (D); $1\ \text{D} = 10^{-18}$ esu cm $= 3.33 \times 10^{-30}$ C m. For example, the molecules ClI, HCl, and KCl have dipole moments of 0.65, 1.03, and 10.48 D, respectively.

3.5. SPINS IN MAGNETIC FIELD

In this section, we are concerned with the effect of an external magnetic field. The energy of a magnetic moment $\boldsymbol{\mu}$ in a magnetic field $\mathbf{B}$ is $-\boldsymbol{\mu} \cdot \mathbf{B}$ where $\boldsymbol{\mu}$ is proportional to angular momentum, so the expectation value over the rotational wave function ψ_{JM} is proportional to M. The magnetic moment associated with the rotating nuclei of a molecule is of size μ_n (nuclear magneton) $= e\hbar/2m_p c$ (m_p = proton mass); the contribution of the electron cloud is of the same size, as is that of the nuclear spins. (All these are much smaller than the contribution of the orbital or spin angular momentum of the electrons, if any, since the Bohr magneton is 1870 times greater than the nuclear magneton.) Writing the magnetic moment as $g_r \mu_n \sqrt{J(J+1)}$, we have for the energy of a rotator in a magnetic field (Zeeman energy) $-g_r \mu_n MB$.

More generally, we consider a collection of independent systems, each with a magnetic moment $\boldsymbol{\mu}$ associated with an angular momentum $\mathbf{J}$. In the absence of an external magnetic field, the average of $\boldsymbol{\mu}$ along any space-fixed direction is zero. The operator for the energy of one moment in a magnetic field B in the z direction is $-g \mu_m J_z B/\hbar$, with eigenvalues $-g \mu_m MB$ where $M = -J, -J+1, \ldots, J$ and μ_m is the appropriate magneton. We may refer to these systems as spins.

Let us first consider a collection of N independent spins $\frac{1}{2}$ (which could be nuclear spins of protons or magnetic ions) in a magnetic field B, at temperature T. The spins are distinguishable because they do not change places, and each has an associated magnetic moment G. Since there are two values of M, $+\frac{1}{2}$ and $-\frac{1}{2}$, there are two spin orientations: for spin "up" and magnetic moment in the direction of field, the energy is $-\frac{1}{2}GB$; for spin "down" and magnetic moment opposite to field direction, the energy is $+\frac{1}{2}GB$.

The states of the system as a whole are specified by giving the number of spins up, U. The number of spins down is $D = N - U$ and the total magnetic energy is $-\frac{1}{2}GBU + \frac{1}{2}GB(N - U)$. The magnetization I is defined as the total magnetic moment in the field direction per unit volume: $I = G(U - D)/2V$. We expect I to become 0 for $B = 0$ (no net orientation of spins), to approach $NG/2V$ for very large positive B, and to approach $-NG/2V$ for B large and negative. The work required to increase B by dB when the intensity of magnetization is I is $-IV\,dB$. Neglecting P–V work, we may write the first and second laws of thermodynamics as $dE = T\,dS - IV\,dB$ so that

$$dA = dE - T\,dS - S\,dT = -S\,dT - IV\,dB \tag{3.31}$$

Thus $IV = -(\partial A/\partial B)_{T,V}$ and $S = -(\partial A/\partial T)_{B,V}$ where the Helmholtz free energy A is $-\beta^{-1} \ln Q$.

To calculate the system partition function Q, note that the possible energies are $-\frac{1}{2}GBU + \frac{1}{2}GB(N - U)$ and the number of ways of choosing U spins to be up is $N!/[U!(N - U)!]$. Then

$$Q = \sum_{U=0}^{N} \frac{N!}{U!(N-U)!} \left[e^{-\beta\frac{1}{2}GB}\right]^{N-U} \left[e^{\beta\frac{1}{2}GB}\right]^{U}$$

where $\beta = 1/kT$. We recognize this as a binomial expansion:

$$Q = \left[e^{-\beta\frac{1}{2}GB} + e^{\beta\frac{1}{2}GB}\right]^N = \left(2\cosh \beta\tfrac{1}{2}GB\right)^N$$

[The hyperbolic cosine is defined as $\cosh x = \frac{1}{2}(e^x + e^{-x})$, the hyperbolic sine as $\sinh x = \frac{1}{2}(e^x - e^x)$.] Then $A = -\beta^{-1}N \ln(2\cosh \beta\frac{1}{2}GB)$ and

$$I = \beta^{-1}\frac{N}{V}\frac{\beta\frac{1}{2}G \sinh \beta\frac{1}{2}GB}{\cosh \beta\frac{1}{2}GB} = (N/V)\tfrac{1}{2}G \tanh \beta\tfrac{1}{2}GB$$

When $B = 0$, $I = 0$; when $B \to \infty$, $I = (N/V)\frac{1}{2}G$ (all dipoles oriented up); when $B \to -\infty$, $I = -(N/V)\frac{1}{2}G$ (all dipoles oriented down).

If x is not too large, $e^x \simeq 1 + x$, so that $\tanh x \simeq x$, and I may be approximated as $(\frac{1}{2}NG/V)(\beta\frac{1}{2}GB)$, so that $I/B = (N/V)(\frac{1}{2}G)^2/kT$. Usually, the magnetic susceptibility χ_m is defined as the ratio of the magnetization to H, where, in a nonmagnetic material, $H = B/\mu_0$ and μ_0 is the permeability of free space, $4\pi \times 10^{-7} NA^{-2}$. The susceptibility of the system of N spins of $\frac{1}{2}$ is therefore

$$\chi_m = \frac{I}{B/\mu_0} = \mu_0 \frac{N(\frac{1}{2}G)^2}{VkT}$$

The inverse proportionality of the magnetic susceptibility to the temperature is an example of Curie's law. This formula is applicable to the conduction electrons of a metal, except that one must remember that most of the conduction electrons have their spins paired. The number of thermally excited electrons with unpaired spins is small, and given by T/T_F where T_F is the Fermi temperature (see Section 5.3), usually between 10^4 K and 10^5 K. Multiplying χ_m by T/T_F shows that the magnetic susceptibility of the conduction electrons in a metal is small and temperature independent.

Evaluation of the system partition function Q for N spins of $\frac{1}{2}$ was simple because the system consists of independent spins. We could have considered the partition function q^s for a single spin and calculated Q as $(q^s)^N$. Here, $q^s = e^{\beta\frac{1}{2}GB} + e^{-\beta\frac{1}{2}GB}$, so $Q = (2\cosh\beta\frac{1}{2}GB)^N$ as before. This approach works for any spin J: there will be $2J+1$ values for M, and the energies in the presence of a field B are $-MGB$. The partition function for one spin is a geometric series:

$$q^s = \sum_{M=-J}^{+J} e^{-(-MGB\beta)} = \frac{e^{\beta(J+1)GB} - e^{-\beta JGB}}{e^{\beta GB} - 1}$$

$$= \frac{e^{\beta(J+\frac{1}{2})}GB - e^{-\beta(J-\frac{1}{2})GB}}{e^{\beta\frac{1}{2}GB} - e^{\beta\frac{1}{2}GB}} \tag{3.32}$$

Let $y = \beta GB = GB/kT$, so that

$$\ln q^s = \ln\left(\frac{\sinh\left[\left(J+\frac{1}{2}\right)y\right]}{\sinh\left[\frac{1}{2}y\right]}\right)$$

The entropy is

$$S = -\frac{\partial A}{\partial T} = \frac{\partial}{\partial T}\left[NkT\ln q^s\right]$$

$$= Nk\ln q^s + NkT\left(\frac{-GB}{kT^2}\right)\left[\left(J+\tfrac{1}{2}\right)\coth\left[\left(J+\tfrac{1}{2}\right)y\right] - \left(\tfrac{1}{2}\right)\coth\left(\tfrac{1}{2}y\right)\right].$$

We have used $(d/dx)\sinh x = \cosh x$, $(d/dx)\cosh x = \sinh x$, and $\coth x = (e^x + e^{-x})/(e^x - e^{-x})$.

Per mole, the spins' contribution to the entropy is

$$S^{\text{spin}} = R\left[\ln\left(\frac{\sinh\left[\left(J+\frac{1}{2}\right)y\right]}{\sinh\left[\frac{1}{2}y\right]}\right) - y\,\frac{2J+1}{2}\coth\left(\frac{2J+1}{2}y\right) + \frac{y}{2}\coth\left(\frac{y}{2}\right)\right] \tag{3.33}$$

To consider the behavior of S^{spin} for large fields, note that, when $x \to \infty$, $\sinh x \to \frac{1}{2}e^x$, $\cosh x \to \frac{1}{2}e^x$, and $\tanh x \to 1$. Thus if $B \to \infty$, $y \to \infty$ and the molar entropy becomes

$$R\{\ln \exp[(J + \tfrac{1}{2})y] - \ln \exp(\tfrac{1}{2}y) - y[J + \tfrac{1}{2} - \tfrac{1}{2}]\} = 0$$

The entropy is at the lowest possible value because all the spins are aligned: perfect order. For $B \to -\infty$, the calculation is identical and the entropy is again zero because all the spins are aligned. For small field, $y \to 0$, $\coth(ay) \to (ay)^{-1}$ and the molar spin entropy is

$$R\left[\ln[(J + \tfrac{1}{2})y] - \ln(\tfrac{1}{2}y) - y\left(\frac{J + \frac{1}{2}}{(J + \frac{1}{2})y} - \frac{\frac{1}{2}}{\frac{1}{2}y}\right)\right]$$

$$= R\left[\ln\left(\frac{J + \frac{1}{2}}{\frac{1}{2}}\right)\right] = R\ln(2J + 1)$$

This is the maximum possible entropy. It corresponds to an entropy of $k\ln(2J + 1)$ for each spin, because there are $2J + 1$ equally probable orientations. See (2.23) with all p_j equal to $(2J + 1)^{-1}$. The molar entropy is shown as a function of y in Figure 3.3 for $J = 2$.

These results are the basis of the method of magnetic cooling to obtain very low temperatures. At low T, of the order of 1 K, a magnetic field is applied to a collection of paramagnetic ions in a solid. The field orients the magnetic moments, reducing their molar entropy from $R\ln(2J + 1)$ to $S^{\text{spin}}(y)$ [3.33, with $y = \beta GB$] and releasing heat from the system, which is held at constant T. Then the system is demagnetized adiabatically; keeping the system insulated, the field is slowly reduced. For reversible change in an isolated system, $\Delta S = 0$. Since the molar entropy of the crystal vibrations is very small compared to that of the spins at these temperatures, this means $\Delta S^{\text{spin}}(y) = 0$. Then $y = GB/kT$ is constant, so a reduction in B should give a proportional reduction in T. If B is reduced to 0, the final temperature should be 0, which violates the third law of thermodynamics. Of course this doesn't happen.

The model of independent angular momenta J cannot be correct at very low temperatures, since it makes $S^{\text{spin}} = R\ln(2J + 1)$ at low T, and the third law requires $S \to 0$ as $T \to 0$. In a real system, two effects become important at low temperatures, leading to a nondegenerate ground state and zero entropy as $T \to 0$. They are the crystal-field splittings and the interaction between the paramagnetic ions. The crystal field at an ion position is the electric field due to surrounding charges. For example, Cr^{3+} in chrome alum, often used in magnetic cooling experiments, is in an almost octahedral environment, and has spin angular momentum $3/2$ and orbital angular momentum $L = 0$, leading to $J = 3/2$. A small trigonal distortion from

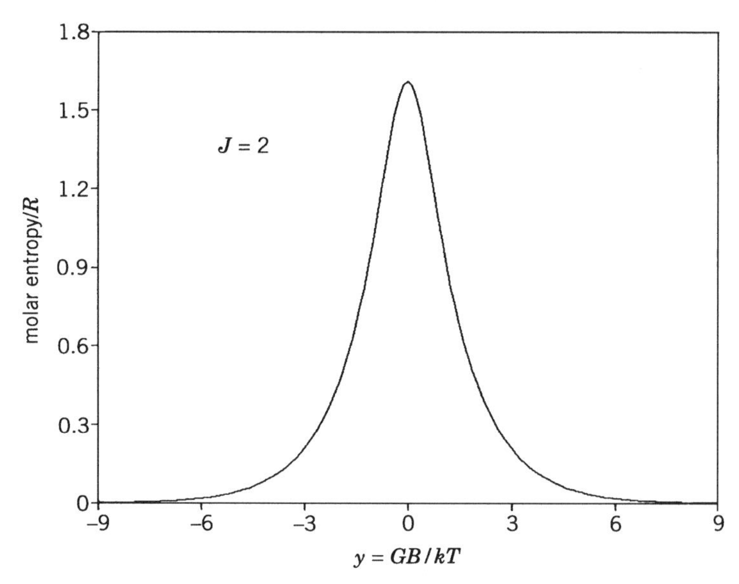

Figure 3.3. Molar entropy of spins $J = 2$ in magnetic field B. The energy levels of a magnetic dipole G are $-MGB$, with $M = -J, \ldots, J$. The maximum molar entropy, occuring at $B = 0$, is $R \ln(2J + 1) = 1.609R$.

octahedral symmetry splits the four-fold degeneracy: the states with $|M_J| = 3/2$ are $\varepsilon = 2.4 \times 10^{-17}$ ergs lower in energy than the states with $|M_J| = \frac{1}{2}$. Thus for temperatures below $\varepsilon/k = 0.17$ K, the effective degeneracy of the ground state is 2 instead of 4. The interaction between the paramagnetic ions becomes important at lower temperatures. The energy of interaction between magnetic dipoles m separated by a distance r is proportional to m^2/r^3. If m is a Bohr magneton, 0.927×10^{-20} ergs/Gauss, and $r = 5 \times 10^{-8}$ cm, $m^2/r^3 = 6.9 \times 10^{-19}$ ergs, equal to kT at $T = 0.0050$ K. At very low temperatures, interactions between the ions are important enough so one can no longer speak of independent angular momenta. The lowest-energy state of the system has the angular momenta ordered, with neighbouring spins aligned. This state is nondegenerate so that S becomes 0.

One can take the magnetic interaction between angular momenta into account approximately, while still maintaining the formalism of independent subsystems, by considering that the magnetic field at an ion is equal to the imposed field B plus the "molecular field" due to neighbouring ions. (Compare the Einstein model for crystal vibrations: the force field in which each atom vibrates is the average of the forces exerted by its neighbors.) The molecular field B_m is defined as the magnetic interaction energy of an ion with its neighbors, divided by the magnetic moment of the ion. The spin entropy of a mole of ions is $S^{\text{spin}}(y')$ where

$$y' = \frac{GJ(B + B_m)}{kT}$$

For very large B, S^{spin} becomes 0, but S^{spin} does not approach $R\ln(2J+1)$ for $B \to 0$ because $y' \to GJB_m/kT$. For an adiabatic demagnetization starting at temperature T_i in which B is reduced from B_i to 0, the constancy of S^{spin} and y' imply that $(B_i + B_m)/T_i = B_m/T_f$. Then, the final temperature T_f is given by $T_i(1 + B_i/B_m)^{-1}$. To make T_f as low as possible, one should make the molecular field as small as possible, which can be done by increasing the distance between the paramagnetic ions, but this means lowering the concentration of these ions in the crystal, which makes it a less efficient magnetic cooler.

In (3.32) and what follows, one can put, for the magnetic moment G, $g\mu_B$ where μ_B is the Bohr magneton and g is of size 1 for electrons and two thousand times smaller for nuclei. Then the Helmholtz free energy for N spins is

$$A = -\beta^{-1} N \ln q^s = -\frac{N}{\beta} \ln\left(\frac{\sinh\left[\left(J + \frac{1}{2}\right)\beta g \mu_{\mathrm{B}} B\right]}{\sinh\left[\frac{1}{2}\beta g \mu_{\mathrm{B}} B\right]}\right) \tag{3.34}$$

The magnetization is the negative derivative of A/V with respect to B at constant temperature and volume:

$$I = \frac{Ng\mu_{\mathrm{B}}}{V}\left\{\left(J + \tfrac{1}{2}\right)\coth\left[\left(J + \tfrac{1}{2}\right)\beta g \mu_{\mathrm{B}} B\right] - \tfrac{1}{2}\coth\left[\tfrac{1}{2}\beta g \mu_{\mathrm{B}} B\right]\right\}$$

This is often written as $(Ng\mu_{\mathrm{B}}J/V)B_J(y)$ where $y = \beta g \mu_{\mathrm{B}} B$ and B_J is called the Brillouin function:

$$B_J(y) = \left(\frac{2J+1}{2J}\coth\left[\left(J + \tfrac{1}{2}\right)y\right] - \frac{1}{2J}\coth\left[\tfrac{1}{2}y\right]\right) \tag{3.35}$$

The Brillouin function is plotted in Figure 3.4 for several values of J. The magnetic energy, $-IVB$, is equal to

$$E^{\text{mag}} = -(Ng\mu_{\mathrm{B}} JB) B_J(y)$$

and the magnetic heat capacity at constant volume is equal to

$$C^{\text{mag}} = \left(\frac{\partial E^{\text{mag}}}{\partial T}\right)_V = -(Ng\mu_B JB)\left(\frac{dB_J}{dy}\right)\left(\frac{dy}{dT}\right)$$

$$= \frac{Ng^2\mu_B^2 B^2}{kT^2}\left(\frac{\left(\frac{1}{2}\right)^2}{\sinh^2\left(\frac{1}{2}y\right)} - \frac{\left(J+\frac{1}{2}\right)^2}{\sinh^2\left[\left(J+\frac{1}{2}\right)y\right]}\right) \tag{3.36}$$

The Brillouin function becomes equal to $(J+1)y/3$ for small y (see Problem 3.24) so that, for small fields, the magnetization is equal to

$$I = \frac{Ng\mu_B J}{V}(J+1)\frac{\beta g\mu_B B}{3} = \frac{N}{V}\frac{g^2\mu_B^2}{3kT}J(J+1)B$$

This is the Curie law, inverse proportionality of the magnetization and the susceptibility to the absolute temperature.

The magnetization is the result of competition between the ordering effect of the magnetic field B and the disordering effect of the temperature T. The heat capacity (3.36) becomes proportional to B^2/T^2 for high temperatures or small fields (small values of y). This is because the magnetic energy changes

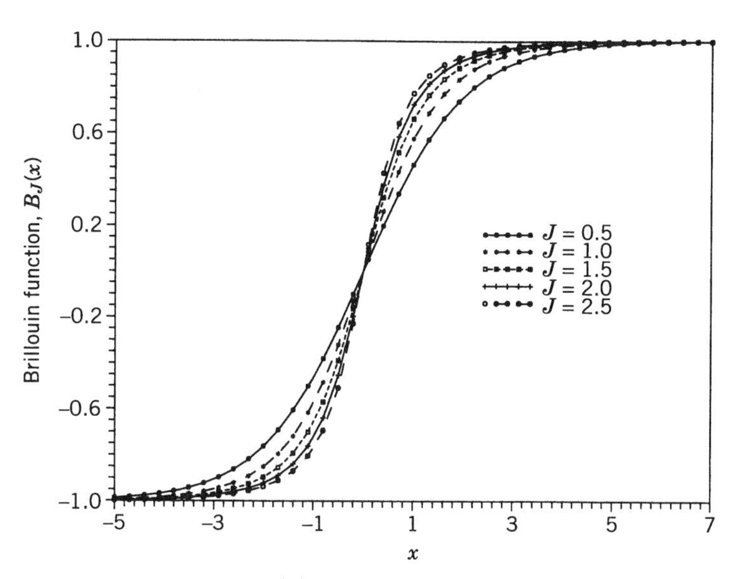

Figure 3.4. Brillouin function $B_J(x)$ vs. $x = \beta g\mu_B B$ for various J-values. Here the magnetic moment is written as $g\mu_B$ instead of G as in Figure 3.3. The magnetization I is $(Ng\mu_B J/V)B_J(x)$.

little with temperature in this regime. The heat capacity also approaches zero for low temperatures or large fields, in this case because there is not enough energy available to cause spins to flip into higher-energy states. (This is analogous to what happens to the vibrational heat capacity of a crystal at low temperatures.)

3.6. THERMODYNAMICS: THIRD LAW

Let us have another look at the laws of thermodynamics, now given a molecular interpretation by statistical mechanics. We have been considering a system of N independent and identical subsystems, held at constant temperature. The system partition function Q is equal to q^N if the subsystems are distinguishable and to $q^N/N!$ if the subsystems are indistinguishable, where q is the subsystem partition function.

$$q = \sum_j e^{-\beta \varepsilon_j}$$

where $\beta = 1/kT$, the sum is over states and the state j has energy ε_j. The probability that any subsystem is in the jth state is $p_j = e^{-\beta\varepsilon_j}/q$. The average energy of a system is therefore $N\sum_j p_j \varepsilon_j$ or

$$(N/q)\sum_j e^{-\beta\varepsilon_j}\varepsilon_j = -N\left(\frac{\partial \ln q}{\partial \beta}\right)_V = NkT^2\left(\frac{\partial \ln q}{\partial T}\right)_V$$

This is identified with the thermodynamic energy E.

According to the first law of thermodynamics for a closed system (N fixed), $dE = đq + đw$ (heat + work). In statistical mechanical terms,

$$dE = Nd\left(\sum_j p_j \varepsilon_j\right) = N\sum_j (p_j d\varepsilon_j + \varepsilon_j\, dp_j) \tag{3.37}$$

The change in ε_j is due to a change in some system parameter, like volume or magnetic field, which can be controlled from the exterior. Denoting these parameters by $\{a_r\}$, we may write

$$d\varepsilon_j = \sum_r \left(\frac{\partial \varepsilon_j}{\partial a_r}\right)_{a_s(s\neq r)} da_r$$

where, for example, $(\partial\varepsilon_j/\partial V)$ is the negative of the pressure for state j, and $(\partial\varepsilon_j/\partial B)$ is the negative of the magnetic moment for state j. The quantities $(-\partial\varepsilon_j/\partial a_r)$ may be thought of as generalized forces corresponding to the parameters a_r. This makes $\sum_j p_j\, d\varepsilon_j$ a sum of work terms $đw$.

Then, if (3.37) is the first law, $\sum_j \varepsilon_j \, dp_j$ is the heat $đq$, which is associated with changes in the distribution of systems over the energy states. According to the second law of thermodynamics, $đq = T\,dS$. Since $\ln p_j = -\beta\varepsilon_j - \ln q$, we can write

$$T\,dS = \sum_j \left(\frac{\ln p_j + \ln q}{-\beta} \right) dp_j = -kT \sum_j (\ln p_j + 1)\, dp_j$$

(note that $\sum_j dp_j = 0$) and

$$dS = -k\, d \sum_j (p_j \ln p_j)$$

We have repeated the derivation of (2.23). Note that only dS is found; S itself is known only to within an additive constant. Writing

$$S = -k \sum_j p_j \ln p_j \tag{3.38}$$

means this constant is chosen as zero. This is a reasonable choice because all the p_j are between 0 and 1, so that every term in the sum is negative or zero and S cannot be less than zero. It becomes zero if one of the p_j is unity, so that all the others vanish.

The smallest entropy possible occurs when the state of the subsystem is known; the probability of finding it in one state is unity, that of finding it in any other state is zero. This is perfect order. Perfect disorder means equal probability of finding the subsystem in any state. (Unless all the accessible states have the same energy, this can occur only if $\beta = 0$ or $T \to \infty$, since p_j is proportional to $e^{-\beta\varepsilon_j}$.) If there are n possible states, $p_j = 1/n$ for all j and S has its maximum value, $-k \ln(1/n) = k \ln n$. If there are an infinite number of states, as for the particle in a box, S has no maximum value. An example of a system with a finite number of states is an angular momentum J in a magnetic field: $n = 2J + 1$.

As the temperature approaches absolute zero, so β becomes infinite, the probability of finding a subsystem in any state with energy above the ground state energy becomes 0. Then only states with the ground-state energy contribute to the sum in (3.38). If the ground state is nondegenerate, $S = 0$; if the ground state has a degeneracy d, $S = k \ln d$. The ground state of a real chemical system is almost always degenerate because of the nuclear spins (a nuclear spin $\mathbf{N}$ has $2N + 1$ degenerate states) and the mixing of nuclei of different isotopes, as we will discuss in the next chapter; chemical energies are not affected by the mass of the nuclei of the nuclear spin states. By the same token, chemical reactions do not affect the nuclei, so that the nuclear contribution to the entropy is the same for reactants and products of a reaction. For this reason, the nuclear contribution to the entropy is conventionally ignored. Then, if there are no other sources of degeneracy, the entropy of any pure substance should approach zero as $T \to 0$.

This is the statistical mechanical derivation of the third law of thermodynamics sometimes known as Nernst's heat theorem. It was found that, for many reactions, the Gibbs free energy change approached the enthalpy change as the temperature decreased. This led Nernst to suggest that, for all reactions,

$$\lim_{T \to 0} \{\Delta_r S\} = 0$$

The reason for this, we understand, is that, as $T \to 0$, the probability of a molecule's being in any state other than the lowest-energy states becomes zero. Furthermore, the lowest-energy state is nondegenerate except for the nuclear states, and the nuclear contribution cancels off in the entropy of reaction $\Delta_r S$.

The so-called third law entropy or absolute entropy of a substance at a temperature T is obtaincd by taking its entropy at absolute zero as 0 and integrating $đq/T$ from 0 to T:

$$S(T) - S(0) = \int_0^T \frac{C_P}{T'}\, dT' \tag{3.39}$$

Measured heat capacities are used down to some low temperature, below which the Debye theory (C_P proportional to T^3; see Section 3.3) is used to extrapolate. Also, if the substance undergoes a phase change at temperature $T_x < T$, contributions $\Delta H_x / T_x$ must be added to the integral. Note that, for the integral to converge, the heat capacity must approach 0 at low temperatures, as we know is the case for the contributions of crystal vibrations (C_P proportional to T^3) and conduction electrons (C_P proportional to T).

Third law entropies agree with entropies calculated from molecular partition functions (see Chapter 4) with a few exceptions. These occur for molecular crystals in which the molecules may have several different orientations of almost equal energy. For example, CO may be oriented as CO or as OC in its crystal, and the two ends of the molecule are so similar that the difference in energy between the two orientations is comparable to kT only at very low T. At such temperatures, the rate at which the molecules can reorient to give the lowest-energy arrangement is negligibly slow. Since each molecule has two possible orientations, a crystal of N molecules has 2^N almost-degenerate states and a "residual entropy" of $k \ln(2^N)$ is locked in at low temperatures. Per mole, $S(0) = R \ln 2 = 0.693R$. In fact, the third law entropy for CO at 300 K, calculated assuming $S(T = 0) = 0$, is less than the statistical mechanical entropy (calculated in Chapter 4) by 4.6 J $= 0.56R$. For NO, the difference is significantly less, only $0.37R$. The explanation for this is that NO in the solid state consists of dimers $(NO)_2$, which can either have two possible orientations or exist in two isomeric forms. Per mole of NO, the entropy at low T is therefore $\frac{1}{2}R \ln 2 = 0.35R$.

For H_2O there appears to be a residual entropy of $0.42R$ at $T = 0$, and for D_2O a residual entropy of $0.37R$. The explanation is again in terms of multiple orientations of the molecules, of almost equal energy. Each molecule in the crystal is surrounded by four others, arranged tetrahedrally. There is one hydrogen on each O—O line, twice as far from one O as from the other (to which it is bonded), so that each O has two near (bonded) hydrogens. The O—H bonds of an H_2O molecule point in two of the tetrahedral directions, and the lone pairs point in the other two, where they form hydrogen bonds with neighboring H_2O's. As the two bonds (and the two lone pairs) are indistinguishable, there would be six possible orientations for an H_2O or D_2O molecule in the crystal, except that an O—H bond cannot point in a direction in which there already is an H from a neighboring molecule. Since the probability of this is $\frac{1}{2}$, the number of orientations is reduced to $6(\frac{1}{2})^2$. The residual entropy is then $R \ln(3/2) = 0.405R$, which agrees with the experimental results.

H_2 and D_2 also show residual entropies, $0.77R$ and $0.36R$, respectively. The explanation for this involves the nuclear spins. H has spin $\frac{1}{2}$, so the nuclear spin quantum number for H_2 can be 1 or 0, referred to as ortho- and para-hydrogen. A nuclear spin angular momentum of 1 has three degenerate states and a nuclear spin angular momentum of 0 only one, so, at high temperatures, H_2 is $\frac{3}{4}$ ortho and $\frac{1}{4}$ para. The nuclear-spin states for ortho-H_2 are symmetric to an exchange of protons, and those for para-H_2 antisymmetric. H being a fermion (see chapter 5.1) the overall wave function for H_2 must be antisymmetric to exchange. Since rotational states of odd(even) J are antisymmetric(symmetric) to exchange, ortho-H_2 has only odd-J rotational states, and para-H_2 only even-J rotational states. In the absence of a suitable catalyst, transitions between nuclear-spin states occur very slowly, so, if H_2 is cooled, it remains $\frac{3}{4}$ ortho and $\frac{1}{4}$ para. For low enough temperatures, the ortho-H_2 is in its lowest rotational state, $J = 1$, and the para-H_2 in *its* lowest rotational state, $J = 0$. Instead of being in a nondegenerate ground state and having no entropy, H_2 would have a molar entropy of

$$\tfrac{3}{4}R \ln 3 + \tfrac{1}{4}R \ln 1 = R \ln 3^{3/4}$$

or $0.824R$, which agrees with what is observed. The situation for D_2 is similar, but a little more complicated (see Problem 3.28).

The residual entropy in the cases discussed above is a consequence of the system's not being in thermal equilibrium at T approaching 0, because the rate at which molecules can change their states becomes very slow at low temperatures. As the system is cooled, the energy that would be released to the surroundings as molecules go from higher-energy to lower-energy states is not released. Then the observed heat capacity is smaller than the true heat capacity, which is for a system remaining in thermal equilibrium. If one measures $S(0)$ by using (3.39), starting with a value for $S(T)$ determined from entropies of reaction or calculated from partition functions, and using the observed heat capacity, one will find a value higher than 0 for $S(0)$.

PROBLEMS

3.1. In a system of N identical indistinguishable particles, the particles suddenly become distinguishable by having numbers painted on their backs. By how much does the energy change? By how much does the entropy change?

3.2. Evaluating the free-particle partition function by an integral, as in (3.6), is a good approximation when the de Brogle wave length, $\Lambda = \sqrt{h^2\beta/2\pi m}$, is small compared to the length of the box in which the particles are contained. Calculate Λ for H atoms at 10 K, 100 K, and 1000 K, assuming $\beta = 1/kT$ with k the Boltzmann constant. Calculate Λ for O_2 molecules at the same three temperatures.

3.3. For H atoms at the lowest temperature of Problem 3.2 (10 K) and a box $10A$ (10^{-7} cm) in length, evaluate the first ten terms of the partition function q. Compare the sum with what one gets by approximating it as an integral,

$$q = \left(\frac{2\pi mL^2}{\beta h^2}\right)^{3/2}$$

with $\beta = 1/kT$ and k = Boltzmann constant.

3.4. The partition function for a particle in a cubic box of side L is

$$q = \sum_{ijk} e^{-\beta h^2(i^2+j^2+k^2)/8mL^2}$$

where the energy levels are $h^2(i^2 + j^2 + k^2)/8mL^2$ and i, j, and k are integers. If the energy levels are closely spaced, this can be approximated as an integral. In Section 3.3 it was shown that the number of sets of three positive integers with sum of squares less than n^2 is $\pi n^3/6$. Thus the number of states with energy less than ε is $(\pi/6)(\varepsilon 8mL^2/h^2)^{3/2}$ and the number of states with energy between ε and $\varepsilon + d\varepsilon$ is

$$g(\varepsilon)d\varepsilon = \left(\frac{\pi}{4}\right)\left(\frac{8mL^2}{h^2}\right)^{3/2} \varepsilon^{1/2}\, d\varepsilon.$$

Now the partition function is

$$q = \left(\frac{\pi}{4}\right)\left(\frac{8mL^2}{h^2}\right)^{3/2} \int_0^{\infty} \varepsilon^{1/2}\, d\varepsilon\, e^{-\beta\varepsilon}$$

Evaluate this and compare the result to the cube of expression (3.6).

3.5. To compare the Einstein theory, with a single vibrational frequency ν_0, to the Debye theory, with a range of frequencies, one might interpret ν_0 as the average frequency of the Debye theory. Calculate this average frequency in terms of ν_{max} of the Debye theory. Express the Einstein temperature, defined as $\Theta_E = h\nu_0/k$, in terms of Θ_D. (To check your result, note that values of Θ_E, from fitting heat capacity curves, are 1364 K and 769 K for diamond and beryllium, corresponding values of Θ_D are 1860 K and 1160 K.)

3.6. The low-temperature form of the Debye heat capacity, equation (3.24), is considered to be a good approximation for $T < \sim 0.1\Theta_D$. We wrote C^{vib} as $3R\mathrm{D}(u)$ with $u = \Theta_D/T$ and approximated the Debye function at low T as

$$\mathrm{D}(u) = \frac{4\pi^4}{5u^3} - \frac{3u}{e^u - 1} - 12\, e^{-u}$$

What is the error in neglecting the second and third terms for $T = 0.1\Theta_D$?

3.7. At very low temperatures, the conduction electrons' contribution to the molar heat capacity is given by γT, where $\gamma = 37.6 \times 10^{-5} R/\mathrm{K}$ for Pb and $80.0 \times 10^{-5} R/\mathrm{K}$ for Pt. For these metals, $\Theta_D = 89$ K and 240 K, respectively. At what temperature, for each metal, does the electronic contribution to the heat capacity become equal to C^{vib}? (Below this temperature, the electronic contribution is larger than the vibrational.)

3.8. The molar heat capacity of aluminum oxide at 10 K is 0.0094 J/K. According to the Debye theory, what is the molar heat capacity at 5 K? What is the Debye temperature? [Use (3.24) instead of the full formula for the Debye heat capacity.]

3.9. Calculate the pressure of a crystal according to the Einstein model, assuming that only ν_0 depends on V, and show it is always positive. Note that an increase in atomic volume, V/N, "loosens up" a crystal and hence decreases ν_0.

3.10. Consider N identical but distinguishable one-dimensional harmonic oscillators, each with $q = e^{-\frac{1}{2}x}/(1 - e^{-x})$, $x = h\nu/kT$ [equation (3.11)].

The number of oscillators in state i with energy $\varepsilon_i = (i + \frac{1}{2})h\nu$, obeys

$$\ln(n_i) = \ln(N) - \varepsilon_i/kT - \ln q$$

and the average energy is $kT^2(\partial \ln q/\partial T)_V$. Show, by differentiating $\ln(n_i/N)$, that an increase in temperature decreases the fractional population n_i/N for states with energies below the average energy and increases n_i/N for states with energies above the average energy. Show that this holds for any collection of N identical but distinguishable particles, regardless of whether they are harmonic oscillators, as long as the above equation for $\ln(n_i)$ is valid.

3.11. Evaluate the rotational partition function (3.28) to an accuracy of 0.0001 by evaluating individual terms and summing, for $T = \theta_r$, where the "rotational temperature" θ_r is defined as hcB/k. Compare with the high-temperature limit, $q_r = kT/hcB$. Repeat the calculation for $T = 2\theta_r$. (*Answer:* at $T = 2\theta_r q_r = 1.05498$ and high-T limit is 2.0.)

3.12. For CO, the rotational temperature (see Problem 3.11) is 2.77 K. What is the moment of inertia of the molecule ($h/8\pi^2 I = Bc$)? What is the average rotational energy at 300 K? What is the partition function and average rotational energy at 3 K? (*Answer:* at 3 K, $q_r = 1.493027$ and $E_{\text{rot}} = 0.983966kT$.)

3.13. The rotational temperature θ_r is 15.37 K for HCl. What l-level has the highest population at temperatures of 100 K, 200 K, 300 K, 500 K, and 1000 K?

3.14. Calculate the rotational entropy of a mole of CO molecules at 300 K and at 3 K (rotational temperature of CO = 2.77 K). (*Answer:* At 3 K, $S = (E - A)/T = 1.384772$ K.)

3.15. For a homonuclear diatomic at high enough T for (3.29) to be valid, the rotational partition function must be divided by the symmetry number $\sigma = 2$. What is the effect of this factor on the energy? What is the effect of this factor on the entropy?

3.16. The rotational temperatures of N_2 and CO are, respectively, 2.86 K and 2.77 K. At 298 K, the standard molar entropies are 191.5 and 197.9 J/K. What would the difference in molar entropies be if (3.29) were used for the rotational partition functions of both molecules? Why is the difference so much greater in reality?

3.17. The moment of inertia of a diatomic molecule is given by $I = m_r R^2$ where m_r, the reduced mass, is the product of the atomic masses divided by their sum. The internuclear distance for O_2 is 1.207×10^{-10} m and the mass of an oxygen atom is 2.66×10^{-26} kg. Calculate the molar rotational entropy and energy at 300 K. Note $h/8\pi^2 I = Bc$.

3.18. The molecule HBr has a rotational constant of 8.473 cm^{-1} and a permanent dipole moment of 0.78 D. When the molecule in its lowest rotational state ($J = 0$) is placed in an electric field of 10^4 V/m at $T = 300$ K, what is the induced dipole moment in the field direction? What is the induced moment for each of the three states with $J = 1$?

3.19. What is the average induced dipole moment for HBr (see preceding problem) in an electric field of 10^4 V/m at $T = 10$ K? Assume that the probability of finding the molecule in a rotational eigenstate characterized by quantum numbers J and M is given by the Boltzmann factor, $e^{-\varepsilon(J,M)/kT}$, where ε is the eigenvalue of rotational energy, and consider all states with fractional populations above 1%.

3.20. The most populated rotational level is for $l = (kT/2Bhc)^{1/2} - \frac{1}{2}$. The intensity of an absorption line in the pure rotational spectrum is proportional to

$$2Bc(l+1)(2l+2)\, e^{-hcBl(l+1)/kT}$$

Show that the most intense line occurs for l approximately $(kT/Bhc)^{1/2} - \frac{3}{4}$. For an emission line, in which the rotational quantum number changes from l to $l-1$, the intensity is proportional to $[2Bc(l+1)]^4(2l)e^{-hcBl(l+1)/kT}$. Show the most intense line is for l approximately equal to $(5kT/2Bhc)^{1/2} - \frac{1}{2}$.

3.21. A system of spins $J = 2$ is to be used for magnetic cooling of a crystal lattice at 20 K. There is one magnetic ion per 20 atoms in the crystal. After complete magnetization, the system is insulated and the magnetic field slowly reduced to zero, increasing the spin entropy from 0 to $0.005R\ln(2J+1)$ and decreasing the vibrational entropy of the crystal by $0.005R\ln(2J+1)$. At these temperatures, the molar (vibrational) heat capacity of the crystal is given by $3R(4\pi^4T^3)/5\Theta_D^3$ where Θ_D, the Debye temperature, is 180 K. Remembering that $\Delta S = \int đq/T$, calculate the final temperature of the system.

3.22. In the preceding problem, one gets a ridiculous answer if a more reasonable (i.e., higher) value is used for Θ_D. This is because crystal fields and interionic magnetic interactions, which split the degeneracy, have been neglected.

(a) If $\Theta_D = 250$ K and all other parameters have the same values as in Problem 3.21, calculate the final temperature of the system. Now suppose that the crystal fields separate the energy levels so that, at low temperatures, only one energy level, with degeneracy 2, is occupied. Calculate the final temperature of the system in this case. (*Answer:* -20.11 K, $+10.17$ K.)

(b) To do the calculation of (a) more correctly, suppose the crystal fields split the $(2J + 1)$-fold degenerate energy level of an ion into a doubly degenerate ground level and a triply degenerate level, $\varepsilon = 6 \times 10^{-16}$ ergs above the ground level, so the ionic partition function is

$$q = 2 + 3\,e^{-\varepsilon/kT}$$

Calculate the spin entropy of 0.05 moles of these ions at temperature T in the absence of a magnetic field. Calculate the temperature T after adiabatic demagnetization at 20 K, by setting the negative of the spin entropy at T in the absence of a field equal to the change in the entropy of the crystal vibrations in going from 20 K to T. (*Answer:* 1.01 K.)

3.23. Derive equation (3.36) for the magnetic heat capacity.

3.24. Show that for small fields B the Brillouin function $B_J(\beta g \mu_B B)$, equation (3.35), is linear in $\beta g \mu_B B$, so that the magnetization, $I = (N g \mu_B J/V) B_J(\beta g \mu_B B)$, is inversely proportional to T (Curie's law). Show that for small fields C^{mag} approaches

$$\frac{N g^2 \mu_B^2 B^2 J(J+1)}{3kT^2}$$

Note that in expanding the hyperbolic functions in power series, one must be consistent about how many terms one keeps in numerator and denominator.

3.25. The heat capacity, equation (3.36) with $y = g\mu_B B/kT$, approaches zero for large y and for small y. Therefore it goes through a maximum. Show that the maximum value of the molar heat capacity is about $0.44R$ for $J = 0.5$ and about $0.64R$ for $J = 1$. For what value of y does the maximum occur in each case?

3.26. The mean magnetic moment of an atomic ion, along the magnetic field, is given by $(g\mu_B J)B_J(y)$ where $y = Bg\mu_B B$ and $\mu_B = 0.927 \times 10^{-20}$ ergs Gauss. In alums, the ions Cr^{3+} and Fe^{3+} are magnetic because of their unpaired electron spins. Their angular momenta are $J = 3/2$ and $5/2$ respectively, and, since the moment is spin-only, $g = 2$. Calculate the mean magnetic moment for each ion at temperatures of 0.5, 1, 1.5, 2, and 3 K for a field of 10^4 Gauss, equivalent to 1 T (Tesla). What fractions of the maximum possible magnetic moment do these represent? (*Answer:* for Cr^{3+} at 0.5 K and 3 K, moment is $2.854\mu_B$ and $1.060\mu_B$, 0.951 and 0.351 of the maximum possible.)

3.27. For the following substances, the third law entropies are smaller than the entropies obtained calorimetrically or by calculation, by the amounts indicated. Estimate the number of almost-equivalent orientations of each molecule in its crystal.

CO	$0.56R$	N_2O	$0.58R$
CH_3	$1.39R$	NO	0.37R

3.28. Deuterium has a nuclear spin of 1, so the nuclear spin for D_2 can be $N = 0$, 1, or 2. With respect to an exchange of nuclei, the nuclear spin wave functions for these are symmetric, antisymmetric, and symmetric respectively. Because the overall wave function for D_2 must be symmetric (D is a boson), $N = 0$ and $N = 2$ go with even values of J and $N = 1$ goes with odd values of J, the rotational quantum number. Remembering that a nuclear spin N implies $2N + 1$ degenerate states, calculate the fraction of D_2 molecules which are in states with even values of J at high temperatures. Assuming that this fraction is preserved as D_2 is cooled down, calculate the residual molar entropy of D_2 at low temperatures. The experimental value is $0.36R$.

CHAPTER 4

ATOMS AND MOLECULES

For chemists, the most important example of a system consisting of independent subsystems is probably a gas of atoms or molecules. The system partition function Q is $q^N/N!$ where q is the partition function for one atom or molecule, $q = \Sigma_i e^{-\varepsilon_i}$ where ε_i is now the energy of the ith atomic or molecular state. The factor of $N!$ comes from the indistinguishability of the subsystems. The wave functions for the atomic or molecular states involve the coordinates of the nuclei and the electrons. Because the motions of nuclei and electrons are treated differently in quantum chemistry (see Section 4.2), an atom (with a single nucleus) is simpler to deal with than a molecule, and we will consider atoms first.

4.1. ATOMS

An atomic Hamiltonian (and a molecular Hamiltonian too) includes kinetic energy operators for all the particles and a potential energy operator which involves the relative coordinates of the particles. This means that, rewritten in terms of the coordinates of the center of mass and internal coordinates, the Hamiltonian will be a sum of the center-of-mass Hamiltonian, with no potential energy operator, and the internal Hamiltonian. All states will be products of center-of-mass states and internal states, the former being free-particle states, so that

$$q = q^{\text{trans}} q^{\text{int}} \tag{4.1}$$

where q^{trans} is the free-particle partition function derived in Chapter 3, equations (3.5) to (3.7):

$$q^{\text{trans}} = \Lambda^{-3}V, \qquad \Lambda = \sqrt{\frac{h^2\beta}{2\pi m}} \tag{4.2}$$

Here, m is the mass of the atom (or molecule) and V the volume in which the gas is contained. In the system partition function, the factor of $N!$ is associated with q^{trans} because it is the translational motion that gives rise to the indistinguishability of the atoms. Therefore

$$Q = \frac{(\Lambda^{-3}V)^N}{N!}(q^{\text{int}})^N$$

where q^{int} is a sum over internal states.

A further separation is possible. To a very high accuracy, the electronic states are independent of the nuclear states, so that q^{int} is itself a product:

$$q^{\text{int}} = q^{\text{en}}q^{\text{nuc}} \tag{4.3}$$

[In (4.3), "en" stands for "extranuclear."] The electronic states which appear in q^{en} involve electronic spins and electronic coordinates relative to the center of mass. In q^{nuc}, the nuclear states involves nuclear coordinates (i.e., coordinates of the components of the nucleus) only. The separation (4.3) is valid for molecules as well as atoms, except for special cases in which one has to consider symmetry to exchange of identical nuclei. This was mentioned in Chapter 3 [discussion after (3.29) and last few paragraphs of Section 3.2], and will be returned to later in this chapter.

The nuclear partition function is quite simple. In the sum over nuclear states,

$$q^{\text{nuc}} = \sum_k e^{-\beta\varepsilon_k} = \sum_k e^{-\varepsilon_k/kT}$$

there are generally very few terms because differences between nuclear energy levels are very large compared to kT. Only the lowest energy level need be considered and $q^{\text{nuc}} = g_N$, where g_N is the degeneracy of this level. The degeneracy contributes $R \ln g_N$ to the molar entropy and nothing to the energy. Furthermore, the nuclear energy levels will be the same whether the nucleus is in an atom or a molecule. The factor g_N will be present in the partition function of any atom or molecule containing this nucleus, and the contribution $R \ln g_N$ to the entropy likewise. In chemistry, we are interested in changes in thermodynamic functions, and nuclei are not changed in chemical reactions. The nuclear entropy contributions will always cancel out (except for the special cases alluded to previously). The convention, which we adopt, is to leave out the nuclear factor in the partition function, thus neglecting nuclear-spin contributions to the entropy.

Now the partition function for an atom is

$$q^{\text{en}} = \sum_j e^{-\beta\varepsilon_j}$$

where the sum is over electronic states, and the energy of the lowest electronic state is taken as zero. The spacing of electronic levels is of the order of electronvolts, where 1 eV $= 1.6 \times 10^{-19}$ J. At $T = 1000$ K, $\beta = (kT)^{-1} = 7.2 \times 10^{19}$ J^{-1}, so $e^{-\beta\varepsilon_j} = 10^{-5}$ if $\varepsilon_j = 1$ eV and less for lower T or higher ε_j. Therefore, for most atoms under normal conditions, the sum in q^{elec} is limited to a few terms because $e^{-\beta\varepsilon_j}$ is small compared to unity.

Usually, the atomic (electronic) partition function is written as a sum over energy levels instead of states:

$$q^{\text{en}} = \sum_k g_k e^{-\beta\varepsilon_k} \tag{4.4}$$

where g_k is the multiplicity (number of degenerate states) of the kth level. The multiplicity is determined by the total angular momentum $\mathbf{J}$ which, for the lighter atoms, is the vector sum of the spin and orbital angular momenta $\mathbf{S}$ and $\mathbf{L}$ [see Section 3.4, equation (3.27)]. For example, the ground state of Li, with electronic configuration $1s^2 2s^1$, has $S = \frac{1}{2}$ and $L = 0$, so $J = \frac{1}{2}$. The first excited electronic configuration, $1s^2 2p^1$, has $S = \frac{1}{2}$ and $L = 1$, so J can be $\frac{1}{2}$ or $\frac{3}{2}$. These terms lie 2.96×10^{-19} J above the ground state, with the $J = \frac{3}{2}$ term only 6.8×10^{-24} J above the other. Since terms from other excited configurations are much higher in energy, (4.4) reduces to

$$q^{\text{en}} = 2 + 6\, e^{-\beta(2.96 \times 10^{-19}\text{J})}$$

unless the temperature is well above 10^4 K.

If several levels result from the ground state electronic configuration, there will be low-lying excited states. Thus the lowest energy configuration of O, $1s^2 2s^2 2p^4$, gives rise to three terms: ^3P ($S = 1$, $L = 1$), ^1D ($S = 0$, $L = 2$), and ^1S ($S = 0$, $L = 0$). The levels from the ^3P term lie lowest, and the lowest of these is for $J = 2$. The next, for $J = 1$, is at 3.15×10^{-21} J; the term for $J = 0$ is at 4.50×10^{-21} J. The next lowest energy level is from the ^1D term, and it is at 2.03×10^{-19} J. Then

$$q^{\text{en}} = 5 + 3\, e^{-\beta(3.15 \times 10^{-21}\text{J})} + e^{-\beta(4.50 \times 10^{-21}\text{J})}$$

is adequate for temperatures below a few thousand degrees.

The partition function for a gas of N identical atoms is now

$$Q = \frac{(\Lambda^{-3}V)^N}{N!}(q^{\text{en}})^N$$

where q^{en} is given by (4.4), Λ is defined in (4.2), and the nuclear-spin factor has been removed. The energy is

$$E = -\left(\frac{\partial \ln Q}{\partial \beta}\right)_V = \tfrac{3}{2}NkT + N\frac{\sum_k g_k \varepsilon_k e^{-\beta\varepsilon_k}}{\sum_k g_k e^{-\beta\varepsilon_k}}$$

The electronic contribution is generally small compared to the translational contribution. The pressure is

$$P = \beta^{-1}\left(\frac{\partial \ln Q}{\partial V}\right)_\beta = \frac{NkT}{V}$$

since electronic energy levels are independent of V. The entropy is

$$S = \frac{E - A}{T} = \tfrac{3}{2}Nk + \frac{3Nk}{2}\ln\left(\frac{2\pi mkT}{h^2}\right) + k\ln\frac{V^N}{N!}$$
$$+ \frac{N}{T}\frac{\sum_k g_k \varepsilon_k e^{-\beta\varepsilon_k}}{\sum_k g_k e^{-\beta\varepsilon_k}} + Nk\ln\left(\sum_k g_k e^{-\beta\varepsilon_k}\right) \quad (4.5)$$

since $A = -kT\ln Q$. Equation (4.5) is sometimes called the Sackur–Tetrode equation. The chemical potential is the derivative of A with respect to number of moles n, where $n = N/\mathcal{N}_A$ and $\mathcal{N}_A$ is Avogadro's number.

$$\mu = \left(\frac{\partial A}{\partial n}\right)_{V,\beta} = \mathcal{N}_A\left(\frac{\partial A}{\partial N}\right)_{V,\beta} = \mathcal{N}_A\left\{\frac{-3kT}{2}\ln\left(\frac{2\pi mkT}{h^2}\right)\right.$$
$$\left. -kT\ln V + kT\ln N - kT\ln\left(\sum_k g_k e^{-\beta\varepsilon_k}\right)\right\} \quad (4.6)$$

The standard state of an ideal gas at temperature T is for $P = NkT/V = 1$ atm, so that

$$\mu = \frac{-3RT}{2}\ln\left(\frac{2\pi mkT}{h^2}\right) + RT\ln\frac{P}{kT} - RT\ln\left(\sum_k g_k e^{-\beta\varepsilon_k}\right)$$
$$= \mu^0 + RT\ln P \quad \text{(in atmospheres)} \quad (4.7)$$

where the chemical potential in the standard state, μ^0, is μ for $P = 1$ atm. (The gas constant R is $k\mathcal{N}_A$.)

4.2. MOLECULES: INTERNAL MOTIONS

The separation of center-of-mass motion goes through for molecules just as for atoms, leading to (4.1). The separation of internal nuclear states from other internal motions also goes through, leading to (4.3). For a molecule, q^{nuc} is the product of partition functions for the individual nuclei, each of which reduces to a degeneracy factor g_N. This makes no contribution to the energy, and contributes $R \ln \prod_N g_N = R \sum_N \ln g_N$ to the entropy. Since internal nuclear states are not affected by chemistry, this contribution would cancel off in any chemical process, so, as mentioned in Section 4.1, we follow convention by dropping it.

We now have to deal with q^{en}, which involves the extranuclear states connected with motions of nuclei and electrons around the center of mass. One may now think of "en" as meaning "electron-nuclear" rather than "extranuclear." In the usual quantum mechanical treatment of molecules, the Born–Oppenheimer separation, electronic motions are treated differently from nuclear motions. Because the electrons are thousands of times lighter than the nuclei, it is a good approximation to drop the nuclear kinetic energy operator from the Hamiltonian, leaving to be solved an electronic Schrödinger equation for fixed nuclei. If the positions of the nuclei are denoted by **R** and electronic coordinates by **r**, this Schrödinger equation is

$$H^{\text{BO}}\psi_i(\mathbf{r};\mathbf{R}) = E_i(\mathbf{R})\,\psi_i(\mathbf{r};\mathbf{R}) \tag{4.8}$$

The Born–Oppenheimer Hamiltonian includes kinetic energy operators for the electrons but not for the nuclei. In addition, it includes all electrostatic interactions between electrons and nuclei; internuclear repulsions, as far as the electrons are concerned, are just constant terms.

In (4.8) the Hamiltonian is different for different nuclear positions. It depends on nuclear positions parametrically (as it depends on nuclear charges); to emphasize this parametric dependence, we have put **R** after a semicolon. For each choice of nuclear positions, one gets a set of electronic eigenfunctions $\psi_i(\mathbf{r})$ and eigenvalues E_i, so the eigenfunctions and eigenvalues are functions of the parameters **R**. The next step in the treatment is to combine $E_i(\mathbf{R})$ with the nuclear kinetic energy operators to form the Schrödinger equation for nuclear motion:

$$\left[T^{\text{nuc}} + E_i(\mathbf{R})\right] \chi_j^{(i)}(\mathbf{R}) = E_j^{(i)}\chi_j^{(i)}(\mathbf{R}) \tag{4.9}$$

The electronic energy acts as the potential energy for motion of the nuclei. The nuclear-motion eigenfunctions and eigenvalues depend on which electronic state was used in (4.9); hence the double indexing on $\chi_j^{(i)}(\mathbf{R})$ and $E_j^{(i)}$.

Within the framework of the Born–Oppenheimer approximation, the wave functions are $\Psi_{i,j}(\mathbf{r}, \mathbf{R}) = \psi_i(\mathbf{r}; \mathbf{R})\chi_j^{(i)}(\mathbf{R})$ and the associated energies are $E_j^{(i)}$. It should be emphasized that $\Psi_{i,j}$ is not an eigenfunction of the molecular Hamiltonian, and $E_j^{(i)}$ is not the expectation value of this Hamiltonian over $\Psi_{i,j}$, but only an approximation to an energy level of the molecule. The partition function is, so far,

$$q^{\text{en}} = \sum_i \sum_j^{(i)} e^{-\beta E_j^{(i)}}$$

where $E_j^{(i)}$ for all j is at least as great as the minimum value of $E_i(\mathbf{R})$. The spacings between the electronic energy levels of molecules are comparable to those in atoms. Thus the sum over i in q^{en} is limited to a few terms.

A further separation is possible in the nuclear coordinates. To specify the positions of the n nuclei in a molecule requires $3n$ coordinates. One can form $3n$ linear combinations of these coordinates such that three of them give the position of the center of mass and the others are relative coordinates. Thus, the nuclear position coordinates $\mathbf{R}$ in (4.9) are $3n - 3$ linear combinations of displacements of individual nuclei. Combinations of these may be chosen so that they include two (for linear molecules) or three (for nonlinear molecules) coordinates which define the orientation of the molecule. The potential energy $E_i(\mathbf{R})$ cannot depend on these rotational coordinates. The remaining $3n - 5$ or $3n - 6$ coordinates, on which $E_i(\mathbf{R})$ *does* depend, are the vibrational coordinates. To a good approximation, $\chi_j^{(i)}(\mathbf{R})$ is a product of rotational and vibrational wave functions and $E_j^{(i)}$ is a sum of rotational and vibrational energies.

We show how this works for diatomic molecules, deriving the energy-level formula (4.12). With $n = 2$, there are six coordinates of nuclei 1 and 2, which we write as x_1, y_1, z_1, x_2, y_2, and z_2. The three coordinates of the center of mass are

$$X \equiv \frac{m_1 x_1 + m_2 x_2}{m_1 + m_2}, \qquad Y \equiv \frac{m_1 y_1 + m_2 y_2}{m_1 + m_2}, \qquad Z \equiv \frac{m_1 z_1 + m_2 z_2}{m_1 + m_2}$$

and the $3n - 3 = 3$ relative coordinates may be taken as

$$x = x_1 - x_2, \qquad y \equiv y_1 - y_2, \qquad z \equiv z_1 - z_2$$

The relative coordinates may be combined to give three new coordinates $\mathbf{R}$:

$$R = \sqrt{x^2 + y^2 + z^2}, \qquad \theta = \cos^{-1}\left(\frac{z}{R}\right), \qquad \phi = \tan^{-1}\left(\frac{y}{x}\right)$$

The first is the internuclear distance and the others give the orientation of the molecular axis in space. The electronic energy $E_i(\mathbf{R})$ can depend only on the internuclear distance, so it appears as $E_i(R)$ in the nuclear-motion Schrödinger equation (4.9). With the nuclear kinetic energy operator written in terms of R, θ, and ϕ, (4.9) becomes

$$-\frac{\hbar^2}{2mR^2}\left[\frac{\partial}{\partial R}\left(R^2\frac{\partial \chi_j^{(i)}}{\partial R}\right)+\frac{1}{\sin\theta}\frac{\partial}{\partial\theta}\left(\sin\theta\frac{\partial \chi_j^{(i)}}{\partial\theta}\right)+\frac{1}{\sin^2\theta}\frac{\partial^2\chi_j^{(i)}}{\partial\phi^2}\right]$$
$$+E_i(R)\chi_j^{(i)}=E_j^{(i)}\chi_j^{(i)}$$

where $\chi_j^{(i)}$ depends on R, θ, and ϕ. Here, m is the reduced mass of the nuclei, $m_1m_2/(m_1+m_2)$. Now let $\chi_j^{(i)}$ be written as $A(\theta,\phi)B(R)/R$ and substitute. After division by AB, one obtains

$$-\frac{\hbar^2}{2mB}\frac{d^2B}{dR^2}+E_i(R)$$
$$-\frac{\hbar^2}{2mAR^2}\left[\frac{1}{\sin\theta}\frac{\partial}{\partial\theta}\left(\sin\theta\frac{\partial A}{\partial\theta}\right)+\frac{1}{\sin^2\theta}\frac{\partial^2A}{\partial\phi^2}\right]=E_j^{(i)} \quad (4.10)$$

The usual (in quantum mechanics) argument of separation of variables is now invoked.

In (4.10), the right-hand side and the first two terms on the left are independent of the angles θ and ϕ, so the second term on the left must also be independent of these angles. This requires

$$\frac{1}{A}\left[\frac{1}{\sin\theta}\frac{\partial}{\partial\theta}\left(\sin\theta\frac{\partial A}{\partial\theta}\right)+\frac{1}{\sin^2\theta}\frac{\partial^2A}{\partial\phi^2}\right]$$

to be independent of angles, that is, to be equal to a constant. Recognizing that this is the Schrödinger equation for spherical symmetry, one knows that the constant must be $-J(J+1)$ where J is a nonnegative integer, and that there are $2J+1$ degenerate wave functions for each value of J; A, which is a spherical harmonic, is the rotational or orientational wave function. Putting the constant into (4.10), we have the vibrational Schrödinger equation:

$$-\frac{\hbar^2}{2m}\frac{d^2B}{dR^2}+\left[E_i(R)+\frac{\hbar^2J(J+1)}{2mR^2}\right]B=E_j^{(i)}B$$

The terms in the square brackets are the potential energy operators, with the second being the centrifugal energy, and B is the vibrational wave function.

Usually, $E_i(R)$ and $1/R^2$ in the centrifugal energy are written as power series in $(R - R_e)$, where R_e is the value of R for which $E_i(R)$ has its minimum, giving

$$-\frac{\hbar^2}{2m}\frac{d^2B}{dR^2} + \left[E_i(R_e) + \frac{k}{2}(R - R_e)^2 + \frac{\hbar^2 J(J+1)}{2mR_e^2}\left\{1 - \frac{2}{R_e}(R - R_e)\right\}\right]B = E_j^{(i)}B \quad (4.11)$$

Here, k is the second derivative of $E_i(R)$ at R_e; the first derivative vanishes because one is expanding about the minimum. Equation (4.11) for $J = 0$ is almost the harmonic oscillator Schrödinger equation in the coordinate $R - R_e$. The "almost" refers to the fact that B is to vanish at $R = 0$ and at $R \to \infty$, whereas the harmonic oscillator functions vanish at $R \to \pm\infty$. It turns out, however, that if B is a harmonic oscillator wave function of $R - R_e$, its value is so small at $R = 0$ that it might as well be zero. If the wave function B is the nth harmonic oscillator function ($n = 0, 1, 2, \dots$), the energy $E_j^{(i)}$ is $E_i(R_e) + (n + \frac{1}{2})h\nu_e$ where $\nu_e = (2\pi)^{-1}\sqrt{k/m}$.

For $J > 0$, two additional terms appear in the Hamiltonian. The first, $\hbar^2 J(J+1)/2mR_e^2$, is a constant, and is added to $E_j^{(i)}$; the effect of the other may be calculated by first-order perturbation theory, which means one takes its expectation value over the harmonic oscillator wave function, and the expectation value of $(R - R_e)$ is zero. Higher terms in the expansion of $1/R^2$ in the centrifugal potential energy make nonzero contributions to the energy, corresponding to vibration–rotation interactions; we shall not include them here. The energy $E_j^{(i)}$ is then the sum of the vibrational (harmonic oscillator) energy and the rotational (rigid rotor) energy. Using the two quantum numbers n and J instead of j to index the energies, we write

$$E_{n,J}^{(i)} = E_i(R_e) + \left(n + \tfrac{1}{2}\right)h\nu_e + \frac{\hbar^2 J(J+1)}{2mR_e^2} \quad (4.12)$$

where ν_e depends on the electronic state i because k is $(d^2E_i/dR^2)_e$. J and n are independent quantum numbers, so that a sum over states is a sum over n and J independently.

For n-atomic linear molecules, there are $3n - 3$ internal coordinates, of which two can be chosen as the orientational angles θ and ϕ. The electronic energy may be expanded in a power series in the deviations of the $3n - 5$ remaining coordinates from their equilibrium values, keeping terms through quadratic. The linear terms vanish because one is expanding about the minimum of electronic energy. The coordinates, which we designate by $\{S_s\}$, can be chosen such that the kinetic energy operator in the vibrational

Schrödinger equation is a sum of $3n - 5$ second derivatives, and all cross terms in the quadratic terms in the electronic energy vanish. These coordinates, which are linear combinations of the deviations of nuclear positions from their equilibrium values, are the normal modes. The vibrational Schrödinger equation for an n-atomic linear molecule looks like

$$-\frac{\hbar^2}{2m}\sum_{s=1}^{3n-5}\frac{\partial^2 B}{\partial S_s^2} + \left[E_i(R_e) + \sum_{s=1}^{3n-5}\left(\frac{\partial^2 E_i}{\partial S_s^2}\right)_e S_s^2 + \frac{\hbar^2 J(J+1)}{2I_e}\right]B = E_j^{(i)}B$$

where $E_i(R_e)$ is the value of the electronic energy at its minimum and I_e is the moment of inertia [compare (4.11)]. The wave function B is then a product of harmonic oscillator wave functions and the energies are

$$E_{\text{linear}} = E_i(R_e) + \sum_{s=1}^{3n-5}(n_s + \tfrac{1}{2})h\nu_s + \frac{\hbar^2 J(J+1)}{2I_e}$$

Here, ν_s is the frequency corresponding to the sth normal mode.

For nonlinear molecules, there are three rotational coordinates: two of them define the orientation of one principal axis of inertia of the molecule and the third measures rotation of the molecule about this axis. A symmetric top molecule, one with two equal moments of inertia, has rotational energies

$$E_{\text{sym}}^{\text{rot}} = \frac{\hbar^2}{2}\left(\frac{J(J+1) - K^2}{I_x} + \frac{K^2}{I_z}\right)$$

Here, I_z designates the unique moment of inertia and I_x one of the two equal ones. There are two quantum numbers in the energy, J and K: J is a nonnegative integer and K takes on integral values from $-J$ to $+J$ ($2J + 1$ values). A third quantum number, M, does not appear in the energy, but may have values from $-J$ to J, leading to a $(2J + 1)$-fold degeneracy of each rotational energy level.

When the proper vibrational coordinates are chosen, the vibrational Schrödinger equation for a symmetric top looks like

$$-\frac{\hbar^2}{2m}\sum_{s=1}^{3n-6}\frac{\partial^2 B}{\partial S_s^2}$$

$$+\left[E_i(R_e) + \sum_{s=1}^{3n-6}\left(\frac{\partial^2 E_i}{\partial S_s^2}\right)_e S_s^2 + \frac{\hbar^2}{2}\left(\frac{J(J+1) - K^2}{I_x} + \frac{K^2}{I_z}\right)\right]B = E_j^{(i)}B$$

and B is again a product of harmonic oscillator wave functions. The energy levels for a symmetric top are then

$$E_{\text{s.t.}} = E_i(R_e) + \sum_{s=1}^{3n-6} (n_s + \tfrac{1}{2})h\nu_s + \frac{\hbar^2}{2}\left(\frac{J(J+1) - K^2}{I_x} + \frac{K^2}{I_z}\right)$$

If the molecule is more complicated than a symmetric top, the last term in the above formula is also more complicated. It is not given here. In all the energy-level formulas given, one can identify three contributions to the energy, which we will refer to as electronic, vibrational, and rotational. Each has its quantum numbers. Note, however, that this is not exact, but only an approximation; we had to neglect higher terms in power-series expansions to get it.

4.3. THERMODYNAMIC PROPERTIES

The calculation of thermodynamic functions requires calculation of partition functions. This means we need to evaluate a sum over states of exponentials, like

$$q^{\text{en}} = \sum_t \exp\left\{\frac{-E_t}{kT}\right\}$$

The sum is over all values taken on by all quantum numbers. Because the energy is a sum of three contributions, each exponential is a product of three exponentials; because the quantum numbers are independent of each other, the sum of all products is equal to a product of sums. For a diatomic molecule,

$$q^{\text{en}} = \sum_i \sum_n \sum_J \exp\left\{-\beta\left[E_i(R_e) + (n + \tfrac{1}{2})h\nu_e + \frac{\hbar^2 J(J+1)}{2mR_e^2}\right]\right\}$$

$$= \sum_i e^{-\beta E_i(R_e)} \sum_n e^{-\beta(n+\frac{1}{2})h\nu_e} \sum_J e^{-\beta\hbar^2 J(J+1)/2mR_e^2}$$

For a polyatomic, q^{en} will also be of this form with a product of harmonic oscillator partition functions. It is important to remember that the vibrational frequencies $\{\nu_s\}$ and the moments of inertia (mR_e^2, I_e, I_x, and I_z) depend on the electronic state, so the vibrational and rotational partition functions are different for different electronic states.

In q^{en}, the sum is over all states with energies low enough for the exponentials to matter: for high energies, the exponentials become negligible.

If the spacing between electronic energies is large enough, only a few electronic states will contribute to the sum. Thus one might have

$$q^{\text{en}} = g_1\, e^{-\beta E_1(R_{e1})} q^{\text{vib},1} q^{\text{rot},1} + g_2\, e^{-\beta E_2(R_{e2})} q^{\text{vib},2} q^{\text{rot},2}$$

(where g_1 and g_2 are the degeneracies of electronic energy levels 1 and 2) if electronic levels E_3, E_4, and so on were high enough for the exponentials of $-\beta E$ to be negligible. Note that the equilibrium internuclear distances differ for different electronic states, just as the vibrational frequencies do.

The partition functions q^{vib} and q^{rot} must now be evaluated for diatomic, linear polyatomic, and symmetric top molecules. Consider the vibrational partition functions first. For a diatomic,

$$q_{\text{diat}}^{\text{vib}} = \sum_{n=0}^{\infty} e^{-\beta h(n+\frac{1}{2})\nu_e} = \frac{e^{-\beta h \nu_e/2}}{1 - e^{-\beta \nu_e}} \tag{4.13}$$

As noted in Section 3.3 [see equation (3.11)], this is a geometric series. For a polyatomic molecule, there is more than one vibration, so

$$q_{\text{poly}}^{\text{vib}} = \sum_{n_1=0}^{\infty} \sum_{n_2=0}^{\infty} \sum_{n_3=0}^{\infty} \cdots e^{-\beta h[(n_1+\frac{1}{2})\nu_1 + (n_2+\frac{1}{2})\nu_2 + (n_3+\frac{1}{2})\nu_3 + \cdots]}$$

$$= \sum_{n_1=0}^{\infty} e^{-\beta h(n_1+\frac{1}{2})\nu_1} \sum_{n_2=0}^{\infty} e^{-\beta h(n_2+\frac{1}{2})\nu_2} \cdots$$

The exponential of a sum is a product of exponentials and the quantum numbers for the different vibrations are independent. Now q^{vib} is a product of harmonic oscillator partition functions:

$$q_{\text{poly}}^{\text{vib}} = \frac{e^{-\beta h \nu_1/2}}{1 - e^{-\beta h \nu_1}} \frac{e^{-\beta h \nu_2/2}}{1 - e^{-\beta h \nu_2}} \cdots \tag{4.14}$$

For a linear molecule, there are $3N - 5$ vibrational frequencies; for a nonlinear molecule, there are $3N - 6$.

The rotational partition function for a linear molecule

$$q_{\text{lin}}^{\text{rot}} = \sum_{J=0}^{\infty} (2J + 1) e^{-\beta \hbar^2 J(J+1)/2I_e}$$

where $2J + 1$ is the degeneracy. For a diatomic, the moment of inertia I_e is equal to MR_e^2 where M is the reduced mass and R_e the equilibrium internuclear distance. This partition function has been evaluated in Section 3.4 by approximating it as an integral over J [see equations (3.28 and 3.29)].

The approximation, a good one when the spacing between energy levels is small compared to kT so that $\beta\hbar^2/I_e$ is small compared to 1, gives

$$q_{\text{lin}}^{\text{rot}} = \frac{2I_e}{\beta\hbar^2} \tag{4.15}$$

For a symmetric top molecule,

$$\begin{aligned} q_{\text{sym}}^{\text{rot}} &= \sum_{J=0}^{\infty} \sum_{K=-J}^{J} (2J+1)\exp\left[-\beta\frac{\hbar^2}{2}\left(\frac{J(J+1)-K^2}{I_x} + \frac{K^2}{I_z}\right)\right] \\ &= \sum_{K=-\infty}^{\infty} \exp\left[\beta\frac{\hbar^2}{2}K^2\left(\frac{1}{I_x} - \frac{1}{I_z}\right)\right] \sum_{J=|K|}^{\infty} (2J+1)e^{-\beta\hbar^2 J(J+1)/2I_x} \end{aligned} \tag{4.16}$$

This partition function can also be evaluated by approximating the sums as integrals. An additional approximation, which is also a good one when the spacing of energy levels is small compared to kT, is to replace the lower limit on J by $J^* = -\frac{1}{2} + \sqrt{\frac{1}{4} + K^2}$, which equals $|K|$ for $K = 0$ and approaches K for $|K|$ much larger than $\frac{1}{2}$. Then

$$\begin{aligned} q_{\text{sym}}^{\text{rot}} &= \int_{K=-\infty}^{\infty} dK \exp\left[\beta\frac{\hbar^2}{2}K^2\left(\frac{1}{I_x} + \frac{1}{I_z}\right)\right] \int_{J^*}^{\infty} dJ(2J+1)e^{-\beta\hbar^2 J(J+1)/2I_x} \\ &= \int_{K=-\infty}^{\infty} dK \exp\left[\beta\frac{\hbar^2}{2}K^2\left(\frac{1}{I_x} - \frac{1}{I_z}\right)\right] \frac{2I_x}{\beta\hbar^2} e^{-\beta\hbar^2 K^2/2I_x} \\ &= \frac{2I_x}{\beta\hbar^2}\sqrt{\frac{2\pi I_z}{\beta\hbar^2}} \end{aligned} \tag{4.17}$$

The rotational partition functions must be divided by a symmetry number if rotation can interchange identical nuclei, as we now explain.

As mentioned in Sections 4.1 and 4.2, one conventionally omits the nuclear-spin factor in the partition function because its contribution to thermodynamic functions cancels off in chemical reactions. There is one complication: the overall wave function for a molecule must be either symmetric or antisymmetric to an exchange of identical nuclei, and identical nuclei may be exchanged by molecular rotation, so only certain nuclear-spin states can occur for a given rotational state. If many rotational states are populated, which occurs when the spacing between them is small compared to β^{-1}, this restriction can be taken into account by dividing the rotational partition function by a symmetry number σ. The use of $\sigma = 2$ for homonuclear diatomics has been discussed in Section 3.4. For polyatomic molecules containing identical nuclei which can be exchanged by a rotation, σ is equal to the number of permutations of identical nuclei which can be obtained by rotation. Thus $\sigma = 2$ for H_2O, $\sigma = 3$ for NH_3, and $\sigma = 12$ for CH_4. For molecules without identical nuclei, $\sigma = 1$.

For diatomic and polyatomic molecules, with the approximations made so far, the internal motion partition functions are sums of terms, each being an electronic factor multiplied by vibrational and rotational partition functions. The electronic factor is usually very small for all electronic states except those of lowest energy, reducing the partition function to a single product. Then the internal motion thermodynamic functions are sums of electronic, vibrational, and rotational contributions, to which the translational contribution must be added.

For a diatomic molecule, assuming only one electronic state (subscript 1) contributes, the partition function is

$$q_{\text{diat}} = g_1 \left(\frac{2\pi M}{\beta h^2} \right)^{3/2} V e^{-\beta E_1(R_e)} \frac{e^{-\beta h \nu_e/2}}{1 - e^{-\beta h \nu_e}} \frac{2I_e}{\sigma \beta \hbar^2} \tag{4.18}$$

Here, $I_e = mR_e^2$ where $m = M_1 M_2/(M_1 + M_2)$, and g_1 is the degeneracy of the electronic state. The Helmholtz free energy for N molecules is

$$\begin{aligned} A = -kT \ln\left(\frac{q^N}{N!} \right) &= -NkT \ln g_1 + \tfrac{3}{2} NkT \ln\left(\frac{h^2}{2\pi MkT} \right) \\ &\quad - NkT \ln \frac{V}{N} - NkT + NE_1(R_e) + \frac{Nh\nu_e}{2} \\ &\quad + NkT \ln(1 - e^{-h\nu_e/kT}) + NkT \ln\left(\frac{\hbar^2 \sigma}{2I_e kT} \right) \end{aligned}$$

and the energy is

$$\begin{aligned} E = kT^2 \left(\frac{\partial \ln[q^N/N!]}{\partial T} \right)_{N,V} &= \tfrac{3}{2} NkT + NE_1(R_e) \\ &\quad + Nh\nu_e \left[\frac{1}{2} + \frac{1}{e^{-h\nu_e/kT} - 1} \right] + NkT \end{aligned} \tag{4.19}$$

The four terms are translational, electronic, vibrational, and rotational contributions. The entropy can be calculated from A and E since $A = E - TS$.

For a linear polyatomic molecule having s vibrations, the free energy has two terms for each vibration. The energy, instead of (4.19), is

$$E = \tfrac{3}{2}NkT + NE_1(R_e) + Nh\sum_{v=1}^{s}\nu_v\left[\frac{1}{2} + \frac{1}{e^{-h\nu_v/kT} - 1}\right] + NkT$$

For a symmetric top molecule, still assuming only one electronic energy level is important, the partition function is

$$q^{\text{sym}} = g_1\left(\frac{2\pi M}{\beta h^2}\right)^{3/2} Ve^{-\beta E_1(R_e)}\prod_{v=1}^{s}\left\{\frac{e^{-\beta h\nu_v/2}}{1 - e^{-\beta h\nu_v}}\right\}\frac{2I_x}{\sigma\beta\hbar^2}\sqrt{\frac{2\pi I_z}{\beta\hbar^2}}$$

so that for N molecules

$$\begin{aligned} A = {} & -NkT\ln g_1 + \tfrac{3}{2}NkT\ln\left(\frac{h^2}{2\pi MkT}\right) - NkT\ln\frac{V}{N} - NkT \\ & + NE_1(R_e) + \sum_{v=1}^{s}\left\{\frac{Nh\nu_e}{2} + NkT\ln(1 - e^{-h\nu_e/kT})\right\} \\ & + NkT\ln\left[\frac{\hbar^3\sigma}{\pi^{1/2}I_xI_z^{1/2}(2kT)^{3/2}}\right] \end{aligned} \tag{4.20}$$

In the above formulas, $E_1(R_e)$ is the ground-state electronic energy when the nuclei are at their equilibrium positions. The energy calculated from (4.20) is

$$E = \tfrac{3}{2}NkT + NE_1(R_e) + Nh\sum_{v=1}^{s}\nu_v\left[\frac{1}{2} + \frac{1}{e^{-h\nu_v/kT} - 1}\right] + \tfrac{3}{2}NkT$$

(The last term is the rotational contribution; there are three rotational degrees of freedom.) The entropy is calculable from $A = E - TS$.

Chemical potentials for a gas of molecules may be calculated by differentiating A with respect to $n = N/\mathcal{N}_A$ (T and V constant). For diatomic molecules,

$$\begin{aligned} \mu_{\text{diat}} = \mathcal{N}_A\Bigg\{ & -kT\ln g_1 + \tfrac{3}{2}kT\ln\left(\frac{h^2}{2\pi MkT}\right) + kT\ln\frac{V}{N} + E_1(R_e) \\ & + \frac{h\nu_e}{2} + kT\ln(1 - e^{-h\nu_e/kT}) + kT\ln\left(\frac{\hbar^2\sigma}{2I_ekT}\right)\Bigg\} \end{aligned} \tag{4.21}$$

For symmetric top molecules,

$$\mu_{\text{sym}} = \mathcal{N}_{\text{A}}\left\{ -kT \ln g_1 + \tfrac{3}{2}kT \ln\left(\frac{h^2}{2\pi MkT}\right) - kT \ln \frac{V}{N} \right.$$

$$+ E_1(R_e) + \sum_{v=1}^{s}\left[\frac{h\nu_e}{2} + kT \ln(1 - e^{-h\nu_e/kT})\right]$$

$$\left. + kT \ln\left[\frac{\hbar^3\sigma}{\pi^{1/2} I_x I_z^{1/2} (2kT)^{3/2}}\right]\right\} \tag{4.22}$$

Note that $\mathcal{N}_{\text{A}}k = R$. The chemical potential for atoms in a perfect gas has already been given in equation (4.7). Like μ_{atom}, μ_{diat} and μ_{sym} may be written as $\mu_0 + RT \ln P$. These formulas apply to atoms and molecules in a perfect gas mixture, as well as to atoms and molecules in a pure perfect gas, so they can be used to discuss chemical equilibrium involving perfect gases.

4.4. CHEMICAL EQUILIBRIUM

Let us first consider the simple dissociation reaction, $AB \rightleftarrows A + B$, with the diatomic AB and the two atoms A and B in the gas phase. The partition function for the mixture of gases is

$$Q = \frac{q_A^{N_A} q_B^{N_B} q_{AB}^{N_{AB}}}{N_A!\, N_B!\, N_{AB}!}$$

To find the numbers of atoms of A and B, and the numbers of molecules of AB, at equilibrium, one can maximize $\ln Q$ with respect to N_A, N_B, and N_{AB}:

$$\delta \ln Q = \delta N_A \ln q_A + \delta N_B \ln q_B + \delta N_{AB} \ln q_{AB} - \delta N_A \ln N_A$$

$$- \delta N_B \ln N_B - \delta N_{AB} \ln N_{AB} = 0 \tag{4.23}$$

Because the changes in mole numbers are the result of the chemical reaction, $\delta N_A = \delta N_B = -\delta N_{AB}$, and the maximization equation becomes

$$[\ln q_A + \ln q_B - \ln q_{AB}]\, \delta N_{AB} - [\ln N_A + \ln N_B - \ln N_{AB}]\delta N_{AB} = 0$$

which requires

$$\ln\left[\frac{q_A q_B N_{AB}}{q_{AB} N_A N_B}\right] = 0.$$

Therefore, the equilibrium condition is that $N_A N_B / N_{AB}$ be equal to $q_A q_B / q_{AB}$, which is the equilibrium constant in terms of partition functions.

It should be realized that, since $-kT \ln Q$ is the Helmholtz free energy, differentiating $\ln Q$ with respect to N_A gives the chemical potential μ_A, and similarly for μ_B and μ_{AB}. Thus (4.23) is really

$$\mu_A \, \delta N_A + \mu_B \, \delta N_B + \mu_{AB} \, \delta N_{AB} = 0$$

and we have simply carried through the thermodynamic manipulations of (1.23) to (1.25) in another notation to show that, at equilibrium, $\mu_A + \mu_B = \mu_{AB}$. We can thus use the general thermodynamic result: for the chemical reaction

$$\sum_i \nu_i A_i = 0$$

the equilibrium condition is

$$\sum_i \nu_i \, \mu_i = 0 \tag{4.24}$$

The chemical potential of species i is given by

$$\mu_i = -kT\mathcal{N}_{\mathrm{A}}(\ln q_i - \ln N_i)$$

if species i is gaseous. Using this in (4.24) leads to

$$\sum_i \nu_i \ln q_i = \sum_i \nu_i \ln N_i$$

or

$$\prod_i (q_i)^{\nu_i} = \prod_i (N_i)^{\nu_i}$$

as the equilibrium condition.

It will be noticed that q_i for any gaseous atom or molecule includes one factor of V. Then the equilibrium condition is written

$$\prod_i \left(\frac{q_i}{V}\right)^{\nu_i} = \prod_i \left(\frac{N_i}{V}\right)^{\nu_i} \tag{4.25}$$

The right-hand side of (4.25) is a product of number concentrations of reactants and products, each raised to the power of its stoichiometric coefficient. Each factor on the left-hand side (which is the value of the equilibrium constant) is independent of volume.

The equilibrium constant of (4.25) is in terms of number densities. If molar concentrations are desired, one can use $c_i = n_i/V = N_i/(\mathcal{N}_A V)$ and rewrite (4.25) as

$$\prod_i \left(\frac{q_i}{\mathcal{N}_A V}\right)^{\nu_i} = \prod_i c_i^{\nu_i} = K_c$$

Another remark: it is usual to calculate the partition function for an atom or molecule taking as zero of energy the energy of the lowest-lying state for that atom or molecule. For a molecule, this might be $E_1(R_e) + \Sigma \frac{1}{2} h\nu_i$. The derivation of (4.23) to (4.25) assumed the same zero of energy for all species. Denoting the usual partition function as q_i^* one has

$$q_i = q_i^* \, e^{-\varepsilon_i^0/kT}$$

with ε_i^0 the lowest energy level for species i. Then

$$K_c = \prod_i \left[\left(\frac{q_i^*}{\mathcal{N}_A V}\right)^{\nu_i} e^{-\nu_i \varepsilon_i^0/kT}\right]$$

and

$$\ln K_c = \sum_i \nu_i \ln\left(\frac{q_i^*}{\mathcal{N}_A V}\right) - \frac{1}{kT} \sum_i \nu_i \varepsilon_i^0$$

The last sum is the energy of reaction at $T = 0$ (all substances are in their lowest-energy states).

As an example of the calculation of equilibrium constants, consider the reaction

$$H_2 + D_2 = 2HD.$$

For this reaction, $\ln K_c$ is equal to $2 \ln q_{HD} - \ln q_{H_2} - \ln q_{D_2}$. Referring to (4.18), one sees a number of terms cancel off, including $2E_{HD}(R_e) - E_{H_2}(R_e) - E_{D_2}(R_e)$ because all three molecules have the same electronic energy. Then

$$\ln K = 3\ln\left(\frac{2\pi M_{\mathrm{HD}}kT}{h^2}\right) - \beta h\nu_{\mathrm{HD}} - 2\ln(1 - e^{-h\nu_{\mathrm{HD}}/kT}) + 2\ln\left(\frac{2I_{\mathrm{HD}}kT}{\hbar^2}\right)$$

$$-\tfrac{3}{2}\ln\left(\frac{2\pi M_{\mathrm{H}_2}kT}{h^2}\right) + \frac{\beta}{2}h\nu_{\mathrm{H}_2} + \ln(1 - e^{-h\nu_{\mathrm{H}_2}/kT}) - \ln\left(\frac{I_{\mathrm{H}_2}kT}{\hbar^2}\right)$$

$$-\tfrac{3}{2}\ln\left(\frac{2\pi M_{\mathrm{D}_2}kT}{h^2}\right) + \frac{\beta}{2}h\nu_{\mathrm{D}_2} + \ln(1 - e^{-h\nu_{\mathrm{D}_2}/kT}) - \ln\left(\frac{I_{\mathrm{D}_2}kT}{h^2}\right)$$

$$= \tfrac{3}{2}\ln\left(\frac{M_{\mathrm{HD}}^2}{M_{\mathrm{H}_2}M_{\mathrm{D}_2}}\right) - \frac{\beta}{2}h(2\nu_{\mathrm{HD}} - \nu_{\mathrm{H}_2} - \nu_{\mathrm{D}_2}) + \ln\left(\frac{4I_{\mathrm{HD}}^2}{I_{\mathrm{H}_2}I_{\mathrm{D}_2}}\right)$$

$$+ \ln\left[\frac{(1 - e^{-h\nu_{\mathrm{H}_2}/kT})(1 - e^{-h\nu_{\mathrm{D}_2}/kT})}{(1 - e^{-h\nu_{\mathrm{HD}}/kT})^2}\right] \tag{4.26}$$

The ratio of masses is about 9/8, so translation does not contribute much to ln K. The moment of inertia for each molecule is equal to the reduced mass multiplied by the square of the equilibrium internuclear distance, and the equilibrium distance is the same for all three molecules. Thus the ratio of moments of inertia is

$$\frac{I_{\mathrm{HD}}^2}{I_{\mathrm{H}_2}I_{\mathrm{D}_2}} = \frac{M_{\mathrm{H}}^2 M_{\mathrm{D}}^2 (2M_{\mathrm{H}})(2M_{\mathrm{D}})}{(M_{\mathrm{H}} + M_{\mathrm{D}})^2 M_{\mathrm{H}}^2 M_{\mathrm{D}}^2}$$

or about 8/9, so the contribution of rotation to ln K is mainly the contribution of the symmetry numbers. On the other hand, the difference of zero-point vibrational energies makes an important contribution: the vibrational frequency is the square root of the force constant divided by the reduced mass, and the force constant is the same for all three molecules.

4.5. GRAND PARTITION FUNCTION: ADSORPTION

We noted in Section 1.2 that the thermodynamic treatment of chemical reactions imagines the system to be open, so that the numbers of molecules of reactants and products can change, but with the constraint that the changes in the numbers of molecules of different species are related by the stoichiometry of the reaction. An open system is represented by the grand canonical ensemble (Section 2.4). It is interesting to see how the condition of chemical equilibrium arises from consideration of the grand canonical ensemble. Then we will consider a phase equilibrium (adsorption), and see that it can be conveniently treated using the grand canonical partition function.

Consider a system of volume V containing molecules of species a, b, c, and so on, at temperature T. If there are N_a molecules of a, N_b molecules of b, and so on, the (canonical) partition function is $Q(N_a, N_b, \ldots, V, T)$. The grand partition function is

$$\Xi = \sum_{N_a} \sum_{N_b} \cdots e^{\Sigma_m \gamma_m N_m} Q(N_a, N_b, \ldots) \tag{4.27}$$

If the molecules are independent (we want to consider a gas),

$$Q(N_a, N_b, \ldots) = \frac{q_a^{N_a}}{N_a!} \frac{q_b^{N_b}}{N_b!} \cdots$$

where q_m is the single-molecule or single-atom partition function. Using this in (4.27),

$$\Xi = \sum_{N_a} \sum_{N_b} \cdots \frac{e^{\gamma_a N_a} q_a^{N_a}}{N_a!} \frac{e^{\gamma_b N_b} q_b^{N_b}}{N_b!} \cdots$$

$$= \sum_{N_a} \frac{(e^{\gamma_a} q_a)^{N_a}}{N_a!} \sum_{N_b} \frac{(e^{\gamma_b} q_b)^{N_b}}{N_b!} \cdots \tag{4.28}$$

Since the sums over molecule numbers are independent, we have replaced a sum of products by a product of sums. The grand partition function is a product of grand partition functions for the separate species since they are independent.

The sums in (4.28) are easily evaluated as they are just the power series for exponentials:

$$\Xi = \exp(e^{\gamma_a} q_a) \exp(e^{\gamma_b} q_b) \cdots$$

and

$$\ln \Xi = e^{\gamma_a} q_a + e^{\gamma_b} q_b + \cdots \tag{4.29}$$

We know that $\ln \Xi$ is PV/kT [Section 2.4, equation (2.47)], and that $\gamma_a = \beta\mu_a/\mathcal{N}_A$ ($\mathcal{N}_A =$ Avogadro's number). It is interesting to compare (4.29) with what we already know about chemical potentials. For the canonical ensemble (remember $A = -kT \ln Q$)

$$\mu_a = \mathscr{N}_{\mathrm{A}} \left(\frac{\partial A}{\partial N_a} \right)_{T,V} = -kT\mathscr{N}_{\mathrm{A}} \left[\frac{\partial}{\partial N_a} \ln \left(\frac{q_a^{N_a}}{N_a!} \right) \right] = -RT \ln \frac{q_a}{N_a}$$

so that (4.29) is

$$\begin{aligned} \frac{PV}{kT} &= e^{\mu_a/\mathscr{N}_{\mathrm{A}}kT} q_a + e^{\mu_b/\mathscr{N}_{\mathrm{A}}kT} q_b + \cdots \\ &= \left(\frac{q_a}{N_a} \right)^{-1} q_a + \left(\frac{q_b}{N_b} \right)^{-1} q_b + \cdots \end{aligned}$$

or $PV = NkT$, where N is the total number of molecules. [It was permissible to combine results from the canonical ensemble (fixed numbers of particles) and the grand canonical ensemble (variable numbers of particles) because the number of particles is essentially constant for the grand ensemble; [see equations (2.51) to (2.53).] The same result may be obtained staying within the framework of the grand ensemble: the average number of molecules of species a, from (2.42), is

$$\langle N_a \rangle = \left(\frac{\partial \ln \Xi}{\partial \gamma_a} \right)_{\beta, \gamma_b, V} = e^{\gamma_a} q_a$$

which shows that $\ln \Xi$ (4.29), which is PV/kT, is the sum of the average numbers of molecules of all species.

Now we consider a chemical reaction using the grand ensemble. Although more involved than the treatment using the canonical ensemble, it adds to our understanding of the meaning of the formulas. Some readers may want to skip it and go to adsorption equilibria, the paragraph of equation (4.33). Consider an open system of volume V and temperature T, containing molecules of species a and b, so that (4.29) is

$$\Xi = \sum_{N_a} \sum_{N_b} e^{-\gamma_a N_a + \gamma_b N_b} Q(N_a N_b VT)$$

Suppose that a and b molecules can react to give c molecules, according to

$$-\nu_a a - \nu_b b \rightarrow \nu_c c$$

where the stoichiometric coefficients ν_a and ν_b are negative, in accord with the usual convention for writing a chemical reaction,

$$\sum_j \nu_j A_j = 0$$

[see discussion preceding equation (1.23)]. $Q(N_a N_b VT)$ is a sum over all states of the system constructed from N_a molecules of a and N_b molecules of b. This includes states for which $|\xi\nu_a|$ molecules of a and $|\xi\nu_b|$ molecules of b have combined to form $\xi\nu_c$ molecules of c. Here, ξ is the degree of advancement of the reaction [(1.23) and (1.24)].

The grand partition function becomes

$$\Xi = \sum_{N_a} \sum_{N_b} e^{\gamma_a N_a + \gamma_b N_b} \left[\sum_{\xi} \frac{q_a^{N_a + \nu_a \xi}}{(N_a + \nu_a \xi)!} \frac{q_b^{N_b + \nu_b \xi}}{(N_b + \nu_b \xi)!} \frac{q_c^{\nu_c \xi}}{(\nu_c \xi)!} \right]$$

where the last sum is over nonnegative values of ξ such that $N_a + \nu_a \xi$ and $N_b + \nu_b \xi$ are also nonnegative. The sum over ξ can be replaced by a sum over nonnegative values of $N_c^\bullet$, the number of actual c molecules in the system. At the same time, $N_a + \nu_a \xi$ is replaced by $N_a^\bullet$, the number of free or uncombined a molecules, $N_b + \nu_b \xi$ is replaced by $N_b^\bullet$, and $\nu_c \xi$ is replaced by $N_c^\bullet$. Note that N_a is the total number of a molecules, $N_a^\bullet$ plus the number present in the c molecules formed:

$$N_a = N_a^\bullet - \frac{\nu_a}{\nu_c} N_c^\bullet \tag{4.30a}$$

Similarly,

$$N_b = N_b^\bullet - \frac{\nu_b}{\nu_c} N_c^\bullet \tag{4.30b}$$

Then

$$\Xi = \sum_{N_a} \sum_{N_b} e^{\gamma_a N_a + \gamma_b N_b} \left[\sum_{N_c^\bullet} \frac{q_a^{N_a^\bullet}}{(N_a^{N_a^\bullet})!} \frac{q_b^{N_b^\bullet}}{(N_b^\bullet)!} \frac{q_c^{N_c^\bullet}}{(N_c^\bullet)!} \right] \tag{4.31}$$

The sums are over all values of N_a and N_b, and over all values of $N_c^\bullet$ consistent with each N_a and N_b; equivalently, one can say that we consider all possible values of $N_a^\bullet$, $N_b^\bullet$, and $N_c^\bullet$.

The grand partition function is, with the use of (4.30) in (4.31),

$$\Xi = \sum_{N_a^\bullet} \sum_{N_b^\bullet} \sum_{N_c^\bullet} \exp\left(\gamma_a N_a^\bullet - \frac{\gamma_a \nu_a}{\nu_c} N_c^\bullet + \gamma_b N_b^\bullet - \frac{\gamma_b \nu_b}{\nu_c} N_c^\bullet \right)$$

$$\times \frac{q_a^{N_a^\bullet}}{(N_a^\bullet)!} \frac{q_b^{N_b^\bullet}}{(N_b^\bullet)!} \frac{q_c^{N_c^\bullet}}{(N_c^\bullet)!} \tag{4.32}$$

where the sums are over all nonnegative values of $N_a^\bullet$, $N_b^\bullet$, and $N_c^\bullet$. Now, instead of considering an open system composed of a and b, which can react to form c, consider an open system at the same V and T, containing the three species a, b, and c. The grand partition function would be

$$\Xi = \sum_{N_a} \sum_{N_b} \sum_{N_c} \exp(\gamma_a N_a + \gamma_b N_b + \gamma_c N_c) \frac{q_a^{N_a}}{(N_a)!} \frac{q_b^{N_b}}{(N_b)!} \frac{q_c^{N_c}}{(N_c)!}$$

This should be identical to (4.32), since it describes the same system at thermal equilibrium. It *will* be identical to (4.32) if

$$\gamma_c = \frac{-\gamma_a \nu_a}{\nu_c} + \frac{-\gamma_b \nu_b}{\nu_c}$$

which, since $\gamma_a = \beta\mu_a / N_{\mathscr{A}}$ and so on, means that $\nu_a \mu_a + \nu_b \mu_b + \nu_c \mu_c = 0$. This is the usual equilibrium condition in terms of the chemical potentials.

For adsorption equilibria, the grand canonical ensemble is more natural and easier to deal with than the canonical ensemble. Consider a surface having S equivalent sites at each of which one molecule of species a may absorb. An adsorbed molecule has an internal partition function q_a, which depends on the temperature T. (An empty site has, of course, no internal states, so its internal partition function is just 1.) Then the grand partition function for the system of molecules on the surface is

$$\Xi = \sum_{N_a=0}^{\infty} e^{\gamma_a N_a} Q(N_a; S) = \sum_{N_a=0}^{\infty} e^{\gamma_a N_a} \frac{S!}{N_a! (S - N_a)!} q_a^{N_a} \tag{4.33}$$

where N_a is the number of molecules adsorbed. The quotient of factorials is the degeneracy, the number of ways of arranging N_a molecules on S distinguishable sites. Note that Ξ describes the monolayer of adsorbed molecules, and not the molecules in the gas or the liquid from which they adsorb. The gas or liquid is in equilibrium with the monolayer, and serves as the "molecule bath" fixing the chemical potential of a.

Since the factorial of a negative integer is infinite, the sum in (4.33) is effectively from $N_a = 0$ to $N_a = S$. Then the sum is a binomial expansion and

$$\Xi = \sum_{N_a=0}^{S} \frac{S!}{N_a!(S-N_a)!}(q_a e^{\gamma_a})^{N_a} = (1 + q_a e^{\gamma_a})^S = (1 + \lambda_a q_a)^S$$

We have introduced a conventional abbreviation λ_a for e^{γ_a} (λ_a is the activity). The average number of adsorbed molecules, which is also the most probable number, is

$$\langle N_a \rangle = \left(\frac{\partial \ln \Xi}{\partial \gamma_a}\right)_T = \lambda_a \left(\frac{\partial \ln \Xi}{\partial \lambda_a}\right)_T = \lambda_a S \frac{q_a}{1 + q_a \lambda_a}$$

The fractional coverage of the surface, $\langle N_a \rangle$ divided by the number of sites, is

$$\theta = \frac{q_a \lambda_a}{1 + q_a \lambda_a} \tag{4.34}$$

where $\ln \lambda_a = \beta\mu_a/\mathcal{N}_A = \mu_a/RT$. The chemical potential μ_a is equal to the chemical potential of a in the gas or liquid with which the surface is in equilibrium. If it is an ideal gas, $\mu_a = \mu_a^0 + RT \ln P_a$ (P = pressure in atmospheres) so that $\ln \lambda_a = \mu_a^0/RT + \ln P_a$ and

$$\theta = \frac{q_a e^{\mu_a^0/RT} P_a}{1 + q_a e^{\mu_a^0/Rt} P_a} \tag{4.35}$$

This is the famous Langmuir adsorption isotherm.

Of course, the Langmuir isotherm (4.35) can also be obtained from the canonical ensemble. Let there be N molecules of a, of which N_a are adsorbed on a surface of S sites and $N - N_a$ remain in the gas phase. The canonical partition function for the N molecules is

$$Q(N_a; N, S, T) = \frac{S!}{N_a!(S-N_a)!} q_a^{N_a} \frac{q_g^{N-N_a}}{(N-N_a)!} \tag{4.36}$$

with q_g representing the partition function of a molecule of a in the gas phase. The volume (gas plus surface) and the temperature are fixed. For equilibrium at fixed V and T, the Helmholtz free energy A must be minimized. Since $A = -kT \ln Q$, we require

$$\frac{\partial \ln Q(N_a; N, S, T)}{\partial N_a} = 0$$

which leads to

$$\frac{N_a V}{(S - N_a)(N - N_a)} = \frac{q_a}{q_g/V} \tag{4.37}$$

The left-hand side is a product of concentrations, as in (4.25), for the reaction of an a molecule in the gas phase with an empty surface site to produce an adsorbed a molecule. Then the right-hand side is the equilibrium constant (remember the partition function for an empty surface site is taken to be unity). In terms of $\theta = N_a/S$, (4.37) may be rewritten as

$$\frac{\theta}{1-\theta} = \frac{q_a(N - N_a)}{q^* V} = \frac{q_a}{q^*}\frac{P_a}{kT}$$

where, since q_g is proportional to V, we have introduced $q^* = q_g/V$; the ideal-gas law has been used in the form $P_a V = (N - N_a)/kT$. The above equation is equivalent to (4.35), with $\ln q^*$ replacing μ_a^0/RT.

Since the approach by the grand canonical ensemble is simpler, we apply it to the problem of competitive adsorption. Suppose molecules of two species, a and b, complete for the S sites on a surface. Let N_a and N_b be the numbers of molecules of the two species adsorbed, so that there are $S - (N_a + N_b)$ empty sites. The canonical partition function for this situation, considering the number of arrangements of the N_a a molecules and the N_b b molecules, is

$$\frac{S!}{N_a!\,N_b!\,(S - N_a - N_b)!}\, q_a^{N_a} q_b^{N_b}$$

and the grand partition function is

$$\Xi = \sum_{N_a}\sum_{N_b} \frac{e^{\gamma_a N_a}\, e^{\gamma_b N_b} S!}{N_a!\,N_b!\,(S - N_a - N_b)!}\, q_a^{N_a} q_b^{N_b} \tag{4.38}$$

The values of N_a and N_b are those for which $S - N_a - N_b \geq 0$; for other values, a factorial is infinite and the term in the sum vanishes. Then the grand partition function is a multinomial expansion:

$$\Xi = [1 + e^{\gamma_a} q_a + e^{\gamma_b} q_b]^S = [1 + \lambda_a q_a + \lambda_b q_b]^S$$

The activities λ_a and λ_b have been introduced in the last member.

The equilibrium number of a molecules adsorbed is the average number:

$$\langle N_a \rangle = \lambda_a \left(\frac{\partial \ln \Xi}{\partial \lambda_a} \right) = S \frac{\lambda_a q_a}{1 + \lambda_a q_b + \lambda_b q_b}$$

and similarly for b molecules. Letting θ_a and θ_b be the fractional coverages of the surface by a molecules and b molecules,

$$\theta_a = \frac{\lambda_a q_a}{1 + \lambda_a q_b + \lambda_b q_b} \quad \text{and} \quad \theta_b = \frac{\lambda_b q_b}{1 + \lambda_a q_b + \lambda_b q_b} \tag{4.39}$$

The ratio of θ_a to θ_b is given by the ratio of $\lambda_a q_b$ to $\lambda_b q_b$. The chemical potentials and the activities are fixed by the properties of the gas or liquid from which the molecules are adsorbing. Assuming an ideal-gas mixture,

$$\mu_a = \mu_a^0 + RT \ln P_a$$

where P_a is the partial pressure of a, and similarly for b. Then $\ln \lambda_a = \gamma_a = \mu_a/RT = \mu_a^0/RT + \ln P_a$. Abbreviating $\exp(\mu_a^0/RT)$ by λ_a^0, we have

$$\theta_a = \frac{\lambda_a^0 q_a P_a}{1 + \lambda_a^0 q_a P_a + \lambda_b^0 q_b P_b}$$

and similarly for θ_b.

The adsorption described in the calculations above is localized: each adsorbed particle is bound to a site on the surface. For delocalized adsorption, the particles are not localized on adsorption sites; one can specify only whether or not a particle is adsorbed on the surface. The canonical partition function for N_a delocalized adsorbed particles is just $q_a^{N_a}/N_a!$. Then the grand partition function is, instead of (4.33),

$$\Xi = \sum_{N_a} \frac{e^{\gamma N_a}}{N_a!} = \exp[e^{\gamma} q_a] \tag{4.40}$$

The average number of adsorbed particles is

$$\langle N_a \rangle = \frac{1}{\Xi} \sum_{N_a} \frac{N_a \, e^{\gamma N_a} q_a^{N_a}}{N_a!} = \frac{\partial \ln \Xi}{\partial \gamma}$$

which, for nonlocalized adsorption, gives

$$\langle N_a \rangle = \frac{\partial [e^{\gamma} q_a]}{\partial \gamma} = e^{\gamma} q_a$$

Since nonlocalized adsorbed particles on a surface constitute a two-dimensional (ideal) gas,

$$q_a = q_a^{\text{int}}\left(\frac{2\pi mkTL^2}{h^2}\right)$$

where q_a^{int} is the internal partition function [see Section 3.2, especially equations to (3.5) to (3.7)]. L^2 is the area of the surface, A, and $\gamma = \beta\mu/N_{\mathscr{A}}$.

To define the fractional coverage θ, we assume that each adsorbed molecule occupies an area α, so

$$\theta = \frac{\langle N_a \rangle \alpha}{A} = e^{\beta\mu/N_{\mathscr{A}}}\frac{q_a^{\text{int}}(2\pi mkT)\alpha}{h^2} \tag{4.41}$$

The chemical potential μ is equal to the chemical potential of a in the gas phase from which it absorbs, $\mu^0 + RT \ln P$. Then (4.41) shows that θ is proportional to

$$e^{(\mu^0 + RT \ln P)/(N_{\mathscr{A}}kT)} = Pe^{\mu^0/RT}$$

Note the difference between this and the Langmuir isotherm, (4.35). The problem of nonlocalized adsorption can also be handled by the canonical ensemble (see Problem 4.23).

PROBLEMS

4.1. The N atom has a ground state configuration $1s^2 2s^2 2p^3$, giving rise to ^4S, ^2D, and ^2P terms. The lowest-energy term is ^4S, with $J = \frac{3}{2}$. The next, ^2D, splits into two levels, $J = \frac{5}{2}$ at 19,223 cm^{-1} and $J = \frac{3}{2}$ at 19,231 cm^{-1} (to convert cm^{-1} to energy units, multiply by hc, where c is the speed of light). The ^2P term gives rise to levels with $J = \frac{1}{2}$ and $J = \frac{3}{2}$, both at 28,840 cm^{-1}. At 83,340 cm^{-1} above the ground state one finds levels from a ^4P term which derives from the excited configuration $1s^2 2s^2 2p^2 3s$. Calculate the electronic entropy of one mole of N atoms at $T = 50$ K, 100 K, 300 K, 1000 K, 5000 K, and 10,000 K. Calculate the probability that an N atom is in the lowest energy state (^4S) at each of these temperatures. (*Answer:* at 5000 K, $S_{\text{el}} = 8.790$ J K^{-1} mol^{-1}.)

4.2. For the rare gases He, Ne, Ar, Kr, and Xe, the ground state is nondegenerate (^1S term). The first excited states are, respectively, at 159,850 cm^{-1}, 134,044 cm^{-1}, 93,144 cm^{-1}, 79,973 cm^{-1}, and 67,068 cm^{-1}. For each gas, find the temperature at which excited states must be taken into account in calculating the partition function q^{elec} (the temperature at which q^{elec} differs from unity by 1%).

4.3. Calculate the molar entropies of the following monatomic gases at $P = 1$ atmosphere and $T = 298$ K, assuming electronic excitation makes no contribution: He, Ar, Xe, Na, Ag, Hg. Experimental values, corrected in some cases for nonideal behavior, are respectively: 30.13, 36.98, 40.53, 36.72, 41.32, 41.79 cal/K.

4.4. For a mixture of ideal gases containing N_A molecules of A and N_B molecules of B in a volume V and at temperature T, the partition function is [see (4.2) for the definition of Λ]

$$Q = \frac{\left(\Lambda_A^{-3/2} V\right)^{N_A}}{N_A!} \left(q_A^{\text{int}}\right)^{N_A} \frac{\left(\Lambda_B^{-3/2} V\right)^{N_B}}{N_B!} \left(q_B^{\text{int}}\right)^{N_B}$$

Derive an expression for the entropy of this system and compare it to the sum of the entropies of (1) N_A molecules of A in a volume $N_A V/(N_A + N_B)$ and (2) N_B molecules of B in a volume $N_B V(N_A + N_B)$, all at the same temperature T. The choice of volumes corresponds to having the same pressure in (1), (2), and the combined system.

4.5. The ground state term of atomic nitrogen is $^4S_{3/2}$ (4-fold degenerate), with an energy of -109.22713 Ry, where 1 Rydberg $= 109{,}737$ cm^{-1}. Other terms arising from the ground state configuration $1s^2 2s^2 2p^3$ are ^2D, including ten states with average energy -109.05192 Ry, and ^2P, including six states with average energy -108.96431 Ry. These energies are relative to the energy of seven electrons infinitely separated from the nucleus and each other.

(a) Evaluate the electronic partition function at 500 K, assuming terms from other configurations are too high in energy to matter.

(b) Usually, atomic partition functions are evaluated taking the energy of the lowest state to be zero. What is the electronic partition function for N at 500 K if this is done?

4.6. The lowest electronic configuration of NO gives a $^2\Pi$ term, from which arise two energy levels. The lower one is $^2\Pi_{1/2}$ and is doubly degenerate; the upper is $^2\Pi_{3/2}$, fourfold degenerate, and 121.1 cm^{-1} above the lower one. The next electronic state is at 43,966 cm^{-1}. Calculate the electronic partition function, electronic energy, and electronic entropy at 300 K and 1000 K. (*Answer:* at 1000 K, $q = 4.36$, $E = 1.849 \times 10^{-14}$ ergs/molecule, $S = 2.218$ ergs K^{-1} molecule^{-1}.)

4.7. The harmonic oscillator model for vibrations is applicable only to the lower vibrational states because it neglects terms higher than the quadratic term in the expansion of $E_i(R)$, equation (4.11). The potential $E_i(R_e) + \frac{1}{2}k(R - R_e)^2$ is clearly wrong for $R - R_e \to \infty$ (the fact that molecules dissociate means E_i approaches a finite value for $R - R_e \to \infty$). The effect of higher terms in the expansion may be represented by writing the vibrational energy level formula as $\varepsilon_n = (n + \frac{1}{2})h\nu_e - \alpha(n + \frac{1}{2})^2 h\nu_e$, where α, the anharmonicity constant, is positive and less than unity. The spacing of vibrational energy levels decreases with n.

(a) According to this formula, what is the dissociation energy?

(b) Evaluate the vibrational partition function, assuming that α is small enough so $e^{-\alpha x}$ can be approximated by $1 - \alpha x$. What is the effect of anharmonicity on the variation of vibrational energy and heat capacity with temperature?

4.8. The rotational partition function for a polyatomic molecule which has three different moments of inertia is

$$q^{\text{rot}} = \frac{\pi^{1/2}}{\sigma}\left(\frac{2}{\beta\hbar^2}\right)^{3/2}\sqrt{I_x I_y I_z}$$

(a) Show that this is the symmetric-top partition function when two moments of inertia are equal.

(b) Calculate the rotational energy for this partition function.

4.9. A linear molecule is a symmetric top with the unique moment of inertia I_z equal to zero. Show that (4.16) gives the linear-molecule rotational partition function in this case. (What happens to the sum over K?)

4.10. CO and N_2 are very similar molecules. Their "rotational temperatures," $\Theta_r = \hbar^2/2I_e k$, are 2.77 K and 2.86 K, respectively, and their "vibrational temperatures," $\Theta_v = h\nu_e/k$, are 3070 K and 3340 K, respectively. Calculate the rotational and vibrational partition functions for both molecules at 300 K. Calculate the rotational and vibrational energies for both molecules at 300 K. Compare the vibrational energy for each molecule with what it would be at 0 K.

4.11. For H_2, Θ_r and Θ_v (see preceding problem) are 85.3 K and 6215 K, respectively. Calculate Θ_r and Θ_v for HD and for D_2. Note that the force constant, $k = (d^2E_i/dR^2)_e$ is the same for all three molecules.

4.12. Using the data in Problem 4.11, evaluate the terms in (4.26) at 298 K and 383 K, and calculate the equilibrium constant for $H_2 + D_2 = 2HD$ at these temperatures. The experimental values are 3.28 and 3.50. Experimentally measured equilibrium constants pass through a maximum and decrease at higher temperatures. According to (4.26), is this possible?

4.13. Equation (4.22) gives the chemical potential of a symmetric-top molecule in terms of V, N, and molecular parameters. The chemical potential is usually written in terms of the chemical potential in the standard state, $\mu = \mu^0 + RT \ln P$. Give an expression for μ^0.

4.14. The triatomic H_2O molecule, being nonlinear, has $3N - 6$ or 3 vibrations. They are the symmetric stretch of both bonds, with frequency $\nu_1 = 3652\ \text{cm}^{-1}$; the bend vibration, with frequency $\nu_2 = 1595\ \text{cm}^{-1}$; and the asymmetric bond stretch, with frequency $\nu_3 = 3756\ \text{cm}^{-1}$. (Multiply by the speed of light to get frequencies in sec^{-1}.) Calculate the contribution of each vibration to the heat capacity of H_2O at 600 K, and the total vibrational heat capacity at this temperature. What is the limiting value of the vibrational heat capacity as T approaches infinity (assuming purely harmonic vibrations)?

4.15. The thermal ionization of atoms or molecules in the gas phase can be represented by the equation

$$A \rightleftarrows I^+ + e^-$$

and treated as a dissociation reaction. The reactant is the atom or molecule A, and the products are the ion I^+ and the free electron e^-. The internal-state partition function for the electron is simply its degeneracy, 2, and the energy difference between I^+ and e^- in their lowest states, and A in its lowest state, is the ionization potential Φ.

(a) Derive an equation for the equilibrium constant K, where

$$K = \left(\frac{N_e}{V}\right)\left(\frac{N_I}{V}\right)\left(\frac{N_A}{V}\right)^{-1},$$

in terms of the internal-state partition functions for A and I^+.

(b) If the internal-state partition functions for A and I^+ are approximately equal (because these two species have similar rotational, vibrational, and electronic states), derive the "Saha equation"

$$\ln K = -1.16 \times 10^4 \frac{\Phi}{T} + \tfrac{3}{2} \ln T + 36.1$$

where N_i/V is in atoms per cm^3, T is in K, and Φ is in electron-volts.

4.16. Diatomic oxygen has a triply degenerate ground electronic state. Its vibrational frequency $\nu_e = 1580.36\ \text{cm}^{-1}$, and its rotational constant $B_e = 1.44567\ \text{cm}^{-1}$, where $hcB_e = \hbar^2/2I$. (The moment of inertia I is $M_{\text{red}} R_e^2$, so what is the internuclear distance?) Calculate the molar entropy and heat capacity at 298 K and 1 atm (1.0133×10^6 dyne cm^{-1}) pressure. Experimental values are $S = 205.138\ \text{J K}^{-1}\ \text{mol}^{-1}$ and $21.041\ \text{J K}^{-1}\ \text{mol}^{-1}$.

4.17. The linear symmetrical triatomic CO_2 has four vibrational frequencies (in cm^{-1}): 667.3, 1285.5, 1388.3, and 2349.3. The rotational constant B, where $hcB = \hbar^2/2I$, is 0.3895 cm^{-1}. (For a linear molecule, the moment of inertia is given by $\Sigma_i m_i x_i^2$ where the sum is over nuclei and x_i is the distance of nucleus i from the center of mass. What is the C—O distance in CO_2?) Calculate the molar entropy and the molar heat capacity of CO_2 at 298 K and 1 atm pressure. Experimental values are $S = 213.74$ J K^{-1} mol^{-1} and $C_V = 28.80$ J K^{-1} mol^{-1}.

4.18. This, be warned, is a big calculation. Cl_2 has a nondegenerate ground electronic state, $\Theta_r = 0.351$ K and $\Theta_v = 808$ K (see Problem 4.10). HCl has a nondegenerate ground state, $\Theta_r = 15.02$ K and $\theta_v = 4227$ K. The data for H_2 are given in Problem 4.11. Calculate the equilibrium constant for $H_2 + Cl_2 = 2HCl$ at 300 K, 400 K, and 500 K. The slope of a plot of ln K vs. $1/T$ can be shown in thermodynamics to be $-\Delta H^0/R$. From your calculated equilibrium constants, get ΔH^0 and compare it to the standard enthalpy of reaction obtained from thermodynamic tables.

4.19. For the grand partition function (4.31), show that the average value of $N_c^\bullet$ is given by $(\partial \ln \Xi/\partial \ln q_c)$, and that the average value of $N_a^\bullet$ is given by $(\partial \ln \Xi/\partial \ln q_a)$.

4.20. Consider the grand partition function Ξ for a reacting gas-phase system, given by the equation just before (4.30), with the reaction being $a + 2b \rightarrow c$. What are the allowed values of ξ, the degree of reaction? Express $N_a^\bullet$ in terms of N_a and $N_c^\bullet$.

4.21. Show that maximizing Q for the adsorbate-gas system, equation (4.36), with respect to N_a leads to [see (4.37)]

$$\frac{N_a}{(S - N_a)(N - N_a)} = \frac{q_a}{q_g}$$

and that $N_a/(S - N_a) = \theta/(1 - \theta)$ where θ is the fractional coverage of the surface.

4.22. Another way to obtain the Langmuir isotherm is to equate the chemical potentials of the adsorbed molecules and the gas molecules. The chemical potential of the adsorbed molecules equals $\mathcal{N}_A(\partial A_a/\partial N_a)_T$ where $A_a = -kT \ln Q_a$ and

$$Q_a = \frac{S!}{N_a!\,(S - N_a)!}\, q_a^{N_a}$$

The chemical potential of the gas molecules is $(\partial A_g/\partial N_g)_T$ where $A_g = -kT \ln Q_g$ and $Q_g = q_g^{N_g}/N_g!$. Noting that q_g is proportional to V, derive the Langmuir isotherm (4.35).

4.23. To treat nonlocalized adsorption by the canonical partition function, suppose that there are N molecules, of which N_a are adsorbed and $N - N_a$ are in the gas phase. Then the partition function is

$$Q = \frac{q_a^{N_a}}{N_a!} \frac{q_g^{(N-N_a)}}{(N-N_a)!}$$

where q_a and q_g are the partition functions for an adsorbed a molecule and an a molecule in the gas phase, respectively. The equilibrium situation corresponds to minimum free energy or maximum Q.

(a) Maximize Q with respect to N_a and find an expression for $N_a/(N - N_a)$, the ratio of adsorbed to nonadsorded molecules.

(b) We know that $q_a = q_a^{\text{int}}(2\pi mkT/h^2)A$ with A the area of the surface, and

$$q_g = q_g^{\text{imt}}(2\pi mkT/h^2)^{3/2}V$$

Assuming that each adsorbed molecule occupies an area α, calculate the fractional coverage $\theta = N_a\alpha/A$ and show it is proportional to the pressure of the gas phase. Compare equation (4.41) and give the expression for μ^0, the chemical potential of the gas at standard pressure.

4.24. Hydrogen (and probably other gases) exists as diatomic molecules in the gas phase, but adsorbs on many surfaces as atoms, so that the adsorption reaction is $H_2(g) = 2H(ads)$. Let θ_a be the fractional coverage of a surface by H so that [see (4.39)]

$$\theta_a = \frac{\lambda_a q_a}{1 + \lambda_a q_a + \Sigma_j \lambda_j q_j}$$

where the sum over j is a sum over all other species that may adsorb on the surface, in competition with H. At equilibrium, $\mu_{H_2} = 2\mu_a$, where μ_{H_2} is calculated in terms of q_{H_2} from the grand partition function for a gas of H_2 molecules and μ_a is calculated in terms of q_a from the grand partition function for adsorbed H atoms. Show that

$$\frac{\theta_a}{1 - \theta_a - \Sigma_j \theta_j} = \lambda_a q_a = \left(\frac{P_{H_2}}{kTq_{H_2}}\right)^{1/2}$$

so that one has the Langmuir equation in terms of the square root of the pressure instead of the pressure.

CHAPTER 5

QUANTUM STATISTICAL MECHANICS

5.1. INDISTINGUISHABILITY

For a system of N noninteracting indistinguishable particles (atoms, molecules, etc.), we took the indistinguishability into account by dividing the partition function by $N!$. (For N_a indistinguishable particles of species a, N_b indistinguishable particles of species b, and so on, we would divide the partition function by the product $N_a!\,N_b!\ldots$) The argument for doing this is that, if one cannot tell whether two indistinguishable particles have switched states, one should consider all system states that differ only by interchanges of identical particles as a single state. By writing the partition function for the system as a product of single-particle partition functions, one counts all system states which differ by interchanges of identical particles as separate states. To correct for this, the partition function must be divided by the number of states that differ by interchanges of identical particles, that is, by $N!$ (see Section 3.1).

This argument is correct only when there are so many accessible states that very few states have more than one particle in them. Interchanging particles that are in the same state changes nothing, so the number of states differing by interchange of identical particles is less than $N!$ if several particles are in the same state. As noted in Section 3.1, the number of accessible states is much larger than the number of particles, except at low temperatures. To treat low-temperature situations, we need to take the indistinguishability of particles into consideration more correctly. For this, we must go back to the quantum mechanical origin of the problem: a many-particle wave function must be either symmetric or antisymmetric with respect to interchange of identical particles. Whether symmetry or antisymmetry is required depends on what the particles are.

Particles for which wave functions must be antisymmetric to interchange are called fermions, and are said to obey Fermi–Dirac statistics. They include electrons, protons, He^3 nuclei, and particles made from an odd number of elementary fermions. For fermions, an N-particle wave function must satisfy

$$\mathscr{I}_{ij}\psi(1,2,3,\ldots,N) = -\psi(1,2,3,\ldots,N) \tag{5.1}$$

In (5.1), $\mathscr{I}_{ij}$ is an interchange operator, which interchanges the labels of particles i and j in the wavefunction ψ. Particles requiring symmetric wave functions are called bosons, and are said to obey Bose–Einstein statistics. They include photons, alpha particles (helium nuclei), and particles composed of an even number of fermions. For bosons,

$$\mathscr{I}_{ij}\psi(1,2,3,\ldots,N) = \psi(1,2,3,\ldots,N) \tag{5.2}$$

Note that any permutation of particle labels can be accomplished by a succession of interchanges. Thus for bosons ψ must be unchanged by any permutation $\mathscr{P}$ of particle labels. For fermions,

$$\mathscr{P}\psi(1,2,3,\ldots,N) = \pm\psi(1,2,3,\ldots,N)$$

where the sign is + or − depending on whether $\mathscr{P}$ requires an even or an odd number of interchanges.

For *independent* identical particles, either fermions or bosons, many-particle states are built from single-particle states with wave functions ψ_j, where j represents one or more quantum numbers. The N-particle wave functions ψ are sums of products of single-particle wave functions. For bosons, it is easy to see that a properly symmetric N-particle wave function is

$$\psi_{\text{sym}} = \sum_{\mathscr{P}} \mathscr{P}\{\psi_k(1)\,\psi_l(2)\cdots\,\psi_p(N)\}$$

The sum is over permutations $\mathscr{P}$, each of which permutes the particles among the single-particle wave functions. For fermions, a properly antisymmetric N-particle wave function is

$$\psi_{\text{anti}} = \sum_{\mathscr{P}} (-1)^{\mathscr{P}} \mathscr{P}\{\psi_k(1)\,\psi_l(2)\cdots\,\psi_p(N)\}$$

In ψ_{anti} the factor $(-1)^{\mathscr{P}}$ is ± 1, according to whether $\mathscr{P}$ involves an even or odd number of interchanges. In fact, ψ_{anti} is a determinant, with quantum numbers indexing the rows and particle numbers indexing the columns.

Because the particles are indistinguishable, it is meaningless to specify which particle is in which state. One can specify only occupation numbers: how many particles are in each state. Let s_j be the occupation number of state j. For bosons, s_j can be any nonnegative integer, but for fermions s_j can be only 0 or 1. If more than one particle is in any one-particle state, ψ_{anti} vanishes, since it is a determinant with two identical columns. This is the Pauli exclusion principle.

Specifying the occupation numbers $\{s_j\}$ specifies the many-particle state. The energy of the many-particle state is

$$E(\{s_j\}) = \sum_j s_j E_j \tag{5.3}$$

where E_j is the energy (eigenvalue) associated with ψ_j. The sum of the occupation numbers is equal to the total number of particles:

$$N = \sum_j s_j \tag{5.4}$$

The equilibrium state of the system, assuming constant N, V, and T, is described by the canonical partition function,

$$Q(NVT) = \sum_{s_1} \sum_{s_2} \cdots e^{-\sum_j \beta E_j s_j} \tag{5.5}$$

The sum over states is replaced by a sum over occupation numbers. All values of the $\{s_j\}$ are to be considered, with the restriction (5.4). The exponential is a sum of exponentials so Q is a product:

$$Q(NVT) = \sum_{s_1} e^{-\beta E_1 s_1} \sum_{s_2} e^{-\beta E_2 s_2} \cdots$$

In spite of this simplification, evaluation of Q is not simple, because of the restriction (5.4) on the sum of the occupation numbers.

However, if one goes over to the grand canonical distribution, the problem becomes simple again. The grand partition function is

$$\begin{aligned} \Xi &= \sum_N e^{\gamma N} Q(NVT) \\ &= \sum_N \left[\sum_{s_1} e^{-\beta E_1 s_1} \sum_{s_2} e^{-\beta E_2 s_2} \cdots \right] e^{\gamma \sum_j s_j} \end{aligned}$$

In the bracket, one has to sum over all values of all the $\{s_j\}$ consistent with the restriction (5.4), but since one is summing over all N in any case, one can drop the restriction and write

$$\Xi = \sum_{s_1} e^{(\gamma - \beta E_1)s_1} \sum_{s_2} e^{(\gamma - \beta E_2)s_2} \cdots \tag{5.6}$$

which is truly a product of independent sums.

All of the sums in (5.6) are the same. For fermions, each sum consists of two terms only ($s_j = 0$ or 1). Writing $\gamma = \beta\mu$ (for chemical species, we usually write $\beta\mu/\mathcal{N}_A$; in this chapter μ is the chemical potential per particle) we have

$$\Xi = \prod_j \sum_{s_j=0}^{s_j=1} e^{\beta(\mu - E_j)s_j} = \prod_j [1 + e^{\beta(\mu - E_j)}]$$

where the product is over particle states. For bosons, each sum in (5.6) is a geometric series:

$$\Xi = \prod_j \sum_{s_j=0}^{\infty} e^{\beta(\mu - E_j)s_j} = \prod_j \left[\frac{1}{1 - e^{\beta(\mu - E_j)}}\right]$$

Both fermions and bosons are conveniently considered together by writing

$$\Xi = \prod_j \xi_j \tag{5.7}$$

where

$$\xi_j = [1 \pm e^{\beta(\mu - E_j)}]^{\pm 1} \tag{5.8}$$

The upper sign is for fermions, the lower sign for bosons. The logarithm of Ξ, which will be needed to calculate thermodynamic properties, is a sum of terms.

The average number of particles may be calculated as:

$$\langle N \rangle = \frac{\Sigma_N N e^{\beta\mu N} Q(NVT)}{\Sigma_N e^{\beta\mu N} Q(NVT)} = \frac{1}{\beta}\left(\frac{\partial \ln \Xi}{\partial \mu}\right)_{T,V} \tag{5.9}$$

Alternatively, one may note that the number of particles is the sum of the numbers of particles in all states so that

$$\langle N \rangle = \sum_j \langle s_j \rangle$$

and the average occupation number of state j is

$$\langle s_j \rangle = \frac{\sum_{s_j=0}^{\infty} s_j \, e^{\beta(\mu - E_j)s_j}}{\sum_{s_j=0}^{\infty} e^{\beta(\mu - E_j)s_j}} = \frac{1}{\beta}\left(\frac{\partial \ln \xi_j}{\partial \mu}\right)_{V,T} \tag{5.10}$$

For fermions, (5.10) gives

$$\langle s_j \rangle = \frac{1}{\beta} \frac{\partial \ln[1 + e^{\beta(\mu - E_j)}]}{\partial \mu} = \frac{e^{\beta(\mu - E_j)}}{1 + e^{\beta(\mu - E_j)}} \tag{5.11}$$

which, obviously, can never be greater than 1. The Pauli exclusion principle prevents any state being occupied by more than one particle. For bosons, (5.10) gives

$$\langle s_j \rangle = \frac{1}{\beta} \frac{\partial \ln[1 - e^{\beta(\mu - E_j)}]^{-1}}{\partial \mu} = \frac{e^{\beta(\mu - E_j)}}{1 - e^{\beta(\mu - E_j)}} \tag{5.12}$$

which has no upper limit. There is no limit to the number of particles in any state.

The two results (5.11) and (5.12) can be written together:

$$\langle s_j \rangle = \frac{1}{e^{\beta(E_j - \mu)} \pm 1} \tag{5.13}$$

(upper sign for fermions, lower for bosons). When $\beta(E_j - \mu) \gg 1$, $\langle s_j \rangle$ becomes $\exp[-\beta(E_j - \mu)]$ for either fermions or bosons, so that all $\langle s_j \rangle$ are much less than 1. This happens when there are many more available states than particles. This, we said in Section 3.1, is when it is justified to correct for indistinguishability by dividing the partition function by $N!$. We are now in a position to prove this statement, independently of whether the particles are fermions or bosons.

When $\beta(E_j - \mu) \gg 1$, $\exp[-\beta(E_j - \mu)]$ becomes very small, and the logarithm of the grand partition function of equations (5.7) and (5.8) becomes

$$\begin{aligned} \ln \Xi &= \ln \prod_j [1 \pm e^{-\beta(E_j - \mu)}]^{\pm 1} \\ &= \pm \sum_j \ln[1 \pm e^{-\beta(E_j - \mu)}] \simeq \sum_j e^{-\beta(E_j - \mu)} \end{aligned} \tag{5.14}$$

We have approximated $\ln(1 \pm x)$ by $\pm x$, which is valid when x is small. The sum is $e^{\beta\mu}$ multiplied by q, the partition function for a single particle, and $\ln \Xi$ is supposed to be equal to βPV, so

$$PV = kTe^{\beta\mu}q \tag{5.15}$$

We may also use (5.14) in (5.9) to get

$$\langle N \rangle = \frac{1}{\beta}\left(\frac{\partial \ln \Xi}{\partial \mu}\right)_{\beta, V} = \sum_j e^{-\beta(E_j - \mu)} = e^{\beta\mu}q \tag{5.16}$$

Combining this with (5.15) gives the ideal gas law, $PV = \langle N \rangle kT$. To justify dividing the partition function by $N!$, we have to go further.

Because fluctuations are small, $\langle N \rangle$ for the open system can be identified with N for the closed system, and μ can be identified with the chemical potential (per particle) for the closed system, which is [see (2.32)]

$$\mu = \frac{-1}{\beta}\left(\frac{\partial \ln Q}{\partial N}\right)_{T, V}$$

According to (5.16), $\beta\mu$ is $\ln(N/q)$, so that

$$\left(\frac{\partial \ln Q}{\partial N}\right)_{T, V} = \ln\left[\frac{q}{N}\right] \tag{5.17}$$

We integrate (5.17) from 1 to some large value of N. Since $Q = q$ when $N = 1$,

$$\begin{aligned}\ln Q_N - \ln q &= \int_1^N [\ln q - \ln N]\, dN \\ &= (\ln q)(N - 1) - (N \ln N - N) - 1\end{aligned}$$

The 1's may be neglected when N is much larger than 1.

$$\ln Q_N \simeq N \ln q - \ln N!$$

and $Q_N = q^N/N!$. This correction for indistinguishability is valid as long as all occupation numbers $\langle s_j \rangle$ are small compared to 1. When this holds, we refer to the statistics as "corrected Boltzmann statistics" and the particles (which may in reality be fermions or bosons) as "corrected boltzons."

5.2. DISTRIBUTIONS FOR FERMIONS AND BOSONS

We now have occupation number formulas for three kinds of particles, of which "corrected boltzons" are fictional.

$$\text{fermions: } \langle s_j \rangle = [1 + e^{-\beta(\mu - E_j)}]^{-1}$$

$$\text{corrected boltzons: } \langle s_j \rangle = [e^{-\beta(\mu - E_j)}]^{-1}$$

$$\text{bosons: } \langle s_j \rangle = [-1 + e^{-\beta(\mu - E_j)}]^{-1}$$

These three expressions are compared in Figure 5.1. The average occupation numbers have been plotted against energy; in order to have a dimensionless parameter, the x-coordinate has been taken as $\beta E_j - \mu$). Fermions have the lowest occupation number (it approaches 1 as $\beta(E_j - \mu)$ approaches $-\infty$), corrected boltzons next, and bosons highest (it approaches ∞ as $E_j - \mu$ approaches 0). In all cases, $\langle s_j \rangle$ is a monotonic function of E_j, its value approaching 0 as E_j becomes very large.

The sum of the average occupation numbers is the total number of particles. The chemical potential μ must be chosen to fix this number, according to

$$\sum_j \{\pm 1 + e^{\beta(E_j - \mu)}\}^{-1} = N \tag{5.18}$$

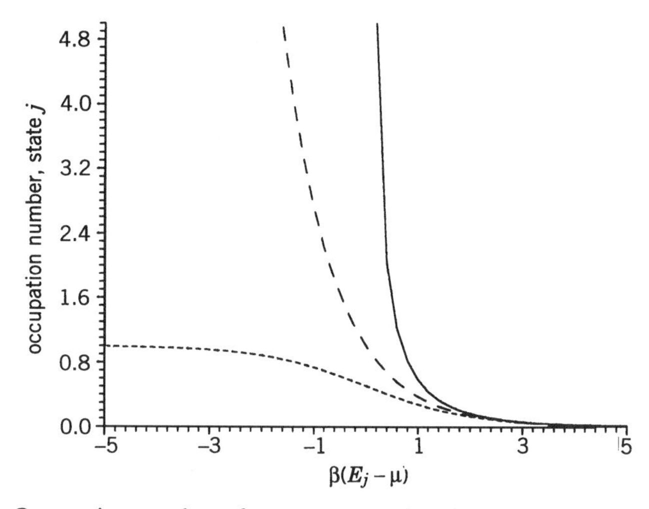

Figure 5.1. Occupation number of a state j as a function of energy E_j. Curves from left to right (or bottom to top) refer to fermions, boltzons, and bosons. The occupation numbers are plotted vs. the dimensionless parameter $\beta(E_j - \mu)$.

For fermions μ is the value of the energy for which $\langle s_j \rangle = \frac{1}{2}$; $\langle s_j \rangle$ is less than $\frac{1}{2}$ for $E_j > \mu$ and greater than $\frac{1}{2}$ for $E_j < \mu$. As $T \to 0$ or $\beta \to \infty$, $\langle s_j \rangle \to 0$ for $E_j > \mu$ and $\langle s_j \rangle \to 1$ for $E_j < \mu$, so $\langle s_j \rangle$ as a function of E_j becomes a step function. Each state with energy below μ is occupied by one particle and each state with energy above μ is completely unoccupied. When $T > 0$, $\langle s_j \rangle$ is smoother (see Figure 5.2), with partial occupation of states with energies above and below μ. For bosons $\langle s_j \rangle$ becomes infinite for $E_j = \mu$, so that μ must lie below the energies of all occupied states. As $T \to 0$, $\langle s_j \rangle \to 0$ for $E_j > \mu$ (as for fermions); this means that the value of μ approaches the lowest energy level, which is the only one occupied. Note that there is no limit on the number of particles which can be in any one state in Bose–Einstein statistics. All the particles go into the lowest-energy states (see Section 5.5).

Using the occupation number formulas, the (average) total energy is

$$E = \sum_j \langle s_j \rangle E_j = \sum_j \frac{E_j}{e^{\beta(E_j - \mu)} \pm 1} \tag{5.19}$$

The pressure may be calculated from the equation of state (see 5.14):

$$PV = kT \ln \Xi = \pm kT \sum_j \ln[1 \pm e^{\beta(\mu - E_j)}] \tag{5.20}$$

The energies E_j are functions of the volume or other external parameters of the system.

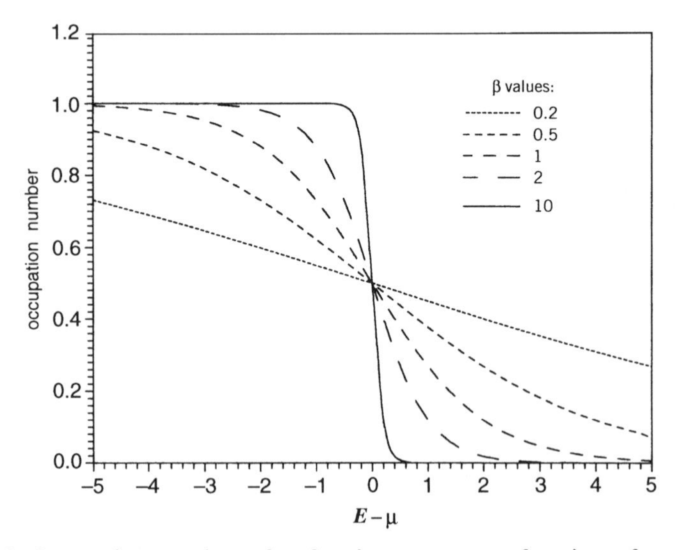

Figure 5.2. Occupation number of a fermion state as a function of energy. The occupation number is plotted vs. $E - \mu$ for various values of $\beta = 1/kT$; it approaches a step function for $\beta \to \infty$.

We now consider a gas of N particles (fermions or bosons) of mass m in a volume $V = L^3$. The one-particle states are the states of a particle in a box, with energies

$$E_{klm} = \frac{h^2(k^2 + l^2 + m^2)}{8mL^2} \tag{5.21}$$

with the quantum numbers k, l, and m taking on positive integer values. The grand partition function is given by (5.7) and (5.8), so that $\ln \Xi$ is a sum of terms, one for each state. Since the number of occupied states is very large, and the energy levels of (5.21) are very close together, the sum may be written as an integral over energy. Let $n(E)$ be the number of states with energy less than or equal to E, so that $dn(E) = (dn/dE)\,dE$ is the number of states with energy between E and $E + dE$. Then

$$\ln \Xi = \int_0^\infty dE \, \frac{dn(E)}{dE} \ln \xi(E) \tag{5.22}$$

where

$$\xi(E) = [1 \pm e^{\beta(\mu - E)}]^{\pm 1}$$

The problem is to calculate $n(E)$ from (5.21).

For this purpose, we imagine three perpendicular coordinate axes labeled by k, l, and m, and planes perpendicular to each axis with unit spacing between them, dividing space into cubical cells of unit volume. Each intersection of three planes corresponds to a state of (5.21), so each state is associated with one cubical cell, with unit volume. Equation (5.21), rewritten as

$$\frac{8mL^2}{h^2} E = k^2 + l^2 + m^2$$

is the equation of a sphere of radius $(8mL^2E/h^2)^{1/2}$. The number of states with energy less than or equal to E is the volume of this sphere, divided by the volume per state (which is 1), except that, since k, l, and m must be positive, we are limited to one-eighth of the sphere. Then

$$n(E) = \frac{1}{8}\,\frac{4\pi}{3}\left(\sqrt{8mL^2E/h^2}\right)^3 = \frac{2^{7/2}\pi(mE)^{3/2}V}{3h^3}$$

and

$$\frac{dn(E)}{dE} = \frac{2^{5/2}\pi m^{3/2}E^{1/2}V}{h^3}$$

The calculation of $n(E)$ is equivalent to that of $g(\nu)$ in the Debye theory of crystal vibrations, (3.14) and following.

If the particles have internal, as well as translational, states, $n(E)$ and dn/dE are multiplied by a degeneracy factor D; for electrons $D = 2$, arising from electron spin. Now equation (5.22) for the partition function becomes

$$\ln \Xi = \frac{2^{5/2}\pi m^{3/2} V}{h^3} D \int_0^\infty dE\, E^{1/2} \ln\left[1 \pm e^{\beta(\mu - E)}\right]^{\pm 1} \tag{5.23}$$

The expression for the total number of particles is written as an integral over energy instead of a sum over states.

$$\begin{aligned} \langle N \rangle &= \int_0^\infty dE \frac{dn(E)}{dE} \langle s(E) \rangle \\ &= \frac{2^{5/2}\pi m^{3/2} DV}{h^3} \int_0^\infty dE \frac{E^{1/2}}{e^{-\beta(\mu - E)} \pm 1} \end{aligned} \tag{5.24}$$

In principle, (5.24) gives the chemical potential μ as a function of N. Unfortunately, the integral cannot be evaluated in closed form. The same goes for the energy:

$$\begin{aligned} \langle E \rangle &= \int_0^\infty dE \frac{dn(E)}{dE} \langle s(E) \rangle E \\ &= \frac{2^{5/2}\pi m^{3/2} DV}{h^3} \int_0^\infty dE \frac{E^{3/2}}{e^{-\beta(\mu - E)} \pm 1} \end{aligned} \tag{5.25}$$

The last expression may be rewritten and integrated by parts, with the boundary terms vanishing:

$$\begin{aligned} \langle E \rangle &= \frac{2^{5/2}\pi m^{3/2} DV}{h^3} \int_0^\infty dE \frac{E^{3/2} e^{\beta(\mu - E)}}{\pm e^{\beta(\mu - E)} + 1} \\ &= \frac{2^{5/2}\pi m^{3/2} DV}{h^3} \int_0^\infty dE \frac{\pm 3}{2\beta} E^{1/2} \ln\{\pm e^{\beta(\mu - E)} + 1\} \end{aligned}$$

Comparing this with the expression (5.23) for $\ln \Xi$ and noting that $\ln \Xi$ is PV/kT,

$$\langle E \rangle = \frac{3}{2\beta} \frac{PV}{kT} = \tfrac{3}{2} PV \tag{5.26}$$

This relation also holds for the corrected Boltzmann gas, with $\langle E \rangle = \frac{3}{2}NkT$ and $PV = NkT$.

However, the equation of state for a gas of fermions or bosons is more complicated than the ideal-gas law. At $T = 0$, the energy of a gas of fermions is not zero since all the particles cannot get into the lowest energy state; the pressure, as will be seen later, is not zero either. For high temperatures, the deviation of the equation of state from that of the ideal-gas law may be derived as a series. The algebra for doing this occupies the rest of this section.

If $L = e^{\beta(\mu - E)}$ is not too large, the integrands in (5.24) and (5.25) can be written as power series in L. For simplicity, we abbreviate the collection of constants in front of the integrals in (5.24) and (5.25) by CV and $\exp(\beta\mu - \beta E)$ by L. On multiplying the numerator and denominator of (5.24) by $e^{\beta\mu}$, we have

$$\langle N \rangle = CV \int_0^\infty dE \frac{E^{1/2} e^{\beta\mu}}{e^{\beta E} \pm e^{\beta\mu}} = CV \int_0^\infty dE \frac{E^{1/2} e^{\beta\mu}}{e^{\beta E}} \frac{1}{1 \pm L}$$

$$= CV \int_0^\infty dE\, E^{1/2} L \sum_{j=0}^{\infty} (\mp L)^j = CV \sum_{j=0}^{\infty} (\mp 1)^j \int_0^\infty dE\, E^{1/2} [e^{\beta(\mu - E)}]^{j+1}$$

Using

$$\int_0^\infty dE\, E^{1/2} e^{-(j+1)\beta E} = \frac{\sqrt{\pi}}{2} [(j+1)\beta]^{-3/2}$$

we have

$$\langle N \rangle = CV \sum_{j=0}^{\infty} (\mp 1)^j e^{(j+1)\beta\mu} \frac{\sqrt{\pi}}{2} [(j+1)\beta]^{-3/2} \tag{5.27}$$

Similarly,

$$\langle E \rangle = CV \int_0^\infty dE \frac{E^{3/2}}{e^{-\beta(\mu - E)} \pm 1}$$

$$= CV \sum_{j=0}^{\infty} (\mp 1)^j \int_0^\infty dE\, E^{3/2} [e^{\beta(\mu - E)}]^{j+1}$$

$$= CV \sum_{j=0}^{\infty} (\mp 1)^j e^{(j+1)\beta\mu} \frac{3\sqrt{\pi}}{4} [(j+1)\beta]^{-5/2} \tag{5.28}$$

Both (5.27) and (5.28) are power series in $e^{\beta\mu}$, converging more quickly for smaller $\beta\mu$ (higher T). The difference between fermions and bosons is whether or not terms in the series alternate in sign.

If $\beta\mu$ is very small, so that it suffices to take the first term in each of the series, we get

$$\langle N \rangle = CVe^{\beta\mu} \frac{\sqrt{\pi}}{2} \beta^{-3/2}$$

and

$$\langle E \rangle = CVe^{\beta\mu} \frac{3\sqrt{\pi}}{4} \beta^{-5/2}$$

Then $\langle E \rangle / \langle N \rangle$, the energy per particle, is $3/(2\beta)$ or $\frac{3}{2}NkT$ for either fermions or bosons. Using (5.26), we recover the ideal-gas equation of state $PV = NkT$. For small values of β, it may suffice to take two terms in the series in $e^{\beta\mu}$, that is,

$$\langle N \rangle = CV \frac{\sqrt{\pi}}{2} \left[\frac{e^{\beta\mu}}{\beta^{3/2}} \mp \frac{e^{2\beta\mu}}{(2\beta)^{3/2}} \right] \tag{5.29}$$

$$\langle E \rangle = CV \frac{3\sqrt{\pi}}{4} \left[\frac{e^{\beta\mu}}{\beta^{5/2}} \mp \frac{e^{2\beta\mu}}{(2\beta)^{5/2}} \right] \tag{5.30}$$

Now E is not $\frac{3}{2}NkT$; furthermore, fermions and bosons behave differently. Equation (5.29) is a quadratic equation for $\lambda = e^{\beta\mu}$ and determines the chemical potential as a function of N, V, and β (or T). Solving for λ and substituting in (5.30) gives E/V as a function of N/V and β. With (5.26), this gives the equation of state.

If one needs more than two terms in the series, the procedure is to use (5.27) to obtain λ as a power series in

$$X = \frac{2\langle N \rangle}{\sqrt{\pi}CV} \beta^{3/2}$$

and substitute the series into (5.28). Suppose $\lambda = bX + cX^2 + \cdots$. Then (5.27) is

$$\begin{aligned} X &= \lambda \mp \lambda^2 2^{-3/2} + \cdots \\ &= bX + cX^2 + \cdots \mp \frac{b^2X^2 + 2bcX^3 + 2bdX^4 + c^2X^4 \cdots}{2^{3/2}} \end{aligned}$$

Equating coefficients of X^1 on the left- and righthand sides, we find $b = 1$, and from X^2 and X^3

$$0 = c \mp 2^{-3/2} b^2$$
$$0 = d \mp 2^{-1/2} bc + 3^{-3/2} b^3$$

One can now find c, and then d, and so on. For illustration, we will show a two-term approximation. With $c = \pm 2^{-3/2}$, $\lambda \simeq X \pm 2^{-3/2} X^2$, to be substituted into (5.30):

$$\begin{aligned}
\langle E \rangle &= \frac{CV}{\beta^{5/2}} \frac{3\sqrt{\pi}}{4} \left[X \pm 2^{-3/2} X^2 \mp \frac{X^2 \pm 2^{-1/2} X^3}{2^{5/2}} \right] \\
&= \frac{CV}{\beta^{5/2}} \frac{3\sqrt{\pi}}{4} \left[\frac{2\langle N \rangle}{\sqrt{\pi} CV} \beta^{3/2} \pm \frac{4\langle N \rangle^2 \beta^3}{\pi CV^2} (2^{-3/2} - 2^{-5/2}) \right] \\
&= \frac{3N}{2\beta} \pm \frac{3N^2}{4CV} \sqrt{\frac{\beta}{2\pi}} \qquad (5.31)
\end{aligned}$$

The equation of state is then

$$PV = \frac{N}{\beta} \pm \frac{N^2}{2CV} \sqrt{\frac{\beta}{2\pi}}$$

showing the first correction to the ideal-gas law. This is the beginning of a virial expansion (see Section 7.1).

To calculate the equation of state of low temperatures, a different approach is required. One wants an expansion in powers of kT rather than β. Fermions and bosons have to be treated separately.

5.3. FERMI GAS: ELECTRONS IN METALS

To develop the equation of state for fermions, we start from $T = 0$. At $T = 0$, $\beta \to \infty$, the occupation number of all states of energy above μ is zero, and the occupation number of all states of energy below μ is unity. Equations (5.24) and (5.25) become

$$\frac{\langle N \rangle}{V} = \frac{2^{5/2} \pi m^{3/2} D}{h^3} \int_0^\mu dE\, E^{1/2} = \frac{2^{5/2} \pi m^{3/2} D}{h^3} \frac{\mu^{3/2}}{3/2} \qquad (5.32)$$

$$\frac{\langle E \rangle}{V} = \frac{2^{5/2} \pi m^{3/2} D}{h^3} \int_0^\mu dE\, E^{3/2} = \frac{2^{5/2} \pi m^{3/2} D}{h^3} \frac{\mu^{5/2}}{5/2} \qquad (5.33)$$

Dividing the second equation by the first gives

$$\frac{\langle E \rangle}{\langle N \rangle} = \frac{3\mu}{5}$$

so the chemical potential, which is the energy of the highest-energy occupied state, equals $\frac{5}{3}$ times the average energy per particle. The chemical potential at $T = 0$ is referred to as the Fermi energy and abbreviated by ε_F. (Some authors use ε_F to refer to the chemical potential at any given temperature.) Since $\langle E \rangle = \frac{3}{2}PV$, $PV = \frac{2}{5}N\mu = \frac{2}{5}N\varepsilon_F$. Note that PV is not zero, even though $T = 0$; the exclusion principle prevents all the particles from occupying the lowest energy state.

The most important use we will make of Fermi–Dirac statistics is to describe the conduction electrons of a metal. We will abbreviate by C the collection of constants that multiplies the integrals in (5.32) and (5.33). For electrons, with $D = 2$ and $m = 9.1 \times 10^{-31}$ kg,

$$C = 1.06 \times 10^{50}\ \mathrm{J}^{-3/2}\ \mathrm{cm}^{-3}$$

so that

$$\varepsilon_F = 5.8 \times 10^{-34}\left(\frac{N}{V}\right)^{2/3}\ \mathrm{J\ cm}^2 = 4.2 \times 10^{-18}\bar{V}^{-2/3}\ \mathrm{J\ cm}^2$$

where $\bar{V}$ is the volume per mole. Suppose a metal has a density of 10 g/cm^3 and an atomic mass of 160 g/mol, and that each metal atom has two valence electrons to contribute to the conduction band. Then $\bar{V} = 8$ cm^3/mol and $\varepsilon_F/k = 7.5 \times 10^4$ K, so that, even for temperatures in the thousands of Kelvins, ε_F/kT is very large. Thus such temperatures are more appropriately considered as low temperatures than high temperatures. The large size of ε_F for electrons means the pressure is high:

$$P = \frac{2}{5}\frac{\varepsilon_F}{V} = 3.1 \times 10^4\ \mathrm{J\ cm}^{-3} = 3.1 \times 10^5\ \mathrm{atm}$$

for the typical metal with $\bar{V} = 8$ cm^3/mol. The electron cloud has a great tendency to expand; it is held in by the electrostatic attraction of the ionic cores, as discussed later.

To calculate properties like E, PV, and μ for $T > 0$, we take advantage of the fact that μ is hundreds of times kT at room temperature, so an expansion in powers of kT/μ will converge rapidly. Integrating (5.24) by parts gives

$$\frac{\langle N \rangle}{CV} = \frac{2\beta}{3}\int_0^\infty dE\, E^{3/2}\, \frac{e^{-\beta(\mu - E)}}{[e^{-\beta(\mu - E)} + 1]^2}$$

The last factor is equal to $\frac{1}{4}$ when $E = \mu$ and decreases rapidly as $|\mu - E|$ increases. If μ/kT is 250, as we calculate for conduction electrons in a typical metal at 300 K, and E is 1% bigger or 1% smaller than μ, this factor is 0.07. Thus only values of E near μ contribute to the integral, so one can expand $E^{3/2}$ in a power series in $E - \mu$ and extend the lower limit on the integral to $-\infty$. Then, changing the integration variable to $x = E - \mu$,

$$\frac{\langle N \rangle}{CV} = \frac{2\beta}{3} \int_{-\infty}^{\infty} dx \left\{ \mu^{3/2} + \frac{3}{2} \mu^{1/2} x + \frac{3}{8} \mu^{-1/2} x^2 + \cdots \right\} \frac{e^{\beta x}}{[e^{\beta x} + 1]^2} \tag{5.34}$$

Equation (5.34) must now be transformed to give an expression for μ in terms of $\langle N \rangle / CV$ or ε_F. This is done in the next two paragraphs, with the result being equation (5.37).

The integral

$$\int_{-\infty}^{\infty} dx x^n \frac{e^{\beta x}}{[e^{\beta x} + 1]^2} = \beta^{-(n+1)} \int_{-\infty}^{\infty} dy \frac{y^n e^y}{[e^y + 1]^2}$$

vanishes for n odd because the integrand changes sign when y is replaced by $-y$, and is seen to equal β^{-1} for $n = 0$. For higher even powers of n, the integrand is expanded in a power series in y and integrated term by term, giving

$$\int_{-\infty}^{\infty} dx x^n \frac{e^{\beta x}}{[e^{\beta x} + 1]^2} = -2n!\, \beta^{-(n+1)} \sum_{m=1}^{\infty} \frac{(-1)^m}{m^n}$$

The sum is $-\pi^2/12$ for $n = 2$ and $-7\pi^4/720$ for $n = 4$. Putting the values for the integrals into (5.34) and substituting $\frac{2}{3}\varepsilon_F^{3/2}$ for $\langle N \rangle / CV$, we have

$$\varepsilon_F^{3/2} = \beta \left(\frac{\mu^{3/2}}{\beta} + \frac{\pi^2 \mu^{-1/2}}{8\beta^3} + \frac{7\pi^4 \mu^{-5/2}}{640} \cdots \right)$$

or

$$\left(\frac{\mu}{\varepsilon_F} \right)^{-3/2} = \left(1 + \frac{\pi^2}{8} \frac{1}{[\beta\mu]^2} + \frac{7\pi^4}{640} \frac{1}{[\beta\mu]^4} + \cdots \right) \tag{5.35}$$

What we really want, however, is an expression for μ in terms of ε_F.

To find it we write μ/ε_F as a power series in the small quantity $z = (\beta\varepsilon_F)^{-1}$.

$$\mu = \varepsilon_F[1 + az + bz^2 + cz^3 + dz^4 + \cdots] \tag{5.36}$$

When this is substituted for μ wherever it appears in (5.35) and coefficients of z, z^2, and so on, on left- and right-hand sides of the equation are set equal, one sees that odd powers of z vanish from (5.36). What remains of (5.35) is

$$-\frac{3}{2}(bz^2 + dz^4 + \cdots) + \frac{15}{8}(bz^2 + dz^4 + \cdots)^2$$

$$= \frac{\pi^2 z^2}{8}\left[1 - 2(bz^2 + dz^4 + \cdots) - 3(bz^2 + dz^4 + \cdots)^2 \cdots\right]$$

$$+ \frac{7\pi^4}{640} z^4(1 + \cdots)$$

Equating the coefficients of z^2, one gets $b = -\pi^2/12$; equating the coefficients of z^4 gives

$$-\frac{3}{2}d + \frac{15}{8}b^2 = -\frac{\pi^2 b}{4} + \frac{7\pi^4}{640}$$

or $d = -\pi^4/80$. Now (5.36) is

$$\frac{\mu}{\varepsilon_F} = 1 - \frac{\pi^2}{12(\beta\varepsilon_F)^2} - \frac{\pi^4}{80(\beta\varepsilon_F)^4} + \cdots \tag{5.37}$$

Equation (5.25) for the energy $\langle E \rangle$ may be treated the same way as that for $\langle N \rangle$, (5.24):

$$\frac{\langle E \rangle}{CV} = \frac{2\beta}{5}\int_{-\infty}^{\infty} dE\, E^{5/2} \frac{e^{-\beta(\mu - E)}}{[e^{-\beta(\mu-E)} + 1]^2}$$

$$= \frac{2\beta}{5}\int_{-\infty}^{\infty} dx \left\{\mu^{5/2} + \frac{5}{2}\mu^{3/2}x + \frac{15}{8}\mu^{1/2}x^2 + \cdots\right\} \frac{e^{\beta x}}{[e^{\beta x} + 1]^2}$$

$$= \frac{2\beta}{5}\left[\mu^{5/2}\beta^{-1} + 0 + \frac{15}{8}\mu^{1/2}\frac{\pi^2}{3\beta^3} + 0 - \frac{15}{384}\mu^{-3/2}\frac{7\pi^4}{15} + \cdots\right]$$

Using (5.37), we find

$$\frac{\langle E \rangle}{CV} = \frac{2}{5}\varepsilon_{\mathrm{F}}^{5/2}\left[1 + \frac{5\pi^2}{12\beta^2\varepsilon_{\mathrm{F}}^2} - \frac{\pi^4}{16\beta^4\varepsilon_{\mathrm{F}}^4} + \cdots\right]$$

Then, since $\varepsilon_{\mathrm{F}}^{3/2} = \frac{3}{2}N/(CV)$,

$$E = \frac{3}{5}N\varepsilon_{\mathrm{F}}\left[1 + \frac{5\pi^2}{12\beta^2\varepsilon_{\mathrm{F}}^2} - \frac{\pi^4}{16\beta^4\varepsilon_{\mathrm{F}}^4} + \cdots\right] \tag{5.38}$$

(the thermodynamic energy E is identified, as usual, with the average energy $\langle E \rangle$).

All of the thermodynamic functions may now be obtained in terms of ε_{F} and β. The Gibbs free energy G is $N\mu$, with μ given by (5.37). The pressure P is $\frac{2}{3}E/V$, with E given by (5.38). Then the Helmholtz free energy A is $G - PV$ and the enthalpy H is $E + PV = \frac{5}{3}E$. The entropy may be obtained from $TS = H - G$. The heat capacity C_V is obtained by differentiating (5.38):

$$C_V = \left(\frac{\partial E}{\partial T}\right)_V = \tfrac{3}{5}Nk\left[\frac{5\pi^2}{6\beta\varepsilon_{\mathrm{F}}} - \frac{\pi^4}{4\beta^3\varepsilon_{\mathrm{F}}^3} + \cdots\right] \tag{5.39}$$

According to (5.39), the heat capacity of an electron gas becomes proportional to T when $T \to 0$. This means that, at low enough temperature, the contribution of the conduction electrons to the heat capacity of a metal crystal exceeds the contribution of the crystal vibrations, which is proportional to T^3 [see Section 3.3, equation (3.24)].

It may seem strange that one can model the conduction electrons as a gas. A gas implies independent or noninteracting particles, and there are strong electrostatic repulsions between the electrons. The model becomes more reasonable when one remembers that each electron is subject to the electrostatic attraction of the positively charged ion cores as well as to the repulsion of the other conduction electrons. Furthermore, the exclusion principle, which requires that conduction-electron wave functions be orthogonal to inner-shell orbitals of the cores, keeps the conduction electrons away from the nuclei, providing a repulsive "pseudopotential" which cancels off much of the attraction of the nuclei. The net result is that the potential "seen" by each electron is essentially constant inside the metal, and rises sharply at the boundaries. A first approximation to this potential (in one dimension) is shown in Figure 5.3. If the height of the walls of this potential well is much greater than the energy of the highest-energy occupied state (approximately ε_{F}), the potential may be considered to be infinite outside the metal. This justifies considering the conduction electrons as particles in a box, that is, as a gas.

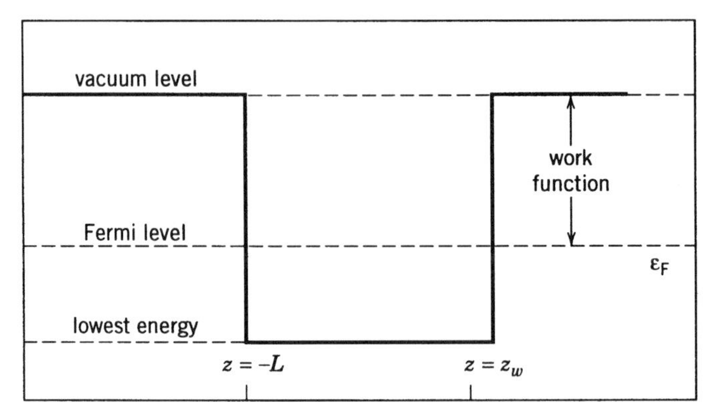

Figure 5.3. Model potential for conduction electrons in a metal. The metal extends from $z = -L$ to $z = z_w$. The lowest electronic energy level inside the metal is lower than the lowest level outside (vacuum level). The work function Φ is the energy difference between the vacuum level and the most energetic electron in the metal, which is at the Fermi level.

The pressure, calculated as $\frac{2}{3}E/V$, is

$$P = \frac{2}{5}\frac{N}{V}\varepsilon_F\left[1 + \frac{5\pi^2}{12\beta^2\varepsilon_F^2} - \frac{\pi^4}{16\beta^4\varepsilon_F^4} + \cdots\right] \tag{5.40}$$

but typically only the first term is important. With $\bar{V} = 8\ \text{cm}^3/\text{mol}$, we found P to be several times 10^5 atm. This P is the contribution of the kinetic energy to the pressure. It is overcome by the contribution of the potential energy, the forces exerted on the conduction electrons by the ion cores and the other electrons, which is extremely large and negative. The result is that the electron gas cannot expand, and electrons do not spontaneously escape from the metal. However, the potential is not infinitely high at the walls, so that an electron can escape if given enough energy.

The minimum energy required is the work function Φ, typically several eV for metals. For example, Na with 2.5×10^{22} electrons per cm^3 has $\Phi = 2.3$ eV, and Cu with 8.5×10^{22} electrons per cm^3 has $\Phi = 4.4$ eV. As shown in Figure 5.3, Φ is the difference between the energy of the most energetic electron inside the metal (Fermi level) and the energy of an electron far outside the metal (vacuum level). It may be considered to consist of two contributions, one a bulk-metal property and one a property of the surface. The latter is the electron charge multiplied by the surface potential, the electrostatic potential difference between just inside the metal and just outside. The bulk-metal contribution includes kinetic energy, short-range interactions, correlation energy, and so on. The kinetic energy part is what we have been calculating. The rest of the bulk-metal contribution and the surface contribution make the lowest electronic energy level inside the metal lower than outside. The surface potential varies more from metal to metal than does the work function itself, being, for example, 5 eV greater for Cu than for Na. The size of the surface potential relative to Φ is larger for polyvalent than for monovalent metals.

The reason for the surface potential is that, like the valence-electron density in atoms or molecules, the conduction-electron density in metals is more delocalized than the positive-ion density. The "spreading out" of these electrons to the exterior of the metal creates a dipolar layer, with the negative side to the outside of the metal, and the surface potential. From the Poisson equation, it is

$$V(\infty) - V(-\infty) = \frac{e}{\varepsilon_0} \int_{-L}^{\infty} dz\, z[\rho_+(z) - \rho(z)] \tag{5.41}$$

where the surface is perpendicular to the z-axis and we later let $L \to \infty$. The charge density due to ion cores is represented by $e\rho_+(z)$ and the electron charge density by $-e\rho(z)$ $(e > 0)$. Since the latter is much more diffuse than the former, ρ_+ is assumed to be a step function, $\rho_+(z) = \rho_b$ for $z < 0$ and 0 for $z \geq 0$, where $e\rho_b$ is the charge density of ion cores in bulk metal (see Figure 5.4).

We use Fermi–Dirac statistics to calculate $\rho(z)$. The electron density $\rho(z)$ must equal $-\rho_b$ in bulk metal $(z \to -\infty)$ and approach 0 outside the metal $(z \to \infty)$. Assume $\rho(z)$ becomes 0 for $z = z_w$, where $z_w > 0$, because of the attraction of the ion cores. This attraction is modeled by an infinite wall at $z = z_w$. The wave functions for electrons (orbitals) are particle-in-box wave functions with the box having length L in the x- and y-directions and length $L + z_w$ in the z-direction. The electron density is obtained by putting two electrons in each orbital and adding together the squares of the wave functions for all energies from 0 to ε_F (the Fermi–Dirac distribution for $T = 0$ is being used).

$$\rho = 2 \sum_{n_x, n_y, n_z} \left(\frac{2}{L}\right)^2 \left(\frac{2}{L + z_w}\right) \sin^2\left(\frac{n_x \pi x}{L}\right) \sin^2\left(\frac{n_y \pi y}{L}\right) \sin^2\left(\frac{n_z \pi (z - z_w)}{L + z_w}\right)$$

Since $L \to \infty$, $L + z_w$ may be replaced by L. The sum is over all the values of the quantum numbers such that

$$\frac{\left(n_x^2 + n_y^2 + n_z^2\right) h^2}{8mL^2} \leq \varepsilon_F \equiv \frac{N_F^2 h^2}{8mL^2} \tag{5.42}$$

where N_F is defined to simplify notation. We will approximate the sums by integrals.

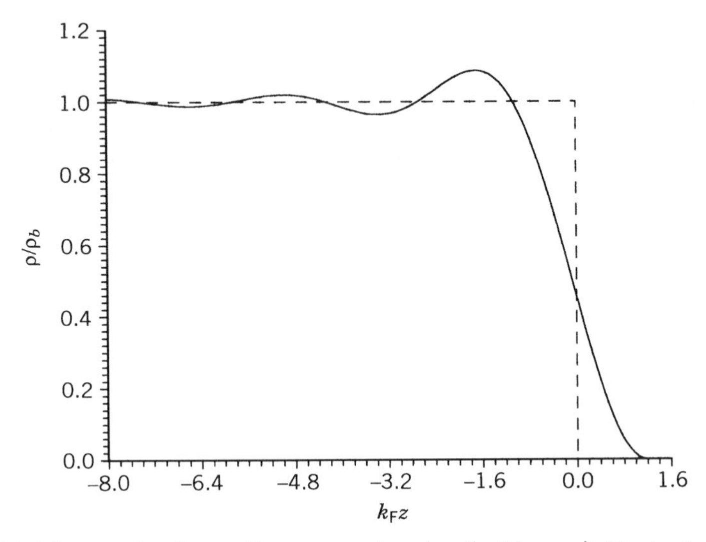

Figure 5.4. Electron density profile at a metal surface (solid curve). The broken curve, a step function, is the profile of positive (ionic) charge density. The electron density and ionic charge density both approach the bulk value, ρ_b, far inside the metal. The electron density drops to zero at $z = z_w = 3\pi/8k_F$ where k_F is the Fermi momentum.

In the integration over n_x, the phase $a_x = n_x \pi x/L$ varies from 0 to *many* times 2π, so that $\sin^2(a_x)$ may be replaced by its average value, $\frac{1}{2}$. The same goes for $\sin^2(a_y)$ where $a_y = n_y \pi y/L$, but *not* for $a_z = n_z \pi(z - z_w)/L$ when z is close to z_w. The integrations may be carried out, giving

$$\rho = 2\left(\frac{2}{L}\right)^3 \int_0^{N_F} dn_z \int_0^{\sqrt{N_F^2 - n_z^2}} dn_y \int_0^{\sqrt{N_F^2 - n_y^2 - n_z^2}} dn_x \frac{1}{4} \sin^2\left(\frac{n_z \pi (z - z_w)}{L}\right)$$

$$= \frac{\pi}{L^3} \frac{N_F^3}{3} \left[1 + 3\frac{\cos[2k_F(z - z_w)]}{[2k_F(z - z_w)]^2} - 3\frac{\sin[2k_F(z - z_w)]}{[2k_F(z - z_w)]^3}\right]$$

where $k_F = \pi N_F/L$. Since $k_F^2 = 8m\pi^2 \varepsilon_F/h^2$ according to (5.42), $\varepsilon_F = \hbar^2 k_F^2/2m$, and $\hbar k_F$ is called the Fermi momentum. Using (5.32),

$$\varepsilon_F = \frac{3^{2/3}(N/V)^{2/3} h^2}{8\pi^{2/3} m}$$

with $V = L^3$, so that (5.42) gives $N_F^3 = 3N/\pi$. Since $N/V = \rho_b$,

$$\rho = \rho_b \left[1 + 3\frac{\cos[2k_F(z - z_w)]}{[2k_F(z - z_w)]^2} - 3\frac{\sin[2k_F(z - z_w)]}{[2k_F(z - z_w)]^3}\right] \tag{5.43}$$

To determine z_w, we invoke the requirement of electroneutrality:

$$\int_{-L}^{z_w} dz\, \rho(z) = \int_{-L}^{\infty} dz\, \rho_+ = \int_{-L}^{0} dz\, \rho_b \tag{5.44}$$

$L \to \infty$) which, on insertion of (5.43), leads to $z_w = 3\pi/8k_F$. The density (5.43) is plotted, along with ρ_+, in Figure 5.4.

Finally, using (5.41), we find for the surface potential

$$\Delta V = V(\infty) - V(-\infty) = \frac{ek_F}{4\pi^2\varepsilon_0}\left(\frac{3\pi^2}{32} - 1\right) \tag{5.45}$$

(the integrals are somewhat messy). To calculate the magnitude of ΔV, we evaluate k_F for our typical metal, with $\varepsilon_F/k = 7.5 \times 10^4$ K. We find $k_F^2 = 8m\pi^2\varepsilon_F/h^2 = 1.70 \times 10^{16}$ cm^{-2} and $\Delta V = -0.446$ V. More realistic models for the conduction electrons and their interaction with the ion cores predict surface potentials slightly larger in magnitude, but concur with (5.45) in predicting a decrease in $|\Delta V|$ with increasing bulk electron density ρ_b. It should be noted that ΔV is not measurable experimentally (it is an axiom of electrochemistry that one cannot measure a difference in electrical potential between points in chemically different phases). One can measure the work function, which is $-e\,\Delta V$ plus the chemical potential of an electron in bulk metal, but not the two contributions separately.

5.4. DENSITY OF STATES AND SEMICONDUCTORS

Another application of Fermi–Dirac statistics is to the valence electrons in a semiconductor. In either a conductor, such as the metal we have been modeling, or a semiconductor, the valence-shell atomic orbitals are combined into delocalized metal wave functions $\{\psi_j\}$, representing states which can be occupied by the valence electrons. The energies of the $\{\psi_j\}$ are close together, and are considered to form a continuum or band, with $\mathcal{N}(\varepsilon)\,d\varepsilon$ wave functions with energies between ε and $\varepsilon + d\varepsilon$; $\mathcal{N}(\varepsilon)$ is called the density of states. In a metal, the number of electrons that are to occupy the $\{\psi_j\}$ is much less than the number of states in the band so that the Fermi level is in the middle of the band. At 0 K there are many unoccupied states with energies very close to occupied states. However, in an insulator or an intrinsic semiconductor, there are enough valence electrons to completely fill the valence band, so there are no empty states with energies just above the highest occupied level. If ε_v is the top of the valence band, $\mathcal{N}(\varepsilon) = 0$ for $\varepsilon_v \le \varepsilon \le \varepsilon_c$; at $\varepsilon = \varepsilon_c$, $\mathcal{N}(\varepsilon)$ becomes nonzero, as a new band, usually called the conduction band, begins. The difference $\varepsilon_c - \varepsilon_v$ is called the band gap. $\mathcal{N}(\varepsilon)$ for these bands is indicated in Figure 5.5 (solid curves).

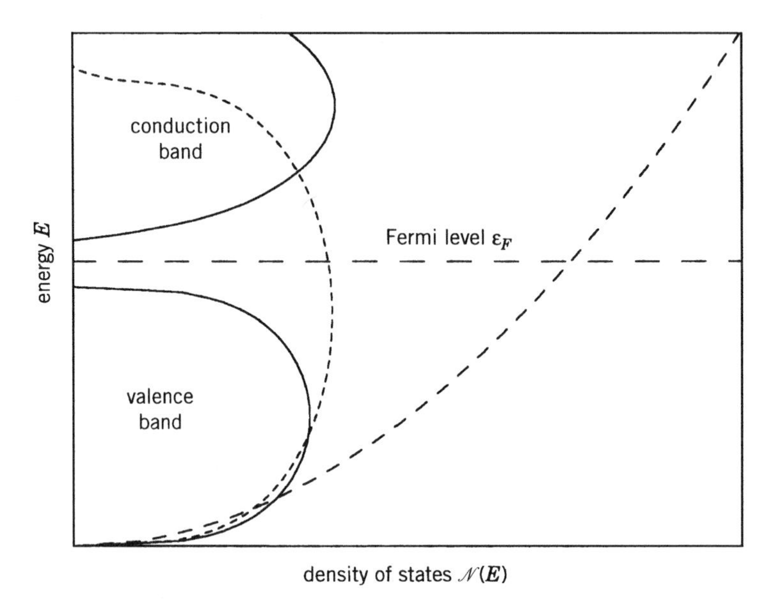

Figure 5.5. Densities of states for bands. The solid curves show $\mathcal{N}(E)$ for the valence and conduction bands of a semiconductor; the Fermi level is shown. The parabola (long dashed curve) is for free particles: $\mathcal{N}(E)$ proportional to $\sqrt{\varepsilon - \varepsilon_0}$ where ε_0 is the bottom of the band. The short dashed curve shows a typical energy band; note the parabolic behavior near the bottom and top.

If the band gap is too large for electrons to be excited from states in the valence band to states in the conduction band when the temperature is increased, one has an insulator. For small enough band gap, thermal excitation can occur, each excited electron leaving a hole (unoccupied state) in the valence band. One then has an *intrinsic* semiconductor, since the current-carrying species, electrons and holes, are created by thermal excitation. In an *extrinsic* semiconductor, there are holes in the valence band and/or electrons in the condition band because of heteroatoms (doping). See Figure 5.6.

The total number of states in a band, $\int \mathcal{N}(\varepsilon)\, d\varepsilon$, is finite (it is in fact proportional to the volume of the system). For particle-in-box wave functions [see (5.22) and (5.23)],

$$\mathcal{N}(\varepsilon) = \frac{dn(\varepsilon)}{d\varepsilon} = D\,\frac{2^{5/2}\pi m^{3/2}\varepsilon^{1/2}V}{h^3}$$

This $\mathcal{N}(\varepsilon)$ is unrealistic because $\int \mathcal{N}(\varepsilon)\, d\varepsilon$ is infinite. It could be used for modeling the conduction electrons in a metal because, in a metal, there are many empty states above the Fermi level, and because the density of states near the bottom of a band can be shown to be proportional to $(\varepsilon - \varepsilon_0)^{1/2}$, where ε_0 is the lowest energy in the band. The value of the heat capacity of an electron gas reflects $\mathcal{N}(\mu)$, the density of states at an energy equal to the chemical potential. We now show [equation (5.51)] how this comes about for a general $\mathcal{N}(\varepsilon)$. The bottom of the band is taken as 0.

Instead of (5.24) and (5.25), the total number of electrons is given by

$$N = \int_0^\infty \mathcal{N}(\varepsilon) \frac{1}{1 + e^{\beta(\varepsilon - \mu)}} \, d\varepsilon \tag{5.46}$$

and the total energy by

$$E = \int_0^\infty \mathcal{N}(\varepsilon) \frac{\varepsilon}{1 + e^{\beta(\varepsilon - \mu)}} \, d\varepsilon \tag{5.47}$$

Integrating (5.46) by parts,

$$N = \beta \int_0^\infty \frac{e^{\beta(\varepsilon - \mu)}}{[1 + e^{\beta(\varepsilon - \mu)}]^2} F(\varepsilon) \, d\varepsilon$$

where

$$F(\varepsilon) = \int_0^\varepsilon \mathcal{N}(\varepsilon') \, d\varepsilon'$$

is the total number of energy levels with energies between 0 and ε. As discussed previously, the fraction in the integrand of N is large only for ε close to μ, so one can extend the lower limit of integration to $-\infty$ and expand $F(\varepsilon)$ in a power series about $\varepsilon = \mu$. Then, with $x = \varepsilon - \mu$,

$$N = \beta \int_{-\infty}^\infty \frac{e^{\beta x}}{[1 + e^{\beta x}]^2} \left\{ F(\mu) + x \left[\frac{dF}{d\varepsilon} \right]_{\varepsilon = \mu} + \frac{1}{2} x^2 \left[\frac{d^2F}{d\varepsilon^2} \right]_{\varepsilon = \mu} + \cdots \right\}$$

We truncate the series after the quadratic term. Since [see (5.35)] the integrals involving x^0, x^1, and x^2 are, respectively, β^{-1}, 0, and $\pi^2/(3\beta^3)$,

$$N \simeq F(\mu) + \frac{\pi^2}{6\beta^2} \left[\frac{d^2F}{d\varepsilon^2} \right]_{\varepsilon = \mu} \tag{5.48}$$

If (5.47) for the energy is treated similarly to (5.46), we get

$$E \simeq G(\mu) + \frac{\pi^2}{6\beta^2} \left[\frac{d^2G}{d\varepsilon^2} \right]_{\varepsilon = \mu} \tag{5.49}$$

where

$$G(\varepsilon) = \int_0^{\varepsilon} \mathcal{N}(\varepsilon')\varepsilon'\, d\varepsilon'$$

Since $F(\varepsilon_F) = N$ and μ is not very different from ε_F,

$$F(\mu) \simeq F(\varepsilon_F) + (\mu - \varepsilon_F)\left[\frac{dF}{d\varepsilon}\right]_{\varepsilon = \varepsilon_F} = N + (\mu - \varepsilon_F)\mathcal{N}(\varepsilon_F).$$

and

$$G(\mu) \simeq \int_0^{\varepsilon_F} \mathcal{N}(\varepsilon')\varepsilon'\, d\varepsilon' + (\mu - \varepsilon_F)\mathcal{N}(\varepsilon_F)\varepsilon_F$$

Note also that $d^2F/d\varepsilon^2$ is $d\mathcal{N}(\varepsilon)/d\varepsilon$ and that $d^2G/d\varepsilon^2$ is $d(\varepsilon\mathcal{N})/d\varepsilon$. Putting all this into (5.48) and (5.49), we obtain

$$0 = (\mu - \varepsilon_F)\mathcal{N}(\varepsilon_F) + \left[\frac{d\mathcal{N}}{d\varepsilon}\right]_{\varepsilon_F} \frac{\pi^2}{6\beta^2} \tag{5.50}$$

and

$$E = \int_0^{\varepsilon_F} \mathcal{N}(\varepsilon')\varepsilon'\, d\varepsilon' + (\mu - \varepsilon_F)\mathcal{N}(\varepsilon_F)\varepsilon_F + \left[\frac{d(\varepsilon\mathcal{N})}{d\varepsilon}\right]_{\varepsilon_F} \frac{\pi^2}{6\beta^2}$$

Finally, solving (5.50) for $[d\mathcal{N}/d\varepsilon]$ and substituting the result into the preceding equation,

$$E = \int_0^{\varepsilon_F} \mathcal{N}(\varepsilon')\varepsilon'\, d\varepsilon' + \mathcal{N}(\varepsilon_F)\frac{\pi^2}{6\beta^2}$$

The heat capacity C_V is obtained by differentiating E with respect to T. The result is

$$C_V = \frac{\pi^2 k^2 T}{3}\mathcal{N}(\varepsilon_F) \tag{5.51}$$

which shows that from a measurement of the heat capacity of the conduction electrons one finds out about the density of states at the Fermi level. For transition metals, the heat capacity is larger by an order of magnitude than for the simple metals; the former have a high density of electronic states because of the incomplete d- and f-shells.

For intrinsic semiconductors, the density of states has a gap, separating the valence band, which is filled at zero temperature, from the conduction band, empty at zero temperature (Figure 5.5). The total number of electrons is

$$N = \int_0^{\varepsilon_v} \mathcal{N}_v(\varepsilon)\, d\varepsilon = \int_0^{\varepsilon_v} \mathcal{N}_v(\varepsilon) f(\varepsilon)\, d\varepsilon + \int_{\varepsilon_c}^{\infty} \mathcal{N}_c(\varepsilon) f(\varepsilon)\, d\varepsilon \quad (5.52)$$

where

$$f(\varepsilon) = \left[1 + e^{\beta(\varepsilon - \mu)}\right]^{-1}$$

Also, ε_v is the highest energy in the valence band, ε_c is the lowest energy in the conduction band, $\mathcal{N}_v$ is the density of states in the valence band, and $\mathcal{N}_c$ is the density of states in the conduction band. Only the two bands are considered. Equation (5.52) may be rewritten as

$$\int_0^{\varepsilon_v} \mathcal{N}_v(\varepsilon)[1 - f(\varepsilon)]\, d\varepsilon = \int_{\varepsilon_c}^{\infty} \mathcal{N}_c(\varepsilon) f(\varepsilon)\, d\varepsilon$$

The right-hand side is the number of electrons in the conduction band, and the left-hand side the number of holes in the valence band; they must be equal.

There is another interesting equality here. Substituting for $f(\varepsilon)$ in the above equation,

$$\int_0^{\varepsilon_v} \mathcal{N}_v(\varepsilon) \frac{1}{1 + e^{\beta(\mu - \varepsilon)}}\, d\varepsilon = \int_{\varepsilon_c}^{\infty} \mathcal{N}_c(\varepsilon) \frac{1}{1 + e^{\beta(\varepsilon - \mu)}}\, d\varepsilon \quad (5.53)$$

In (5.53),

$$f(\varepsilon) = \left[1 + e^{\beta(\varepsilon - \mu)}\right]^{-1}$$

is the probability of finding an electron at energy ε, while

$$1 - f(\varepsilon) = \frac{e^{\beta(\varepsilon - \mu)}}{1 + e^{\beta(\varepsilon - \mu)}} = \left[1 + e^{\beta(\mu - \varepsilon)}\right]^{-1}$$

is the probability of finding a hole at energy ε. Thus the probability of finding an electron in a conduction-band state at energy $|\varepsilon - \mu|$ above the chemical potential is equal to the probability of finding a hole in a valence-band state at energy $|\varepsilon - \mu|$ below the chemical potential. Both electrons and holes obey Fermi–Dirac statistics, but, for holes, energies are measured downward. As we shall show, μ is near the middle of the band gap (see Figure 5.5). Since $\varepsilon_g = \varepsilon_c - \varepsilon_v$ is of the order of electronvolts, where $1\ \text{eV}/k = 1.2 \times 10^4$ K, most of the electrons in the conduction band are near the bottom, and most of the holes in the valence band are near the top.

It can be shown that, near the bottom of any band, the density of states is given by

$$\mathcal{N}(\varepsilon) = \frac{V}{2\pi^2}\left(\frac{2m_e}{\hbar^2}\right)^{3/2}(\varepsilon - \varepsilon_0)^{1/2}$$

where ε_0 is the lowest energy in the band (band origin). This is of the particle-in-box or free-particle form, except that m_e is an effective mass, not the true electronic mass. Also, near the top of a band, the density of states is given by

$$\mathcal{N}(\varepsilon) = \frac{V}{2\pi^2}\left(\frac{2m_h}{\hbar^2}\right)^{3/2}(\varepsilon_t - \varepsilon)^{1/2}$$

where m_h is an effective mass, of the same size as m_e, and ε_t is the highest energy in the band. (Again note the symmetry between electrons and holes.) Using these in (5.53),

$$m_h^{3/2}\int_{-\infty}^{\varepsilon_v}\frac{(\varepsilon_v - \varepsilon)^{1/2}}{1 + e^{\beta(\mu-\varepsilon)}}\,d\varepsilon = m_e^{3/2}\int_{\varepsilon_c}^{\infty}\frac{(\varepsilon - \varepsilon_c)^{1/2}}{1 + e^{\beta(\varepsilon-\mu)}}\,d\varepsilon \tag{5.54}$$

with the lower limit in the first integral extended to $-\infty$ because the integrand is appreciable only near $\varepsilon = \mu$. If $m_h = m_e$, (5.54) requires that $\mu = \frac{1}{2}(\varepsilon_v + \varepsilon_c)$, that is, the chemical potential is in the center of the band gap. If m_h and m_e are not too different, μ is near the center. Then, since $(\varepsilon_c - \varepsilon_v)/k$ is several times 10^4 K, $e^{\beta(\mu-\varepsilon)} \gg 1$ when ε is in the valence band, and $e^{\beta(\varepsilon-\mu)} \gg 1$ when ε is in the conduction band. The Fermi–Dirac distributions of electrons and holes become Boltzmann distributions, as is appropriate for a dilute gas.

The total number of electrons in the conduction band is then

$$n_e = \int_{\varepsilon_c}^{\infty}\frac{V}{2\pi^2}\left(\frac{2m_e}{\hbar^2}\right)^{3/2}(\varepsilon - \varepsilon_c)^{1/2}e^{-\beta(\varepsilon-\mu)}\,d\varepsilon$$

$$= \frac{V}{2\pi^2}\left(\frac{2m_e}{\hbar^2}\right)^{3/2}e^{-\beta(\varepsilon_c-\mu)}\sqrt{\frac{\pi}{4\beta^3}} \tag{5.55}$$

and the total number of holes in the valence band is

$$n_h = \int_{-\infty}^{\varepsilon_v} \frac{V}{2\pi^2}\left(\frac{2m_h}{\hbar^2}\right)^{3/2} (\varepsilon_v - \varepsilon)^{1/2} e^{-\beta(\mu-\varepsilon)} d\varepsilon$$

$$= \frac{V}{2\pi^2}\left(\frac{2m_h}{\hbar^2}\right)^{3/2} e^{-\beta(\mu-\varepsilon_v)} \sqrt{\frac{\pi}{4\beta^3}} \tag{5.56}$$

Since n_e and n_h are equal to each other,

$$\tfrac{3}{2}\ln\left(\frac{m_e}{m_h}\right) = -\beta(2\mu - \varepsilon_v - \varepsilon_c)$$

which allows the determination of μ. Also, one can multiply (5.55) and (5.56) to get

$$\frac{n_e n_h}{V^2} = \frac{1}{16\pi^3\beta^3}\left(\frac{4m_e m_h}{\hbar^4}\right)^{3/2} e^{-\beta(\varepsilon_c-\varepsilon_v)} \tag{5.57}$$

which is independent of μ, and take the square root to get

$$\frac{n_e}{V} = \frac{n_h}{V} = \frac{(m_e m_h)^{3/4}}{\hbar^3\sqrt{2\pi^3\beta^3}} e^{-\beta(\varepsilon_c-\varepsilon_v)/2} \tag{5.58}$$

Equation (5.57) is like a chemical equilibrium expression, with electrons and holes as products and nothing as the reactant: the lowest energy level of products is higher than the lowest energy level of reactant by the band gap.

An extrinsic semiconductor contains a small number of heteroatoms among the host atoms. The heteroatoms may be electron donors or electron acceptors, forming n-type and p-type semiconductors, respectively. Donor atoms, such as a Group V atom like As in a Group IV semiconductor (e.g., Si or Ge), have extra electrons, in states with energy close to ε_c. Some of these electrons may be promoted to the conduction band. Acceptor atoms, such as Group III atoms (e.g., Ga or B) in a Group IV semiconductor, have fewer electrons than the host atoms, and empty states with energy slightly above ε_v. A valence electron from a host atom can jump into such a state, leaving a hole in the valence band. The energies and densities of states are shown in Figure 5.6. In the n-type semiconductors, it is the electrons that are mobile; in the p-type, it is the holes.

Suppose there are N_d donor atoms, each of which possesses one electronic state with energy ε_d, localized on the atom. Interelectronic repulsion prevents more than one electron from being in any one localized state, but, if there is

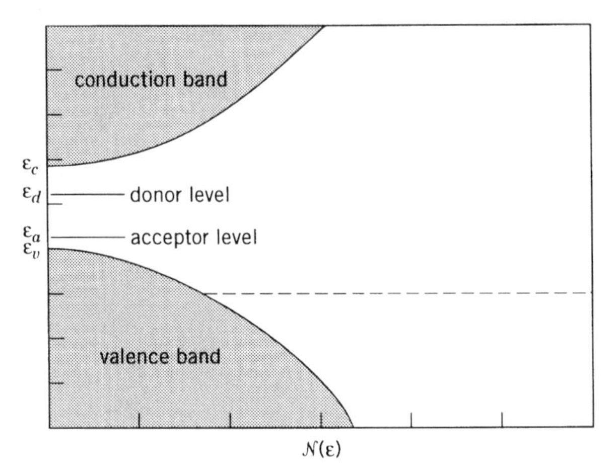

Figure 5.6. Density of states for extrinsic (doped) semiconductor. The top of the valence band ε_v and the bottom of the conduction band ε_c are shown, as well as the discrete levels provided by donor (ε_d) and acceptor (ε_a) atoms.

one electron, it can be of either α or β spin. To describe this, consider N electrons of which n_d go into the N_d donor-atom states and the others into delocalized states. Let the energies of the latter be ε_i, with degeneracy g_i ($g_i = 2$ for spin degeneracy), and let there be n_i electrons in the ith level. The equilibrium situation corresponds to the maximum of

$$W = 2^{n_d} \frac{N_d!}{n_d!\,(N_d - n_d)!} \prod_i \frac{g_i!}{n_i!\,(g_i - n_i)!} \tag{5.59}$$

subject to the constraints

$$n_d + \sum_i n_i = N \quad \text{and} \quad n_d \varepsilon_d + \sum_i n_i \varepsilon_i = E \tag{5.60}$$

The factor of 2^{n_d} arises from the two possible spin states for an electron in a donor-atom state. The result of the minimization (see Problem 5.18) is

$$n_i = g_i[1 + e^{\beta(\varepsilon_i - \mu)}]^{-1}$$

which is the Fermi–Dirac distribution and

$$\frac{N_d}{n_d} = 1 + \tfrac{1}{2} e^{\beta(\varepsilon_d - \mu)} \tag{5.61}$$

Note that (5.61) allows no more than one electron per donor-atom state.

Since donor atoms give up their electrons to the conduction band, one is often interested in calculating the number of holes:

$$n_h^d = N_d - n_d = N_d\left[1 + 2e^{-\beta(\varepsilon_d - \mu)}\right]^{-1} \tag{5.62}$$

is the number of holes in donor-atom states. An equation similar to (5.62) gives the number of electrons in acceptor atom states; if there are N_a acceptor atoms,

$$n_a = N_a\left[1 + \tfrac{1}{2}e^{\beta(\varepsilon_a - \mu)}\right]^{-1}$$

The number of electrons in the conduction band is given by (5.55) or

$$n_e = 2\left(\frac{2\pi m_e}{h^2\beta}\right)^{3/2} V e^{\beta(\mu - \varepsilon_c)} \tag{5.63}$$

and the number of holes in the valence band is given by (5.56) or

$$n_h = 2\left(\frac{2\pi m_h}{h^2\beta}\right)^{3/2} V e^{-\beta(\mu - \varepsilon_v)}$$

Electrical neutrality requires

$$n_h + n_h^d = n_e + n_a$$

This equation is used to calculate the chemical potential μ.

We can now examine some special cases. Suppose $N_a = 0$ (no acceptors, so $n_a = 0$) and N_d is relatively large. Since ε_d is close to ε_c, it will be easier to excite electrons from the donor atoms into the conduction band than to create holes in the valence band, and we will have $n_e \gg n_h$ or $n_e \simeq n_h^d$ (all the electrons in the conduction band come from the donor atoms). Equating (5.55) and (5.62),

$$Qe^{\beta(\mu - \varepsilon_c)} = \frac{N_d}{1 + 2e^{-\beta(\varepsilon_d - \mu)}}$$

where $Q = 2(2\pi m_e/h^2\beta)^{3/2}V$. This is a quadratic equation for $e^{\beta\mu}$ with the solution

$$e^{\beta\mu} = \frac{-1 + \sqrt{1 + 8N_d e^{\beta(\varepsilon_c - \varepsilon_d)}Q^{-1}}}{4e^{-\beta\varepsilon_d}}$$

For small N_d or large T, the second term in the square root will be much less than 1 and

$$e^{\beta\mu} \simeq \frac{N_d}{Q} e^{\beta\varepsilon_c}$$

which makes n_e approximately equal to N_d. In this case, all donor atoms are fully ionized. The opposite extreme is when the second term in the square root is large compared to 1; then

$$e^{\beta\mu} \simeq \sqrt{\frac{N_d}{2Q} e^{\beta(\varepsilon_c + \varepsilon_d)}}$$

and

$$n_e \simeq \sqrt{\tfrac{1}{2} Q N_d e^{-\beta(\varepsilon_d + \varepsilon_c)}}$$

In this case (low temperature or large number of donor atoms), the number of electrons in the conduction band is proportional to the square root of the number of donor atoms.

In an n-type semiconductor $N_d \gg N_a$ but $N_a \neq 0$. From (5.63), $e^{\beta\mu} = (n_e/Q)e^{\beta\varepsilon_c}$, so (5.62) becomes

$$n_h^d = \frac{N_d}{1 + 2e^{-\beta(\varepsilon_d - \mu)}} = \frac{N_d}{1 + 2n_e Q^{-1} e^{\beta(\varepsilon_c - \varepsilon_d)}}$$

Almost all the acceptor states will be occupied by electrons, and there will be almost no holes in the valence band, so that $n_h^d \simeq n_e + N_a$. Using the above expression for n_h^d gives an equation which can be rearranged to

$$\frac{2}{Q} \frac{(n_e + N_a)n_e}{N_d - N_a - n_e} = e^{-(\varepsilon_c - \varepsilon_d)/kT} \tag{5.64}$$

A plot of

$$k \ln\left[T^{-3/2} \frac{(n_e + N_a)n_e}{N_d - N_a - n_e}\right]$$

vs. T^{-1} (note that Q is proportional to $T^{3/2}$) should be linear with slope $\varepsilon_c - \varepsilon_d$. This may be used to determine the important energy difference $\varepsilon_c - \varepsilon_d$, if n_e, the number of carriers (conduction electrons), is measured at a series of temperatures. At low temperatures, there is both a complication and a simplification. The complication is that there may be several low-energy donor states. For Sb and As in Ge, there is a triplet at ε_d and a singlet at $\varepsilon_d - \delta\varepsilon_d$, where $\delta\varepsilon_d/k = 6.6$ K for Sb and 46.4 K for As. Then $e^{\beta(\varepsilon_c - \varepsilon_d)}$

must be replaced by

$$\tfrac{1}{4}e^{\beta(\varepsilon_c - \varepsilon_d + \delta\varepsilon_d)} + \tfrac{3}{4}e^{\beta(\varepsilon_c - \varepsilon_d)}$$

The simplification is that at low temperatures $n_e \ll N_a$ so that (5.64) becomes

$$\frac{2}{Q}\frac{n_e N_a}{N_d - N_a} = \frac{1}{\tfrac{1}{4}e^{\beta(\varepsilon_c - \varepsilon_d + \delta\varepsilon_d)} + \tfrac{3}{4}e^{\beta(\varepsilon_c - \varepsilon_d)}} = \frac{4e^{-\beta(\varepsilon_c - \varepsilon_d)}}{e^{\beta\delta\varepsilon_d} + 3}$$

Since the quantity $D = e^{\beta\delta\varepsilon_d} + 3$ is known as a function of temperature, one can plot the logarithm of $n_e D T^{-3/2}$ vs. $1/T$ to find the value of $\varepsilon_c - \varepsilon_d$.

5.5. BOSE GAS AT LOW TEMPERATURE: PHOTONS

The equation of state of a gas of bosons at low temperatures is quite different from that of a gas of fermions, because there is no exclusion principle preventing all the bosons from going into the lowest-energy state. At high temperatures, the Bose–Einstein gas can be treated along with the Fermi–Dirac gas, as was done in Section 5.2, equations (5.18) through (5.31). According to (5.13), the occupation number of state j is

$$\langle s_j \rangle = \left[\pm 1 + e^{\beta(E_j - \mu)}\right]^{-1}$$

with the lower sign for Bose–Einstein statistics and the upper for Fermi–Dirac. In the former case, $\langle s_j \rangle$ would become infinite if $\beta(E_j - \mu)$ became zero; to avoid this, one requires μ always to lie below E_0, the lowest-energy level. At low temperature ($\beta \to \infty$), $e^{\beta(E_0 - \mu)}$ becomes smaller (and closer to 1) than $e^{\beta(E_j - \mu)}$ for $j > 0$, making the occupation number much larger for the lowest-energy level than for the others. But if one occupation number is much larger than all the others, the replacement of the sum over states by an integral over energy, as in (5.46), is invalid.

The problem appears in a different way for the ideal gas (particles in a box), for which (5.27) was derived. For bosons,

$$\begin{aligned}
\langle N \rangle &= \frac{2^{5/2}\pi m^{3/2} V}{h^3} \int_0^\infty dE \, \frac{E^{1/2} e^{\beta(\mu - E)}}{1 - e^{\beta(\mu - E)}} \\
&= \frac{2^{5/2}\pi m^{3/2} V}{h^3} \int_0^\infty dE \, E^{1/2} e^{\beta(\mu - E)} \sum_{j=0}^{\infty} e^{j\beta(\mu - E)} \\
&= \frac{2^{3/2}\pi^{3/2} m^{3/2} V}{h^3} \sum_{j=0}^{\infty} \frac{e^{(j+1)\mu\beta}}{[(j+1)\beta]^{3/2}}
\end{aligned}$$

We have integrated the series term by term and put the degeneracy D equal to 1. Note that the lowest-energy level is at 0, so μ must be negative. We rewrite this as:

$$\frac{\langle N\rangle}{(2\pi m/\beta h^2)^{3/2}V} = \sum_{j=0}^{\infty} \frac{e^{(j+1)\mu\beta}}{(j+1)^{3/2}} = \sum_{j=1}^{\infty} \frac{\lambda^j}{j^{3/2}} \tag{5.65}$$

where $\lambda = e^{\beta\mu}$. When $\beta\mu$ is very large (and negative), λ is very small and the series reduces to its first term. This corresponds to the ideal gas, equation (5.16), with $\langle N\rangle/V$ proportional to λ. For larger λ (less negative $\beta\mu$), $\langle N\rangle/V$ increases more rapidly with μ, as shown in Figure 5.7. Since $\beta\mu$ cannot be greater than 0, λ cannot exceed 1, and there appears to be a maximum value of $\langle N\rangle$, which will be designated by N^*:

$$\frac{N^*}{V} = \left(\frac{2\pi m}{\beta h^2}\right)^{3/2} \sum_{j=1}^{\infty} \left(\frac{1}{j}\right)^{3/2} \simeq 2.6124\,\frac{q^{\text{tr}}}{V} \tag{5.66}$$

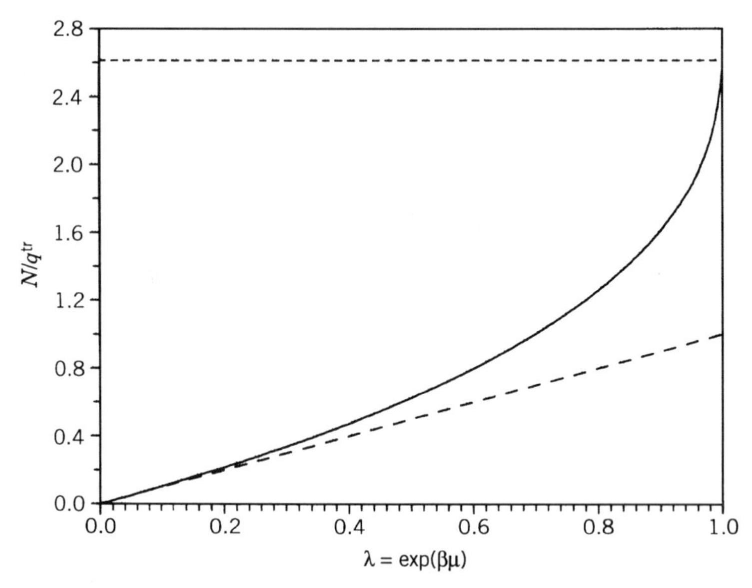

Figure 5.7. Total number of particles N divided by $q^{\text{tr}} = (2\pi mkT/h^2)^{3/2}$ V, as a function of $\lambda = e^{\beta\mu}$ according to (5.65). Since λ cannot exceed 1, N/V is predicted to have a limiting value of $\cong 2.6124 q^{\text{tr}}/V$. For small λ, $N/V = \lambda q^{\text{tr}}/V$, which corresponds to the classical ideal gas. This is represented by the dashed line.

where

$$q^{\text{tr}} = \left(\frac{2\pi mkT}{h^2}\right)^{3/2} V$$

is the translational partition function [see equations (3.6) and (3.7)]. There is no physical reason for a maximum $\langle N \rangle$. This incorrect result comes from not having given the ground state special treatment.

Much of the error may be corrected by an approximate treatment of the problem, in which the ground state is singled out and the sum over all the other states is approximated as an integral. This treatment is not completely rigorous, but the exact analysis is much more complex. The grand partition function is written

$$\begin{aligned} \ln \Xi &= \sum_j \ln[1 - e^{\beta(\mu - E_j)}]^{-1} \\ &\simeq -\ln[1 - e^{\beta\mu}] - \frac{2^{5/2}\pi m^{3/2}}{h^3} V \int_0^\infty dE\, E^{1/2} \ln[1 - e^{\beta(\mu - E)}] \end{aligned} \tag{5.67}$$

The integral is integrated by parts and, in the resulting integral, the series

$$[1 - e^{\beta(\mu - E)}]^{-1} = \sum_{j=0}^{\infty} e^{j\beta(\mu - E)}$$

is inserted and integrated term by term. The result is

$$\ln \Xi = -\ln[1 - e^{\beta\mu}] + q^{\text{tr}} \sum_{j=0}^{\infty} \frac{e^{(j+1)\beta\mu}}{(j+1)^{5/2}} \tag{5.68}$$

Then the total number of particles is

$$N = kT\left(\frac{\partial \ln \Xi}{\partial \mu}\right) = \frac{1}{e^{-\beta\mu} - 1} + q^{\text{tr}} \sum_{j=0}^{\infty} \frac{e^{(j+1)\beta\mu}}{(j+1)^{3/2}} \tag{5.69}$$

The second term in (5.69) is the same as (5.65) and was graphed in Figure 5.7. It now represents the number of particles in states *other* than the ground state. Its value cannot exceed N^*, (5.66).

As μ approaches 0 from below, $\lambda = e^{\beta\mu}$ approaches 1, and the first term in (5.69) approaches infinity. Thus the number of particles in excited states remains finite while the number in the ground state increases without limit. This is called the Bose condensation. In Figure 5.8, N/q^{tr} is plotted against the activity $e^{\beta\mu}$, with q^{tr} taken equal to 10 for clarity (typical values are many

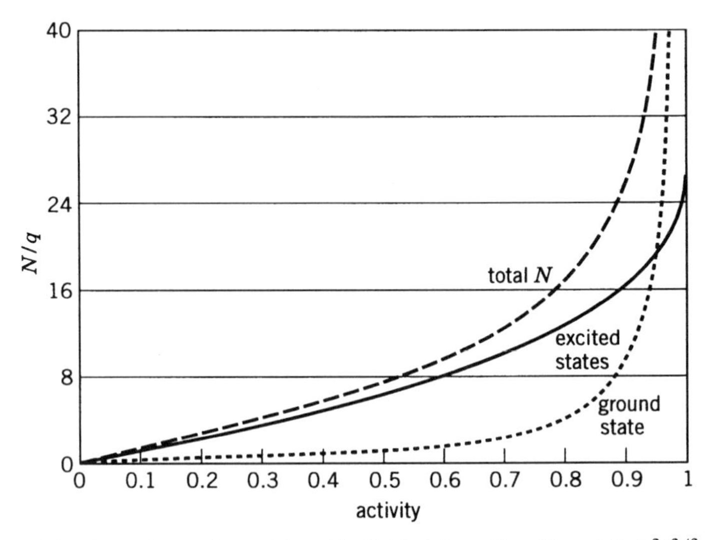

Figure 5.8. Total number of particles N divided by $q^{\text{tr}} = (2\pi mkT/h^2)^{3/2}$ V, as a function of activity $\lambda = e^{\beta\mu}$ (dashed curve), assuming $q^{\text{tr}} = 10$. The solid curve shows the number of particles in states other than the ground state, which cannot exceed $2.6124q^{\text{tr}}$. The dotted curve shows the number of particles in the ground state, which approaches N as λ approaches 1 (Bose condensation).

orders of magnitude larger). Note that the number of particles in the ground state [first term in (5.69)] becomes larger than the number in all other states, which cannot exceed $2.6124q^{\text{tr}}$. The number of particles in the ground state approaches N when $e^{-\beta\mu}$ differs from 1 by less than $1/N$. Since N is typically 10^{20}, μ is essentially 0 (but negative). Another way of expressing this is to define a critical condensation temperature T_c according to

$$\left(\frac{2\pi mkT_c}{h^2}\right)^{3/2} = \frac{N}{2.6124\text{ V}} \tag{5.70}$$

Since $\beta\mu \sim 0$, the second term of (5.69) is

$$2.6124q^{\text{tr}} = 2.6124\left(\frac{2\pi mkT}{h^2}\right)^{3/2}\text{V}$$

When T falls below T_c this term is less than N, and the occupation number of the ground state begins to approach N.

The calculation of thermodynamic properties is simple for $T/T_c < 1$, because μ is 0. The average energy is

$$\langle E \rangle = \frac{2^{5/2}\pi m^{3/2}\,\mathrm{V}}{h^3} \int_0^\infty dE\, \frac{E^{3/2}}{e^{-\beta(\mu - E)} - 1}$$

$$= \frac{2^{5/2}\pi m^{3/2}\,\mathrm{V}}{h^3} \int_0^\infty dE\, E^{3/2} \frac{e^{\beta(\mu - E)}}{1 - e^{\beta(\mu - E)}}$$

Note that the ground state does not contribute. Using $(1 - x)^{-1} = \Sigma_j x^j$ and integrating term by term, one obtains

$$\langle E \rangle = \frac{3}{2\beta} q^{\mathrm{tr}} \sum_{j=1}^{\infty} \frac{e^{j\beta\mu}}{j^{5/2}} \tag{5.71}$$

With $\mu = 0$, the value of the sum is 1.3413. Dividing by NkT and using (5.70),

$$\frac{E}{NkT} = \frac{3}{2}\left(\frac{T}{T_c}\right)^{3/2} \frac{1.3413}{2.6124} = 0.7702\left(\frac{T}{T_c}\right)^{3/2} \tag{5.72}$$

(The energy is the average energy and the particle number the average particle number.) Since $PV = \frac{2}{3}E$,

$$\frac{PV}{NkT} = 0.5134\left(\frac{T}{T_c}\right)^{3/2} \tag{5.73}$$

The Gibbs free energy $G = N\mu = 0$, so the Helmholtz free energy is

$$A = 0 - PV = -0.5134NkT\left(\frac{T}{T_c}\right)^{3/2} = -\tfrac{2}{3}E$$

The entropy is

$$S = \frac{E - A}{T} = 1.2836Nk\left(\frac{T}{T_c}\right)^{3/2} \tag{5.74}$$

and the heat capacity

$$C_V = 1.9254Nk\left(\frac{T}{T_c}\right)^{3/2} \tag{5.75}$$

All these results, it should be remembered, are for $T < T_c$.

For $T > T_c$, equation (5.69) must be solved to give μ (which is significantly different from 0) as a function of T, and the result substituted into equation (5.71) for the energy. The occupation number of the ground state is small, so (5.69) becomes

$$N = q^{\text{tr}} \sum_{j=1}^{\infty} \frac{e^{j\beta\mu}}{j^{3/2}}$$

or, with (5.70) for N,

$$\sum_{j=1}^{\infty} \frac{e^{j\beta\mu}}{j^{3/2}} = 2.6124\left(\frac{T_c}{T}\right)^{3/2}$$

For any value of T/T_c, this equation can be solved (numerically or algebraically) for $e^{\beta\mu}$. Then the value of $e^{\beta\mu}$ is put into (5.71), which may be written

$$\frac{E}{NkT} = \frac{3}{2(2.6124)}\left(\frac{T}{T_c}\right)^{3/2} \sum_{j=1}^{\infty} \frac{e^{j\beta\mu}}{j^{5/2}}$$

One then has E/NkT as a function of T/T_c for $T > T_c$. It gives the same result as (5.72) for $T = T_c$. In Figure 5.9 we have given E/NkT as a function of T/T_c (dashed line). By differentiating E with respect to T, one may obtain C_V as a function of T/T_c. It is continuous at $T/T_c = 1$, as shown in Figure 5.9 (solid line), but there is a discontinuity in the slope of C_V at $T/T_c = 1$. Such a discontinuity occurs for a phase transition. It is at $T = T_c$ that C_V attains its largest value, $1.92Nk$. Note that $C_V \to 1.5Nk$ for $T \to \infty$.

A problem in applying this formalism to actual systems is that, at low temperatures, the effect of interparticle interactions becomes important (as we will discuss in Chapter 6). One no longer has an ideal gas of noninteracting particles, as has been assumed here. The problem is to disentangle the effects of interparticle interactions and of Bose condensation. It is only very recently that Bose condensation in an atomic system has been displayed unambiguously.

Bose–Einstein statistics may be applied to photons or electromagnetic radiation. The linearity of the equations of electromagnetic theory implies the lack of interaction between photons, so they are appropriately considered as constituting a gas. Electromagnetic radiation in an enclosed volume V with walls maintained at temperature T is called "black-body radiation." Since the photons do not interact with each other, the presence of walls or matter, with which they do interact, is necessary to establish thermal equilibrium. According to electromagnetic theory, the pressure of isotropic radiation is equal to $\frac{1}{3}$

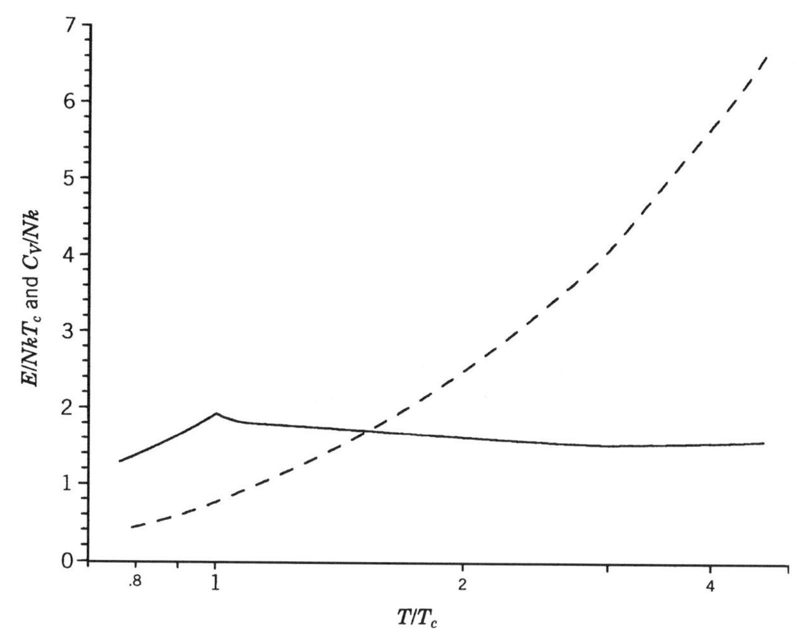

Figure 5.9. Energy (dashed curve) and heat capacity (solid curve) of Bose gas as functions of reduced temperature T/T_c. T_c is the critical temperature, defined by equation (5.70). The heat capacity has a discontinuity in slope at $T/T_c = 1$.

the energy density U/V so

$$dU = 3(PdV + VdP)$$

The energy density can be shown, by thermodynamic arguments, to depend on T, and not on the volume or shape of the container, so that

$$dS = \frac{dU + PdV}{T} = \frac{3V}{T}dP + \frac{4P}{T}dV = \frac{3V}{T}\frac{dP}{dT}dT + \frac{4P}{T}dV$$

Since dS is an exact differential [see equations (1.15) and following],

$$\left(\frac{\partial}{\partial T}\left(\frac{4P}{T}\right)\right)_V = \left(\frac{\partial}{\partial V}\left(\frac{3V}{T}\frac{dP}{dT}\right)\right)_T$$

Using $(\partial T/\partial V)_P = 0$, one derives

$$\frac{dP}{dT} = \frac{4}{3}\frac{P}{T}$$

which integrates to

$$P = \tfrac{1}{3}aT^4 \tag{5.76}$$

where a is a constant of integration. Since $U = \tfrac{3}{2}PV$ by equation (5.26),

$$\frac{U}{V} = aT^4 \quad \text{and} \quad S = \tfrac{4}{3}aVT^3$$

The first of these two equations is known as Stefan's law.

Now we turn to a "microscopic" description of the radiation field. The electric field of electromagnetic waves in the volume V satisfies the wave equation, just as do the vibrations of an elastic continuum. Solutions obeying the proper boundary conditions are standing waves of frequencies ν_i. Associating these with normal modes of vibration, which are quantized harmonic oscillators, we write the energy of the radiation field as a sum of terms $n_i h\nu_i$. The zero-point energy is omitted (this constitutes a redefinition of the zero of energy). Then the canonical partition function is

$$Q = \sum_{n_1=0}^{\infty} \sum_{n_2=0}^{\infty} \sum_{n_3=0}^{\infty} \cdots e^{-\Sigma_i n_i h\nu_i / kT}$$

This sum of products is equal to a product of sums, one for each normal mode. The ith sum is

$$\sum_{n_i=0}^{\infty} e^{-n_i h\nu_i / kT} = \left[1 - e^{-h\nu_i / kT}\right]^{-1}$$

so

$$Q = \prod_i \left[1 - e^{-\beta h\nu_i}\right]^{-1} \tag{5.77}$$

The manipulations just carried out resemble those used to get the grand partition function in equations (5.6) to (5.8). We return to this point later. To get the thermodynamic properties of the radiation, we need to evaluate ln Q, which is a sum of terms, one for each mode. The frequencies are so close together that the sum can be replaced by an integral:

$$\ln Q = \int_0^{\infty} d\nu \frac{dn(\nu)}{d\nu} \ln\left[1 - e^{-h\nu/kT}\right]^{-1}$$

Here, $dn(\nu)/d\nu$ is the number of frequencies per unit frequency range.

Determining $dn(\nu)/d\nu$ is exactly the same as determining the number of particle-in-box states per unit energy range [see equations (5.22) and (5.23)] or the number of crystal vibrations per unit frequency range in the Debye theory (Section 3.3). The wavelength for frequency ν is c/ν, where c is the wave velocity, which in the present application is the velocity of light in a vacuum. In three dimensions, the frequency ν is related to the wavelengths in the x-, y-, and z-directions by $\nu^2 = \nu_x^2 + \nu_y^2 + \nu_z^2$ where $\nu_x = c/\lambda_x$, and so on. The length of the box must be an integral multiple of half-wavelengths: $L = n_x \lambda_x/2$ with n_x an integer, and so on. Then

$$\nu^2 = \frac{c^2}{4L^2}\left(n_x^2 + n_y^2 + n_z^2\right)$$

The number of allowed frequencies less than ν is the number of sets of positive integers such that

$$n_x^2 + n_y^2 + n_z^2 < \left(\frac{2\nu L}{c}\right)^2$$

and this is one-eighth of the volume of a sphere of radius $2\nu L/c$. Then the number of frequencies between ν and $\nu + d\nu$ is

$$\frac{d}{d\nu}\left[\frac{1}{8}\,\frac{4\pi}{3}\left(\frac{2\nu L}{c}\right)^3\right] = \frac{4\pi\nu^2 L^3}{c^3}$$

Differences between electromagnetic waves and crystal vibrations include: (a) there is no upper limit on frequency, because the medium is really continuous, and not composed of atoms; (b) electromagnetic waves are transverse and not longitudinal, so there are two directions of vibration (more correctly, polarization) for each wave, whereas there were three for crystal vibrations. (Because there are an infinite number of frequencies, the zero-point energies we threw away in writing Q actually add up to an infinite quantity.)

One obtains for the logarithm of the partition function (5.77)

$$\ln Q = -\frac{8\pi}{c^3} V \int_0^\infty d\nu\, \nu^2 \ln[1 - e^{-h\nu/kT}]^{-1} \tag{5.78}$$

with $V = L^3$. The Helmholtz free energy A is $-kT \ln Q$, and the energy is

$$U = -T^2 \frac{\partial(A/T)}{\partial T} = \frac{8\pi V}{c^3} \int_0^\infty d\nu \frac{h\nu^3}{e^{h\nu/kT} - 1} \tag{5.79}$$

One can now obtain Stefan's law and even find the explicit value of the constant a in (5.76). Putting $x = h\nu/kT$,

$$\frac{U}{V} = \frac{8\pi}{c^3}\left(\frac{kT}{h}\right)^4 \int_0^\infty dx\, \frac{x^3}{e^x - 1} = \frac{8\pi^5 k^4}{15h^3c^3} T^4$$

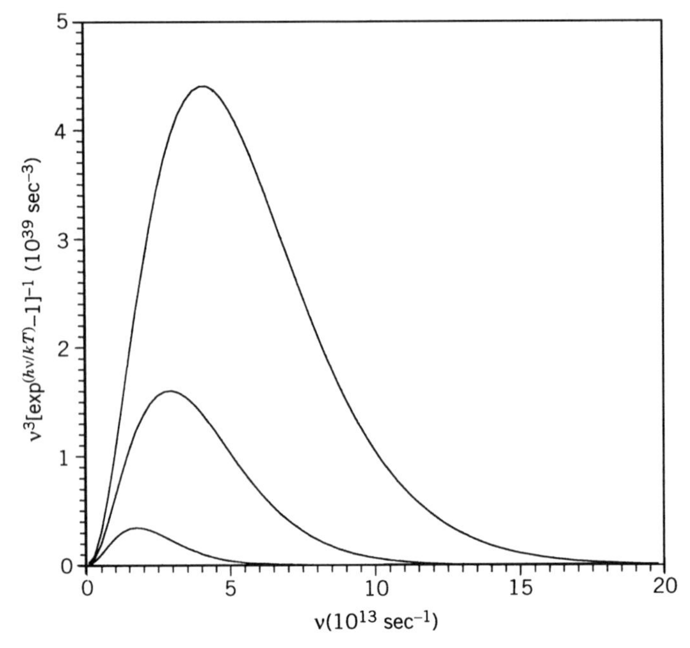

Figure 5.10. Planck distribution function for $T = 300$ K, 600 K, and 900 K (bottom to top). The function plotted must be multiplied by $8\pi Vh/c^3$ to give the number of photons with frequency between ν and $\nu + d\nu$. The total number of photons (area under a curve) increases with T.

and the T^4-dependence is proved. The constant a is $8\pi^5k^4/15h^3c^3$. From (5.79), the energy density of radiation with frequency between ν and $\nu + d\nu$ is

$$U_\nu = \frac{8\pi V}{c^3}\,\frac{h\nu^3}{e^{h\nu/kT} - 1}$$

This is the Planck distribution law, which played a pivotal role in the development of quantum mechanics. The Planck distribution function is shown in Figure 5.10 for several temperatures.

For small frequencies, the denominator in U_ν is $h\nu/kT$ and U_ν is approximately $8\pi VkT\nu^2/c^3$, the Rayleigh–Jeans law. For large frequencies, U_ν is proportional to $\nu^3e^{-h\nu/kT}$. The maximum in U_ν occurs at $\nu = \nu_m$, where

$$1 + \frac{h\nu_m}{3kT}\left[e^{-h\nu_m/kT} - 1\right]^{-1} = 0 \tag{5.80}$$

Since the equation, $1 - e^{-x} = \frac{1}{3}x$, has as root $x = 2.8214$, $\nu_m = 2.8214kT/h$.

Thus the frequency of maximum energy density is proportional to T. A measurement of ν_m at known temperature can give the value of h, assuming k is known.

The form of (5.77) suggests that the radiation, instead of being thought of as waves, can be considered to consist of photons. The normal modes of vibration then correspond to photon states, the state j having wave function

$$\psi_j = e^{2\pi i(p_{xj}x + p_{yj}y + p_{zj}z)/h}$$

and energy $h\nu_j = (p_{xj}^2 + p_{yj}^2 + p_{zj}^2)/2m$. The possible values of p_x are $n_x h/L$ with n_x an integer. The number of photons in state j is n_j so the total energy is $\sum n_j h\nu_j$. The grand partition function is

$$\begin{aligned}\Xi &= \left[\sum_{n_1=0}^{\infty} e^{\beta\mu n_1} e^{-\beta n_1 h\nu_1}\right]\left[\sum_{n_2=0}^{\infty} e^{\beta\mu n_2} e^{-\beta n_2 h\nu_2}\right]\cdots \\ &= \frac{1}{1 - e^{-\beta(h\nu_1 - \mu)}} \frac{1}{1 - e^{-\beta(h\nu_2 - \mu)}} \cdots\end{aligned}$$

which for $\mu = 0$ is identical to (5.77); Q is really a grand partition function with the chemical potential $\mu = 0$. [If Q is a canonical partition function, the meaning of $\mu = 0$ is that the number of photons N is determined so as to minimize A, since $\mu = (\partial A/\partial N)_T$.]

Interpreting the black-body radiation as photons, one gets the thermodynamic properties of the photon gas from (5.78). On integrating by parts and using the integral

$$\int_0^{\infty} \frac{x^3}{e^x - 1}\, dx = \frac{\pi^4}{15}$$

one gets

$$\ln Q = \ln \Xi = -\frac{8\pi^5 V(kT)^3}{45(hc)^3} \tag{5.81}$$

Since $\ln \Xi$ is PV/kT,

$$P = \frac{8\pi^5 (kT)^4}{45(hc)^3}$$

We will find this result from gas kinetic theory in the next chapter (Problem 6.11). Note that the pressure of this gas is independent of the volume at a given temperature. This is because the number of photons is not fixed: an increase in volume is accompanied by the creation of more photons, since

N/V is a function of temperature only. The energy according to (5.79) is

$$U = \frac{8\pi hV}{c^3}\left(\frac{kT}{h}\right)^4 \int_0^\infty dx\, \frac{x^3}{e^x - 1}$$

which is three times PV, in accord with the electromagnetic theory result cited at the beginning of our discussion of photons. In connection with Stefan's law, $U/V = aT^4$, $\sigma = ac/4$ is called the Stefan's law constant or the Stefan–Boltzmann constant. From the value of a,

$$\sigma = \frac{2\pi^5 k^4}{15h^3c^2} = 5.67 \times 10^{-8}\ \mathrm{J\ m^{-2}\ sec^{-1}\ K^{-4}} \tag{5.82}$$

Finally, since $\mu = 0$, $G = 0$, and $A = -PV$. Then the entropy S may be found as $-(\partial A/\partial T)_V$; it is proportional to T^3, as found from thermodynamic reasoning.

For a gas of N particles moving at speed c within a volume V, the rate at which particles collide with unit area of the wall is $\frac{1}{4}cN/V$, as will be discussed in the next chapter. If there is a hole of area $\mathscr{A}$ in the confining wall, the rate at which particles escape is $\frac{1}{4}c\mathscr{A}N/V$. For photons, each particle of frequency ν carries energy $h\nu$ so the rate at which energy of frequency ν escapes is $(c\mathscr{A}/4V)(N_\nu h\nu)$. The frequency distribution of energy in the emitted radiation is the same as in the radiation inside V; the distribution is in fact measured on the emitted radiation. The total energy emitted is

$$\frac{c\mathscr{A}}{4V}\int_0^\infty d\nu\, N_\nu h\nu = \frac{c\mathscr{A}h}{4}\,\frac{8\pi}{c^3}\int_0^\infty d\nu\, \frac{\nu^3}{e^{h\nu/kT} - 1} = \sigma\mathscr{A}T^4$$

where σ is Stefan's constant, given in (5.82).

PROBLEMS

5.1. For fermions, the occupation number of a state with energy E is $s(E) = \{e^{\beta(E-\mu)} + 1\}^{-1}$, so that it is $\frac{1}{2}$ for $E = \mu$. For states with energy E near μ, $s(E) = \frac{1}{2} + A(E - \mu) + B(E - \mu)^2 + \cdots$. Calculate A and B by evaluating the first and second derivatives of $s(E)$ with respect to E at $E = \mu$. Compare your result with Figure 5.2 showing the Fermi distribution at various temperatures.

5.2. For bosons, the occupation number of a state with energy E is $s(E) = \{e^{\beta(E-\mu)} - 1\}^{-1}$. Suppose that a particle can be in one of only two states, a ground state with energy 0 and an excited state with energy ε, and suppose $\beta = 2/\varepsilon$ (so that kT is $\varepsilon/2$). Calculate the value of the chemical potential for total number of particles $N = 0.5, 1, 2, 5, 10, 20$, and 50. [The chemical potential has to be chosen to give the right N, as in (5.18).] Calculate the occupation numbers for the two states for each value of N, and compare the occupation numbers with what the Boltzmann distribution predicts. (*Answer:* with $x = e^{-2\mu/\varepsilon}$, $x^2e^2N - x(N+1)(e^2+1) + N + 2 = 0$, giving x-values 3.19416, 2.07501, 1.52558, 1.20519, 1.10142, 1.05037, 1.02006 and ground state populations 0.455756, 0.930228, 1.902654, 4.873501, 9.859914, 19.8521, 49.84703.)

5.3. From thermodynamics, $N\mu = G = E + PV - TS$. Show that for an ideal Fermi–Dirac or Bose–Einstein gas,

$$S = -k \sum_j \left\{ s_j \ln s_j \pm (1 \mp s_j) \ln(1 \mp s_j) \right\}$$

in terms of the (average) occupation numbers (upper signs for fermions and lower signs for bosons). This formula recalls the canonical-ensemble result,

$$S = -k \sum_j p_j \ln p_j,$$

where p_j is the probability that the system is in state j ($p_j \leq 1$).

5.4. Show using (5.27) that the first three terms in the power series expansion of $\lambda = e^{\beta\mu}$ in

$$X = \frac{2\langle N \rangle}{\sqrt{\pi}CV} \beta^{3/2}$$

are

$$\lambda = X \pm 2^{-3/2}X^2 + \left(\tfrac{1}{4} - 3^{-3/2}\right)X^3$$

(upper signs for fermions, lower for bosons). Solve (5.29) for λ and show it is equivalent to the first two terms of the above expansion.

5.5. Using the three-term expansion of Problem 5.4 in the first three terms of the energy expression (5.28), prove that

$$PV = \frac{N}{\beta} \pm \frac{3}{2^{3/2}} \frac{N^2\beta^{1/2}}{CV\sqrt{\pi}} + \left(\frac{3}{2} - \frac{8}{3\sqrt{3}}\right) \frac{N^3\beta^2}{C^2V^2\pi} + \cdots$$

With $C = 2^{7/2}\pi m^{3/2}h^{-3}$ (internal degeneracy $D = 2$), calculate the pressure of a gas of fermions of mass 1.67×10^{-24} g and $T = 10^4$ K at number densities $N/V = 10^{23}$, 3×10^{23}, and 10^{24} cm^{-3}. Do the same calculations for a gas of bosons and for a classical (corrected boltzons) gas. (*Answer:* for fermions, P in 10^{11} dyn cm^{-2}: 1.5748, 5.8553, 33.035; for boltzons, 1.3806, 4.1418, 13.806; for bosons, 1.1859, 2.3769, -5.6135.) (Series expansion is failing here.)

5.6. The pressure of an electron gas at $T = 0$ was calculated from $PV = \frac{2}{3}E = \frac{2}{5}N\varepsilon_F$. Alternatively, one could note that $P = -(\partial A/\partial V)_T$ and that the Helmholtz free energy A is equal to E at $T = 0$. Calculate $-(\partial E/\partial V)_T$ and show it is equal to $\frac{2}{5}N\varepsilon_F/V$. Then show that $E + PV = N\mu$. (The Gibbs free energy G is equal to $N\mu$ and $G = E + PV$ at $T = 0$.)

5.7. The number of valence (conduction) electrons per unit volume for Ag is approximately 5.9×10^{22} cm^{-3}. Calculate ε_F and the chemical potential μ for the conduction electrons at 10 K. Then calculate their contribution to the heat capacity at 10 K. In the expansions (5.37) and (5.39), how important are higher terms than the first? (The measured heat capacity contribution is 6.7×10^{-3} J K^{-1} mol^{-1}; for most metals, one gets agreement to within 20% between calculated and experimental heat capacities.) (*Answer:* $\beta\varepsilon_F = 6367$, $C_V = 0.00644$ J mol^{-1} K^{-1}.)

5.8. Silver, Ag, has a Debye temperature of 210 K. Below what temperature is the contribution of conduction electrons to the heat capacity (see Problem 5.7) larger than the contribution of the crystal vibrations? (*Answer:* 1.75 K.)

5.9. Equation (5.43) gives the electron density near a wall at $z = z_w$. The positive-ion density extends from $z = -L$ to $z = 0$, where $L \to \infty$. To assure electroneutrality, (5.44) must be satisfied, with ρ_b, the bulk density, a constant. Show, by substituting (5.43) into (5.44), that electroneutrality requires $z_w = 3\pi/8k_F$.

5.10. The free-electron model of equations (5.41) to (5.45) is to be applied to a charged metal. The charge, produced by adding or removing electrons, will of course lie on or near the surface. Let the surface charge density on the metal be $q^M e$, and assume there is a compensating charge density $-q^M e$ located outside the metal, at $z = z_w + d$, where z_w, the plane at which the electron density vanishes, is determined by

$$q^M = -\int_{-L}^{z_w} dz\, \rho(z) + \int_{-L}^{0} dz\, \rho_b$$

Find z_w as a function of q^M. Then calculate the surface potential $\Delta V = V(\infty) - V(-\infty)$, which should involve two contributions, one from the charge q^M and one from the surface potential of the metal electrons. Finally, calculate the capacitance $C = dq^M/d\,\Delta V$.

5.11. For the gas of electrons in a metal (free-particle states with the electrons obeying Fermi–Dirac statistics), calculate the one-dimensional velocity distribution, the speed distribution, and the average speed of an electron.

$$\left(\textit{Answer:} \quad N(v_x)\, dv_x = (4\pi V m^2 kT/h^3) \ln[e^{(\mu-\varepsilon_x)/kT} + 1]\, dv_x \right)$$

5.12. When modeling the conduction electrons in a metal as a confined electron gas (Section 5.3), it sometimes must be remembered that the walls of the "container" are not infinitely high. If an electron in an electron gas hits a wall of the container (the boundaries of the metal) with sufficient kinetic energy, it can pass out of the metal. This is thermionic emission. The rate at which electrons having velocity in the x-direction v_x strike unit area of surface perpendicular to the x-direction is $v_x N(v_x)/V$ [see Problem 5.11 for the distribution $N(v_x)$]. Let ε_m be the minimum kinetic energy in the x-direction (x-kinetic energy $= \frac{1}{2}mv_x^2$) that an electron must have to overcome the potential of the wall and escape from the metal. Show that the rate of thermionic emission of electrons per unit time and per unit surface area is $CT^2 e^{-\phi/kT}$ with C a constant and $\phi = \varepsilon_m - \mu$; ϕ is the work function. Note $\varepsilon_m - \mu$ is large compared to kT.

5.13. Equation (5.51) gives the heat capacity of the conduction electrons in a metal in terms of $\mathcal{N}(\varepsilon)$, and equation (5.39) gives the heat capacity assuming that the density of states is

$$2\,\frac{2^{5/2}\pi m^{3/2} E^{1/2} V}{h^3}$$

which is appropriate for particles in a box. Calculate the Fermi level ε_F (or μ) in terms of N/V for the latter case and show that (5.39) is a special case of (5.51).

5.14. In electrochemistry, one describes a charged species by giving its electrochemical potential, which has a "chemical" and an "electrical" part, corresponding to short-range and long-range forces. The electrochemical potential of electrons in a metal, $\tilde{\mu}_e$, is the work required to bring a mole of electrons into bulk metal from vacuum infinitely far outside the metal. Thus $\tilde{\mu}_e = (\partial A/\partial N)_{V,T,\phi}$ where A is the Helmholtz free energy and ϕ the electrical potential inside the metal so the lowest electronic energy level is at $-e\phi$. Use $N\mu = G = A + PV$ to derive an expression for A in terms of N, μ, β, and Ξ and differentiate it to find the relation between $\tilde{\mu}_e$ and μ. (Note that the proper independent variables are V, β, and μ, and that μ depends on N for constant V.)

5.15. The concentration of atoms in Si, an intrinsic semiconductor, is 10^{22} cm^{-3}. At 300 K, there are about 3×10^{10} electrons (and 3×10^{10} holes) per cm^{-3}. Estimate the band gap, assuming the effective mass of electrons or holes is essentially equal to the electronic mass. If the temperature is doubled, to 600 K, what is the concentration of electrons?

5.16. In the simplified treatment at the end of Section 5.4, we derived

$$(N_+ N_-)/N_0^2 = e^{-\beta(\varepsilon_c - \varepsilon_v)}$$

Assuming that each atom of Si (see Problem 5.14) contributes four electrons to the valence band (so $N_0 = 4 \times 10^{22}$ cm^{-3}), what is the band gap?

5.17. For some intrinsic semiconductor, the effective electron mass m_e is twice the effective hole mass m_h. Where is the chemical potential μ relative to ε_v and ε_c? (If the effective masses are equal, μ is midway between ε_v and ε_c.)

5.18. Minimize $\ln W$, where W is given by (5.59), with respect to n_d and the $\{n_i\}$, subject to the constraints (5.60). Use the Stirling approximation and Lagrange multipliers, so you are actually making $\delta\{\ln W + \alpha N - \beta E\} = 0$. Identifying the parameter α with $\beta\mu$ and β with the usual $1/kT$, prove $n_i = g_i[1 + e^{\beta(\varepsilon_i - \mu)}]^{-1}$ and (5.61).

5.19. Suppose there is an intrinsic semiconductor with a carrier concentration of 1.1×10^{16} cm^{-3} at 300 K and 5.5×10^{16} cm^{-3} at 400 K. What is the energy gap $\varepsilon_c - \varepsilon_v$? What will be the carrier concentration at 200 K?

5.20. Consider a p-type semiconductor, $N_a \gg N_d$, with the acceptor-atom state at energy ε_a. The conduction band and the donor-atom levels will be essentially empty of electrons and there will be holes in the valence band. Write the expressions for the number of holes n_h and the number of electrons in acceptor levels n_a. Combining them with the electroneutrality condition, show that

$$C' \equiv \frac{n_h(n_h + N_d)}{N_a - N_d - n_h} = 2Q e^{-\beta(\varepsilon_a - \varepsilon_v)}$$

where $Q = 2(2\pi m_h/h^2\beta)^{3/2}V$. This means that $\varepsilon_a - \varepsilon_v$ can be obtained by plotting the logarithm of $T^{-3/2}C'$ vs. $1/T$.

5.21. Equation (5.64) can be used to obtain the energy difference $\varepsilon_c - \varepsilon_d$ by plotting the logarithm of the left-hand side vs. 1/T, if there is only one donor level. As discussed after (5.65), if there are several donor levels and the temperature is low, the temperature dependence of the right side is more complicated. For a triplet and a singlet donor level, at ε_d and at $\varepsilon_c - \delta\varepsilon_c$, $\exp[-(\varepsilon_c - \varepsilon_d)/kT]$ is replaced by

$$D \equiv \left(\tfrac{1}{4}e^{\beta(\varepsilon_c - \varepsilon_d + \delta\varepsilon_d)} + \tfrac{3}{4}e^{\beta(\varepsilon_c - \varepsilon_d)}\right)^{-1} = 4e^{-\beta(\varepsilon_c - \varepsilon_d)}\left(e^{\beta\delta\varepsilon_d} + 3\right)^{-1}$$

Suppose $\delta\varepsilon_d = 9 \times 10^{-23}$ J and $\varepsilon_c - \varepsilon_d = 2 \times 10^{-21}$ J (about right for Sb in Ge). Calculate the above quantity for five values of T from 100 K to 110 K and fit the logarithms of the results to a linear function of $1/T$. Note the quality of the fit and the value of the slope, which should be $-(\varepsilon_c - \varepsilon_d)/k$ if $\delta\varepsilon_d$ is small enough to neglect. Repeat the calculation for five values of T from 5 K to 10 K, again noting the value of $-(\varepsilon_c - \varepsilon_d)/k$ obtained.

5.22. Equation (5.67) gives $\ln \Xi$ for an ideal gas obeying Bose–Einstein statistics. Integrate the second term by parts and insert the expansion given in the equation below (5.67). Then, integrating term by term, derive equation (5.68).

5.23. Calculate the critical condensation temperature for He atoms at a density of 1 mole per liter, and at a density of 100 moles per liter.

5.24. Derive equation (5.71) from the equation for $\langle E \rangle$ preceding it.

5.25. Interpreting black-body radiation as a collection of photons, such that one photon of frequency ν has energy $h\nu$, calculate the frequency for which there is the greatest *number* of photons. Calculate an expression for the total number of photons as a function of temperature. Derive an expression for the average energy of a photon. Note that $\int_0^\infty x^n(e^x - 1)^{-1}\,dx = \pi^4/15$ for $n = 3$ and 2.4041 for $n = 2$.

5.26. From (5.78), derive an expression for the Helmholtz free energy A of a photon gas (black-body radiation). Then derive an expression for the entropy S by differentiating A. Show that TS is four times PV.

5.27. Calculate C_V for the photon gas by differentiating U. Since $(\partial S/\partial T)_V$ should equal C_V/T, integrate C_V/T from $T = 0$ to $T = T$ to derive an expression for the entropy S (this assumes $S = 0$ at $T = 0$). Compare with the expression for S of Problem 5.26.

5.28. Wien's Displacement Law for thermal or black-body radiation is that the energy density per unit range of wavelength λ is equal to λ^{-5} multiplied by a function of λT. Prove this from the Planck distribution.

5.29. An empty enclosed volume of 10 cm^3 is heated to 1200 K so that the radiation inside comes to equilibrium at this temperature. What is the energy of the radiation inside with wavelength between 600 and 610 nm? What is the energy of the radiation with wavelength between 500 and 510 nm?

5.30. The light emitted by the sun follows a Planck distribution, with the peak (highest energy per unit frequency) at a frequency of 3.5×10^{14} sec^{-1}. Estimate the surface temperature. If the sun is considered to be a sphere of diameter 1.4×10^{6} km, how much energy is radiated per second?

5.31. (a) The skin temperature of a person is 33°C, and the emitted radiation follows a black-body (Planck) distribution. What is the frequency of maximum emission, and in what region of the spectrum is it?

(b) Assuming that the earth radiates like a black body at a mean temperature of 288 K, what is the frequency of maximum emission and in what region of the spectrum does it lie?

(c) The earth receives energy from the sun, on average, at $\mathscr{E} = 160$ J sec^{-1} m^{-2}. If the earth emitted as a black body in equilibrium with the sun, so that its radiation rate were equal to $\mathscr{E}$, what would the earth's surface temperature be? (*Answer:* 230 K.)

CHAPTER 6

CLASSICAL STATISTICAL MECHANICS

6.1. PHASE SPACE

Everything we have done so far in statistical mechanics has been based on quantum mechanics: we have been considering the probability that a system, or a particle, is in a particular quantum state, which means it has a particular wave function and energy. In a classical mechanical description, the state of a system is defined by giving the values of all its positions and momenta. If the system contains N particles, there are $3N$ positions and $3N$ momenta. In classical statistical mechanics, we must consider the probability that a system has particular values for all positions and momenta. The state of a system changes with time, but the state at one instant of time determines the state at succeeding instants. When we consider an ensemble of systems, with different particle positions and momenta, the properties of interest will be averages over the ensemble.

It is convenient to visualize the state of an N-particle system in terms of a space of $6N$ dimensions, called the "phase space." The coordinate axes are the $3N$ particle positions and the $3N$ particle momenta. A point in the $6N$-dimensional space represents a state of the system, since it is defined by giving values of the $3N$ positions and $3N$ momenta. As the positions and momenta change, the point representing the system traces out a trajectory in phase space. An ensemble of systems at one instant is represented by a collection of points in phase space. If there are a very large number of systems in the ensemble, this collection of points forms a cloud, characterized by its density as a function of position in phase space. In time, this cloud may change its size and shape. Our assumption, that observation of a single system over time gives the same results as observation of all the systems in the ensemble at one time, means that, if one waits long enough, the trajectory of a single system in phase space resembles the cloud representing the ensemble.

As a simple example, consider a one-dimensional harmonic oscillator, which has only one position, say x, and one momentum, say p. The phase space is then two-dimensional and can be represented as a plane. The state of the system is a point, given by two coordinates. The total energy is the sum of kinetic and potential energy:

$$E = \tfrac{1}{2}kx^2 + \frac{p^2}{2m} \tag{6.1}$$

This is the equation of an ellipse in phase space. Since total energy is conserved, the phase point representing the system always remains on the ellipse. If the system is described by quantum mechanics, the only possible values of E are $(n + \frac{1}{2})h\nu$ where $2\pi\nu = \sqrt{k/m}$ and n is an integer. Figure 6.1 shows, for $k = 2$ and $m = 1$, the ellipses for $n = 9$, 10, 11, 12, and 13. If the system is described by classical mechanics, any value of E is allowed.

The force on the particle is equal to its mass multiplied by its acceleration:

$$F_x = \frac{-d}{dx}\left(\frac{1}{2}kx^2\right) = -kx = m\frac{d^2x}{dt^2} = \frac{d}{dt}\left(m\frac{dx}{dt}\right) \tag{6.2}$$

Since $m(dx/dt) = p$, (6.2) is simply $dp/dt = -kx$. Of course, $dx/dt = p/m$ by definition. According to (6.2),

$$\frac{d^2x}{dt^2} = \frac{-kx}{m}$$

which has the general solution

$$x = x_0 \sin(\omega t + b) \tag{6.3}$$

where $\omega = \sqrt{k/m}$ and x_0 and b are constants of integration. From (6.3),

$$p/m = x_0\omega\cos(\omega t + b) \tag{6.4}$$

As a function of time t, the phase point, represented by (6.3) and (6.4), traces out the ellipse with semiaxes x_0 and $x_0\omega$. The conservation of energy is,

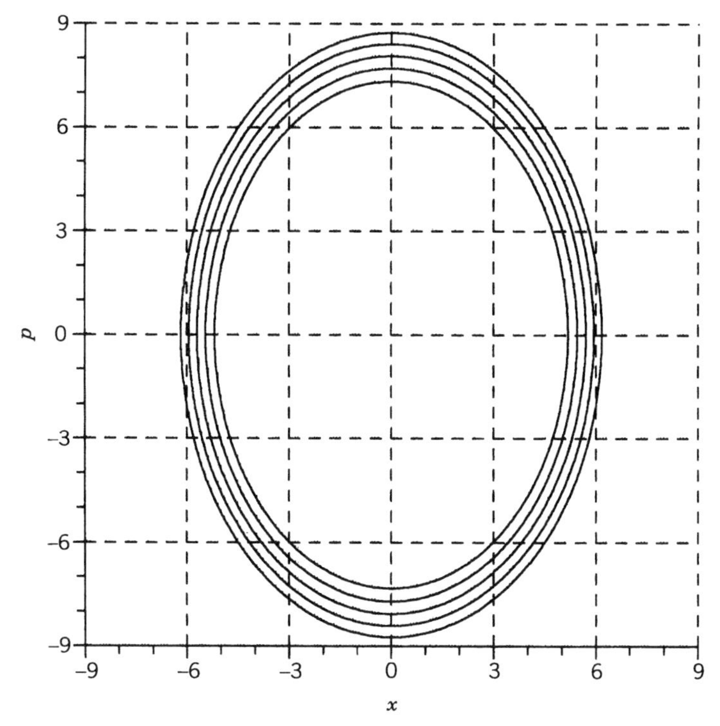

Figure 6.1. Phase space for one-dimensional particle, showing constant-energy surfaces for harmonic oscillator. Any energy is possible in classical mechanics, but a quantum mechanical oscillator must have $E = (n + \frac{1}{2})h\nu$ where $2\pi\nu = \sqrt{k/m}$; the ellipses are (from inside to outside) for $n = 9$, 10, 11, 12, and 13.

using (6.3) and (6.4),

$$E = \tfrac{1}{2}kx^2 + \frac{p^2}{2m} = \tfrac{1}{2}kx_0^2$$

The energy does not depend on b.

If there were a large number of systems with the same energy but different random values of b, their phase points would be randomly distributed over the same ellipse. Since there is no reason for one value of b to be preferred over another, the points would be distributed uniformly on the ellipse. Each point would move around the ellipse with increasing t, the distribution remaining uniform. This is a microcanonical ensemble: all the systems have the same energy. For a canonical ensemble, one would have to include systems of all energies, with phase points all over the diagram, on ellipses of all sizes.

If the system contains N particles in three dimensions, specifying the state of the system means locating a point in the $6N$-dimensional phase space. The canonical ensemble is represented by a large number of points in this space, with the density of points (number of points per unit volume of phase space) being proportional to $\exp(-E/kT)$. Of course, E is a function of the $3N$ positions and $3N$ momenta, which label the $6N$ coordinate axes in the phase space. In mechanics, the expression for the energy as a function of positions and momenta is called the Hamiltonian, and will be written from now on as $H(p,q)$. The values of the positions and the momenta, which define the state of a system, change with time. The way they change with time is given by the canonical equations of motion in the Hamiltonian formulation of mechanics:

$$\frac{dq_i}{dt} = \frac{\partial H(p,q)}{\partial p_i} \quad \text{and} \quad \frac{dp_i}{dt} = -\frac{\partial H(p,q)}{\partial q_i} \tag{6.5}$$

The equations (6.5) therefore describe the trajectory of a system point in phase space.

The ensemble is represented by a swarm or cloud of phase points, with each point moving in time according to (6.5). Because energy is conserved, the trajectory of a given phase point remains on a constant-energy surface in phase space. To show energy is conserved, calculate $dH(p,q)/dt$ along a trajectory, noting that p and q change with time because the system is changing its location in phase space.

$$\frac{dH(p,q)}{dt} = \sum_{i=1}^{3N}\left[\left(\frac{\partial H}{\partial p_i}\right)\frac{dp_i}{dt} + \left(\frac{\partial H}{\partial q_i}\right)\frac{dq_i}{dt}\right] \tag{6.6}$$

Using (6.5) in (6.6), one finds $dH(p,q)/dt = 0$.

A microcanonical ensemble corresponds to a collection of phase points uniformly distributed over a constant-energy surface in phase space. In time, the phase points move, but they always remain on the surface. It will be shown that they remain uniformly distributed. A canonical ensemble is represented by a collection of phase points distributed all over phase space, with the density of phase points proportional to $\exp[-\beta H(p,q)]$ (Boltzmann distribution). We will show that, although the phase points move in time, the density of phase points at any place in phase space remains unchanged. Thus the Boltzmann distribution persists, as an equilibrium distribution should.

Let us denote by $\rho(p,q,t)$ the density of phase points at the position $\{p,q\}$ in phase space at time t. Then $\rho(p,q,t)d^{3N}p\,d^{3N}q$ is the number of phase points in the volume $d^{3N}p\,d^{3N}q = dv$. Consider a volume v in phase space, bounded by a surface s. The change in the number of phase points in v with time is equal to the net rate at which phase points, following their trajectories, enter the volume v through the surface s. Mathematically,

$$-\frac{d}{dt}\int_v dv\,\rho(p,q,t) = \int_s ds\,\mathbf{n}\cdot\mathbf{v}\rho \tag{6.7}$$

where $\mathbf{n}$ is the local normal to the surface and $\mathbf{v}$ is the velocity vector in phase space:

$$\mathbf{v} = \left\{\frac{dp_1}{dt}\,\frac{dp_2}{dt}\cdots\frac{dq_1}{dt}\,\frac{dq_2}{dt}\cdots\frac{dq_{3N}}{dt}\right\}$$

The right-hand side of (6.7) is equal to the volume integral of the divergence, $\nabla\cdot(\mathbf{v}\rho)$. Then, since the volume v is arbitary,

$$-\frac{\partial\rho}{\partial t} = \nabla\cdot(\mathbf{v}\rho) = \sum_{i=1}^{3N}\left[\frac{\partial}{\partial p_i}\left(\frac{dp_i}{dt}\rho\right) + \frac{\partial}{\partial q_i}\left(\frac{dq_i}{dt}\rho\right)\right]$$

The left-hand side has been written as a partial derivative because it is the derivative of ρ with time at a fixed position in phase space. Carrying out the differentiations gives

$$-\frac{\partial\rho}{\partial t} = \sum_{i=1}^{3N}\left[\frac{\partial\rho}{\partial p_i}\frac{dp_i}{dt} + \frac{\partial\rho}{\partial q_i}\frac{dq_i}{dt} + \rho\left(\frac{\partial(dp_i/dt)}{\partial p_i}\right) + \rho\left(\frac{\partial(dq_i/dt)}{\partial q_i}\right)\right]$$

Using (6.5) shows that the last two terms cancel. The result is called Liouville's theorem:

$$\frac{\partial\rho}{\partial t} + \sum_{i=1}^{3N}\left[\frac{\partial\rho}{\partial p_i}\frac{dp_i}{dt} + \frac{\partial\rho}{\partial q_i}\frac{dq_i}{dt}\right] = 0 \tag{6.8}$$

It is important to understand its meaning.

As mentioned above, $\partial\rho/\partial t$ is the rate of change in ρ, the density of phase points, at a particular place in phase space. The sum over i gives the rate of change in ρ due to the changes in positions and momenta, assuming that the positions and momenta change because a point, representing a system, is moving according to the canonical (Hamiltonian) equations of motion (6.5). Together, the terms in (6.8) give the total rate of change in ρ at the position of a moving phase point. This is zero, meaning that, if one sat on a representative phase point moving along its trajectory, one would see no change in the density of phase points. The collection or cloud of phase points behaves like an incompressible fluid. This came from (6.7), which expressed the fact that phase points can neither be created or destroyed.

In the foregoing, we indicated what the density of phase points would be for a microcanonical and for a canonical distribution. In both cases, ρ was a function only of the energy, $H(p, q)$. If ρ depends only on H,

$$\frac{\partial \rho}{\partial p_i}\frac{dp_i}{dt} + \frac{\partial \rho}{\partial q_i}\frac{dq_i}{dt} = \frac{d\rho}{dH}\frac{\partial H}{\partial p_i}\frac{dp_i}{dt} + \frac{\partial \rho}{\partial H}\frac{\partial H}{\partial q_i}\frac{dq_i}{dt}$$

Again using (6.5), one finds that this vanishes, so that (6.8) reduces to $\partial\rho/\partial t = 0$. The density of phase points at any location in phase space remains unchanged through time. A uniform distribution on an energy surface (microcanonical ensemble) remains a uniform distribution on an energy surface. A Boltzmann distribution in phase space (canonical ensemble) remains a Boltzmann distribution. The reason for this is that the distribution of points in phase space (the density of the swarm of points) depends only on the Hamiltonian or energy function. The motion of the phase points, which would be observed as a deformation of the cloud, is governed by the very same Hamiltonian.

6.2. CLASSICAL MECHANICAL PARTITION FUNCTION

The canonical ensemble partition function for classical mechanical systems should be equal to the partition function for quantum mechanical systems in the limit in which quantization (spacing between allowed energy levels) becomes unimportant. The quantum mechanical partition function is a sum over states:

$$q = \sum_j e^{-\beta E_j}$$

In classical mechanics, the energy is written as a function of positions and momenta, and the state of a system is defined by giving values for all positions and momenta. Thus, summing over states must mean integrating over all positions and momenta. Quantization will become unimportant when there are a very large number of states to be summed over; the sum above will then be well approximated by an integral.

For the one-dimensional harmonic oscillator, one would write

$$q = C^{-1}\int_{-\infty}^{\infty} dx \int_{-\infty}^{\infty} dp\, e^{-\beta E} \tag{6.9}$$

where E, given by (6.1), is just the Hamiltonian. For the expression (6.9) to be equal to the sum over quantum states, the integral must be divided by the area of phase space occupied by one quantum state. We have denoted this area by C. To determine C for the one-dimensional harmonic oscillator, refer to Figure 6.1, in which the ellipses are drawn at the energies of the allowed quantum states.

The area of phase space taken up by one quantum state is the area between two adjacent ellipses. The area of an ellipse is πab, where a and b are the lengths of the semiaxes. In our example, $a = x_0$ and $b = m\omega x_0$ according to (6.3) and (6.4), so $\pi ab = \pi m\omega x_0^2$. Since $\frac{1}{2}kx_0^2 = E$, $\pi ab = 2\pi m\omega E/k$. For these ellipses, $E = (n + \frac{1}{2})h\omega/2\pi$, so the area between adjacent ellipses, corresponding to quantum numbers $n + 1$ and n, is

$$\frac{2\pi m\omega}{k}\frac{\left(n + 1 + \frac{1}{2}\right)h\omega}{2\pi} - \frac{2\pi m\omega}{k}\frac{\left(n + \frac{1}{2}\right)h\omega}{2\pi} = h$$

Thus the value of C in (6.9), representing the area of one quantum state in phase space for this one-dimensional problem, is just h.

The partition function (6.9) for the one-dimensional harmonic oscillator, using (6.1) for E, is

$$\begin{aligned} q &= h^{-1}\int_{-\infty}^{\infty} dx \int_{-\infty}^{\infty} dp\, e^{-\beta kx^2/2}\, e^{-\beta p^2/2m} \\ &= h^{-1}\sqrt{\frac{2\pi}{k\beta}}\sqrt{\frac{2\pi m}{\beta}} = \frac{2\pi}{h\omega\beta} \end{aligned} \tag{6.10}$$

The quantum mechanical partition function for a one-dimensional harmonic oscillator was calculated in (3.11):

$$q^{\text{vib}} = \sum_{n=0}^{\infty} e^{-(n+\frac{1}{2})\beta h\omega/2\pi} = e^{-\beta h\omega/2\pi}\left[1 - e^{-\beta h\omega/2\pi}\right]^{-1}$$

At high temperature (small β), $\exp(-\beta h\omega/2\pi)$ approaches 1 and the last factor approaches $\beta h\omega/2\pi$. This partition function then becomes (6.10). As expected, results calculated on the basis of classical mechanics agree with quantum mechanical results when the temperature is high enough for many states to be populated, so sums over states may be approximated as integrals.

For a single-particle problem in three dimensions, the classical mechanical partition function is

$$q = C^{-1}\int dx \int dy \int dz \int dp_x \int dp_y \int dp_z\, e^{-H(x,y,z,p_x,p_y,p_z)/kT} \tag{6.11}$$

The value of C for a three-dimensional problem will be determined by considering a single free particle in a box of volume V. There is no potential energy, so no dependence of H on x, y, and z. The integration over x, y, and z gives V and

$$q = C^{-1}V\int_{-\infty}^{\infty} dp_x\, e^{-p_x^2/2mkT}\int_{-\infty}^{\infty} e^{-p_y^2/2mkT}\int_{-\infty}^{\infty} e^{-p_z^2/2mkT}$$

$$= C^{-1}V(2\pi mkT)^{3/2}$$

Comparing this with the true partition function,

$$q^{\text{trans}} = \left(\frac{2\pi mkT}{h^2}\right)^{3/2} V$$

we conclude that $C = h^3$ for a free particle in three dimensions. (Note that this q^{trans} was derived by replacing the sum over states by an integral, that is, we assumed we were in the high-temperature limit.) For a three-dimensional harmonic oscillator, the Hamiltonian is

$$H(p,q) = \frac{1}{2}kx^2 + \frac{p_x^2}{2m} + \frac{1}{2}ky^2 + \frac{p_y^2}{2m} + \frac{1}{2}kz^2 + \frac{p_z^2}{2m}$$

The partition function (6.11) is a product of six one-dimensional integrals:

$$q = C^{-1}\int dx\int dy\int dz\int dp_x\int dp_y\int dp_z\, e^{-H(x,y,z,p_x,p_y,p_z)/kT}$$

$$= C^{-1}\sqrt{\frac{2\pi}{k\beta}}\sqrt{\frac{2\pi m}{\beta}}\sqrt{\frac{2\pi}{k\beta}}\sqrt{\frac{2\pi m}{\beta}}\sqrt{\frac{2\pi}{k\beta}}\sqrt{\frac{2\pi m}{\beta}}$$

The quantum mechanical partition function, a product of three one-dimensional harmonic oscillator partition functions, approaches $(2\pi/h\omega\beta)^3$ at high temperatures. Again, one finds C equal to h^3.

One can imagine what happens for more than one particle. In general, a state in three-dimensional N-particle phase space takes up a volume of h^{3N}. Then the value of the constant C, which divides the integral over positions and momenta, is h^{3N}, and the N-particle partition function is

$$Q = h^{-3N}\int d^{3N}p\, d^{3N}q\, e^{-\beta H(p,q)} \tag{6.12}$$

In (6.12), q represents all the $3N$ coordinates and p all the $3N$ momenta; $d^{3N}q\, d^{3N}p$ implies an integration over all the momenta and all the coordinates. It may be noted that the value of C is often not important, because most properties (S is a notable exception) are calculated by differentiating

$\ln Q$. The average energy is

$$E = -\left(\frac{\partial \ln Q}{\partial \beta}\right) = \frac{\int d^{3N}p\, d^{3N}q\, H(p,q) e^{-\beta H(p,q)}}{\int d^{3N}p\, d^{3N}q\, e^{-\beta H(p,q)}}$$

The interpretation of this is that

$$P(p,q) = \frac{e^{-\beta H(p,q)}}{h^{3N}Q} \tag{6.13}$$

is the probability of finding values of the momenta in the volume element $d^{3N}p$ and values of the positions in the volume element $d^{3N}q$. To get the average value of any property which can be written in terms of the positions and momenta, say $R(p,q)$, one multiplies $R(p,q)$ by $P(p,q)$ and integrates over phase space.

Equation (6.13) is of course the Boltzmann distribution: the probability is proportional to $\exp[-H/kT]$ and H is the energy. The Boltzmann distribution is for systems in thermal equilibrium at fixed volume and temperature, described by the canonical ensemble. In the canonical ensemble, the density of points (number of points per unit volume of phase space) is proportional to $\exp[-H/kT]$. A number of important general results obtain for the Boltzmann distribution.

One such result is for the expectation value or average value of $q_i(\partial H/\partial q_j)$, with q_i and q_j position variables.

$$\left\langle q_i \frac{\partial H}{\partial q_j}\right\rangle = \frac{\int e^{-\beta H} q_i(\partial H/\partial q_j)\, d\omega}{\int e^{-\beta H}\, d\omega} \tag{6.14}$$

where $d\omega$ is an abbreviation for $d^{3N}p\, d^{3N}q$. Using

$$e^{-\beta H}\frac{\partial H}{\partial q_j} = \frac{-1}{\beta}\frac{\partial(e^{-\beta H})}{\partial q_j}$$

the numerator of (6.14) may be integrated by parts in q_j. The boundary terms vanish because $e^{-\beta H}$ vanishes for large values of $|q_j|$ (otherwise the particles in the system would not be localized) and

$$\left\langle q_i \frac{\partial H}{\partial q_j}\right\rangle = \frac{\int e^{-\beta H}(\partial q_i/\partial q_j)\, d\omega}{\beta \int e^{-\beta H}\, d\omega} = kT\,\delta_{ij} \tag{6.15}$$

We have used the fact that the derivative in the numerator of (6.15) is equal to zero for $i \neq j$ and equal to 1 for $i = j$. Equation (6.15) is sometimes called the general equipartition theorem. It holds for momenta as well.

According to the Hamilton's equations of motion, $\partial H/\partial q_j = -dp_j/dt$, and dp_j/dt, the rate of change of the jth momentum, is the force acting on the jth coordinate. Summing (6.15) over the degrees of freedom,

$$\left\langle \sum_{i=1}^{3N} q_i \frac{dp_i}{dt} \right\rangle = -3NkT \tag{6.16}$$

The quantity $\sum_i q_i(dp_i/dt)$ is called the virial, and (6.16) is called the virial theorem. It will be of interest in connection with imperfect of gases, discussed in section 7.1.

The form of the Hamiltonian has so far not been important. However, when a position or momentum appears quadratically in H, one can derive, from (6.15), something very useful for the expectation value of this "quadratic degree of freedom." Examples of quadratic degrees of freedom are the position and momentum of the harmonic oscillator (equation 6.1) and the momenta of free particles. (It should be noted that the Hamiltonian equations of motion are often formulated in terms of *generalized* positions and momenta, which are combinations of Cartesian coordinates and momenta such as these.) If q_j appears in H as the term aq_j^2 (a is a constant) and nowhere else, $\partial H/\partial q_j = 2aq_j$ and (6.15) reads

$$\langle q_j(2aq_j) \rangle = kT$$

Thus the expectation value of aq_j^2 is $\frac{1}{2}kT$. The result is exactly the same if the quadratic degree of freedom is a momentum. Equation (6.14) would involve the expectation value of $p_i(\partial p_j/\partial t)$, and the manipulations that led to (6.15) would be carried out in the same way. The average value of a quadratic momentum term in the Hamiltonian is $\frac{1}{2}kT$, like that of a quadratic position term.

This result is called the "equipartition of energy." Suppose there are f quadratic degrees of freedom, so that

$$H = \sum_{i=1}^{f} a_i y_i^2$$

with y_i being a coordinate or a momentum, and the $\{a_i\}$ being constants. For a collection of such systems at thermal equilibrium, the average energy of one system, $\langle H \rangle$, is $fkT/2$. The energy is partitioned equally over the degrees of freedom. Examples so far encountered are: the structureless ideal gas, with three translational degrees of freedom and no potential energy, for which the energy per molecule is $3kT/2$; the one-dimensional harmonic oscillator, with energy kT equally divided between kinetic and potential energy; the linear molecule, with two rotational degrees of freedom and no potential energy, giving rotational energy kT; the nonlinear molecule, with three rotational degrees of freedom and rotational energy $3kT/2$.

6.3. GAS OF STRUCTURELESS PARTICLES

Consider N structureless particles in a volume V, with no external fields. The Hamiltonian consists of a kinetic energy part, the sum of the kinetic energies of the N particles, and a potential energy part $\mathscr{V}$, the interactions between the particles. The N-particle partition function (6.12) is then

$$Q = (N!)^{-1} h^{-3N} \prod_i \left\{ \int_V dx_i\, dy_i\, dz_i \int_{-\infty}^{\infty} dp_{xi} \int_{-\infty}^{\infty} dp_{yi} \int_{-\infty}^{\infty} dp_{zi} \right\}$$

$$\times \exp\left\{ -\beta \left[\sum_j \frac{p_{xj}^2 + p_{yj}^2 + p_{zj}^2}{2m} + \mathscr{V} \right] \right\} \tag{6.17}$$

The product over i is over particles, as is the sum over j. The factor of $(N!)^{-1}$ is to correct for the indistinguishability of the particles. If $\mathscr{V}$ is negligible, (6.17) becomes the partition function for an ideal gas, N noninteracting particles (which can also be referred to as molecules) in a volume V. The integration over the coordinates of each particle gives V, and each of the $3N$ momentum integrals is equal to $\sqrt{2\pi m/\beta}$, so that (6.17) becomes

$$Q = V^N \left(\frac{2\pi mkT}{h^2} \right)^{3N/2}$$

which has already been discussed several times.

If $\mathscr{V}$ is a function of the $3N$ particle coordinates, but is assumed not to involve the $3N$ momenta, the exponential factors into a function of coordinates and a function of momenta. The partition function is then a product of an integral over coordinates and an integral over momenta. We shall be concerned with the former integral when we discuss imperfect gases, in the next chapter. Here, we are concerned with the integral over momenta. The average value of some function of the momenta of particle 1, say $F(\mathbf{p}_1)$, is given by

$$\frac{1}{Q} h^{-3N} \prod_{1} \left\{ \int_V dx_i \, dy_i \, dz_i \int_{-\infty}^{\infty} dp_{xi} \int_{-\infty}^{\infty} dp_{yi} \int_{-\infty}^{\infty} dp_{zi} \right\} F(\{\mathbf{p}_1\}) e^{-\beta H}$$

$$= \frac{\int_{-\infty}^{\infty} dp_{x1} \int_{-\infty}^{\infty} dp_{y1} \int_{-\infty}^{\infty} dp_{z1} F(\mathbf{p}_1) \exp\left\{-\beta\left(p_{x1}^2 + p_{y1}^2 + p_{z1}^2\right)/2m\right\}}{\int_{-\infty}^{\infty} dp_{x1} \int_{-\infty}^{\infty} dp_{y1} \int_{-\infty}^{\infty} dp_{z1} \exp\left\{-\beta\left(p_{x1}^2 + p_{y1}^2 + p_{z1}^2\right)/2m\right\}} \tag{6.18}$$

Of course, the same value is obtained for $F(\mathbf{p}_j)$ where j is any particle.

It follows from (6.18) that the probability that a molecule has its x-component of momentum between p_x and $p_x + dp_x$, its y-component of momentum between p_y and $p_y + dp_y$, and its z-component of momentum between p_z and $p_z + dp_z$ is

$$\mathscr{P}(p_x, p_y, p_z) \, dp_x \, dp_y \, dp_z = \frac{\exp\left\{-\beta\left(p_x^2 + p_y^2 + p_z^2\right)/2m\right\}}{(2\pi m/\beta)^{3/2}} \, dp_x \, dp_y \, dp_z$$

This is the Maxwell–Boltzmann distribution. It describes the distribution over momenta for the molecules at each point in space, even in the presence of intermolecular interactions or an external field which leads to a position dependent density.

The Maxwell–Boltzmann distribution may be written in terms of molecular velocities by writing $p_x = mv_x$ and so on, so that

$$\mathscr{P}(v_x, v_y, v_z) \, dv_x \, dv_y \, dv_z = \frac{\exp\left\{-\beta m\left(v_x^2 + v_y^2 + v_z^2\right)/2\right\}}{(2\pi/m\beta)^{3/2}} \, dv_x \, dv_y \, dv_z \tag{6.19}$$

This is a product of three one-dimensional velocity distributions like

$$\mathscr{P}(v_x) \, dv_x = \left(\frac{m\beta}{2\pi}\right)^{1/2} e^{-\frac{1}{2}\beta m v_x^2} \, dv_x$$

The one-dimensional velocity distribution is plotted in Figure 6.2 for several values of $m\beta/2$. Note that $\mathscr{P}$ is normalized so that

$$\int_{-\infty}^{\infty} \mathscr{P}(v_x) \, dv_x = 1$$

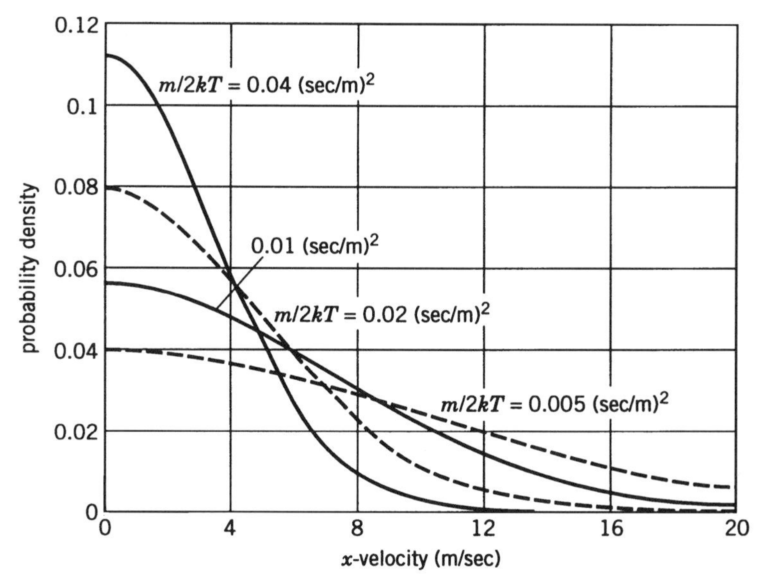

Figure 6.2. One-dimensional Maxwell velocity distribution for various temperatures. The curves give $P(v_x)$ vs. v_x for four different temperatures, where $P(v_x)\,dv_x$ is the probability of finding the x-component of velocity between v_x and $v_x + dv_x$.

Equation (6.19) may also be written in terms of the molecular speed c and the angles θ and ϕ, where $v_x = c \sin\theta \cos\phi$, $v_y = c \sin\theta \sin\phi$, and $v_z = c\cos\theta$, so that $c^2 = v_x^2 + v_y^2 + v_z^2$ and $dv_x\, dv_y\, dv_z = c^2\, dc \sin\theta\, d\theta\, d\phi$.

$$\mathscr{P}(c)\,dc = \int_0^{\pi} \sin\theta\, d\theta \int_0^{2\pi} d\phi\, c^2\, dc \left(\frac{m\beta}{2\pi}\right)^{3/2} e^{-\frac{1}{2}\beta mc^2}\, dc$$

$$= 4\pi \left(\frac{m}{2\pi kT}\right)^{3/2} c^2 e^{-\frac{1}{2}mc^2/kT}\, dc \qquad (6.20)$$

The speed distribution (6.20) is plotted in Figure 6.3 for various values of $m/2kT$. It integrates to unity from $c = 0$ to $c = \infty$.

Average values of functions of the velocity components or the speed are calculated using the velocity and speed distributions. One finds that the average value of any component of the velocity is 0, and that the average speed is

$$\langle c \rangle = \int_0^{\infty} dc\, c\,\mathscr{P}(c) = \sqrt{\frac{8kT}{\pi m}} \qquad (6.21)$$

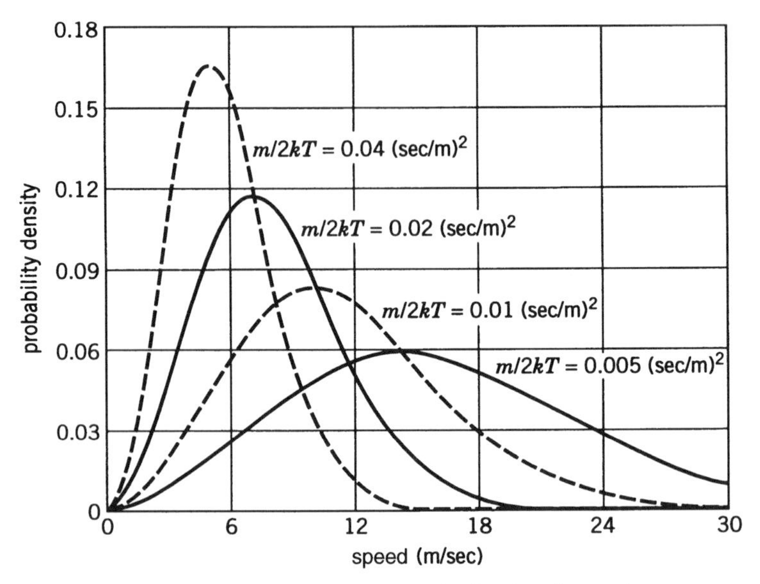

Figure 6.3. Three-dimensional Maxwell speed distribution for various temperatures. The curves give $P(c)$ vs. c, where $P(c)\,dc$ is the probability of finding a speed between c and $c + dc$.

The average square speed, $\langle c^2 \rangle$, may be calculated similarly to $\langle c \rangle$ in (6.21) to give $3kT/m$. This result also follows from the equipartition theorem, since the kinetic energy is

$$E = \frac{mc^2}{2} = \frac{1}{2m}\left(p_x^2 + p_y^2 + p_z^2\right)$$

and the expectation value of each of the three $p^2/2m$ terms is $\frac{1}{2}kT$. The root-mean-square speed is

$$c_{\text{rms}} = \sqrt{\langle c^2 \rangle} = \sqrt{\frac{3kT}{m}} \tag{6.22}$$

which is 1.085 times as big as (6.21). In addition to (6.21) and (6.22), one is often interested in the most probable speed, obtained by maximizing $\mathscr{P}(c)$ with respect to c:

$$c_{\max} = \sqrt{\frac{2kT}{m}} \tag{6.23}$$

which is 0.886 times as big as (6.21).

The velocity distribution (6.19) will now be used to calculate the rate of collisions of molecules with the wall of the container, and from this the

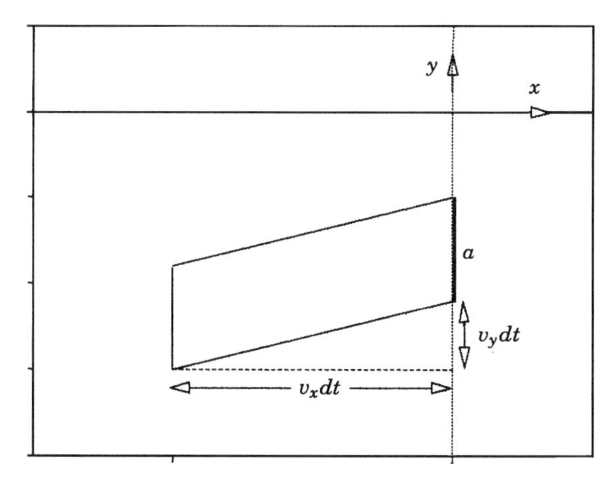

Figure 6.4. Collisions of gas particles with a wall perpendicular to the x-axis. Those molecules with x-component of velocity v_x that will collide with area a of the wall in a time dt lie in a cylinder of height $v_x\,dt$ and area a; regardless of their y- or z-components of velocity, the volume is $av_x\,dt$.

pressure. Suppose the wall, perpendicular to the x-direction, is located at $x = 0$, with the molecules having $x < 0$ (see Figure 6.4). A molecule with a positive x-component of velocity v_x will collide with the wall in the time δt if its x-coordinate is greater than $-v_x\,\delta t$. For the molecule to collide with the area a of the wall, it must lie in a cylinder of height $v_x\,\delta t$ and cross-section a. The angle this cylinder makes with the wall depends on the components of velocity v_y and v_z (see Figure 6.4), but the volume of the cylinder is $av_x\,\delta t$ independently of v_y and v_z.

If there are N molecules in a total volume V, the total number of molecules in the cylinder is $(N/V)av_x\,\delta t$. The number with x-component of velocity equal to v_x which will collide with the area a in the time δt is

$$\begin{aligned}\mathcal{N}(v_x) &= \frac{Na\,\delta t}{V}\int_{-\infty}^{\infty} dv_y \int_{-\infty}^{\infty} dv_z\, v_x \mathcal{P}(v_x, v_y, v_z) \\ &= \frac{Na\,\delta t}{V} v_x \mathcal{P}(v_x) = \frac{Na\,\delta t}{V} v_x \left(\frac{m\beta}{2\pi}\right)^{1/2} e^{-\frac{1}{2}\beta m v_x^2}\end{aligned}$$

The total number of molecules colliding with the area a in time δt is

$$\int_0^{\infty} dv_x \mathcal{N}(v_x) = \frac{Na\,\delta t}{V}\sqrt{\frac{1}{2\pi m\beta}} \tag{6.24}$$

(only positive values of v_x can lead to collision).

The pressure is due to the force exerted on the wall by the molecules during the short time of the collision. It is not necessary to consider the dynamics of the collision in detail, but only to note that, according to Newton's laws,

$$m\left(\frac{dv_x}{dt}\right) = F_x \tag{6.25}$$

where F_x is the force exerted by the wall on the molecule, which is equal and opposite to the force exerted by the molecule on the wall. The collision is assumed elastic, so that, during a collision (when the molecule is in contact with the wall) the force is enough to change the molecule's x-velocity from v_x to $-v_x$. Integrating (6.25) over the duration of a collision shows that $2mv_x$ is equal to the time integral of F_x. Since the force exerted by a molecule on the wall is zero *except* during a collision, the integral may be extended to the interval δt.

Thus the time-average force exerted by a single molecule with x-velocity v_x on the area a is $2mv_x/\delta t$. If $\mathcal{N}(v_x)$ molecules hit the area a during δt, the time average of the total force exerted by them is

$$\mathcal{N}(v_x)\frac{2mv_x}{\delta t} = \frac{Na2m}{V}\left(\frac{m\beta}{2\pi}\right)^{1/2} v_x^2\, e^{-\frac{1}{2}\beta m v_x^2}$$

Integrating over all non-negative values of v_x and dividing by a to get the pressure (force per unit area),

$$P = \frac{Nm}{V}\left(\frac{2m\beta}{\pi}\right)^{1/2} \int_0^\infty dv_x\, v_x^2\, e^{-\frac{1}{2}\beta m v_x^2} = \frac{N}{V\beta} \tag{6.26}$$

This is, of course, the ideal-gas law, with the pressure given a molecular interpretation.

The formula (6.24) for the total number of molecules colliding with an area a during the time δt can be used to calculate the rate of effusion of molecules through a hole, provided that the pressure is low enough for the mean free path [see (6.34)] to be greater than the hole diameter, so that every molecule that hits the hole actually leaves the container. If the hole has area a, the number of molecules hitting the hole (and passing through) per unit time is

$$\frac{Na}{V}\sqrt{\frac{1}{2\pi m\beta}} = \frac{Pa}{\sqrt{2\pi mkT}} \tag{6.27}$$

The rate of effusion is inversely proportional to the square root of the molecular mass. Since the number of molecules inside the container is proportional to the pressure, if the temperature remains constant the pressure inside the container decreases exponentially.

The number of collisions *between* molecules can be calculated by arguments similar to those used to calculate the number of collisions of molecules with the container wall. It is assumed that the molecules are hard spheres with no long-range intermolecular forces, so that a molecule travels in a straight line unless it collides with another molecule. Consider first a molecule with hard-sphere radius r_1, moving at velocity v in a gas of stationary molecules with hard-sphere radius r_2. During a time δt, the moving molecule sweeps out a cylinder of length $v\,\delta t$ and base area πr_1^2, as shown in Figure 6.5. A collision occurs when the distance between the center of the first molecule and the center of a gas molecule is less than $r_1 + r_2$. The situation is equivalent to a molecule of radius $r_1 + r_2$, sweeping out a cylinder of base area $\pi(r_1 + r_2)^2$ in a gas of stationary point particles. The number of gas molecules in this cylinder, and hence the number of collisions the moving molecule makes with gas molecules in time δt, is

$$C' = (N/V)v\,\delta t\,\pi(r_1 + r_2)^2 \tag{6.28}$$

(Of course, collisions cause the molecule's path, and the cylinder, to have kinks or bends in it; giving the cylinder's length as $v\,\delta t$ means that the bends

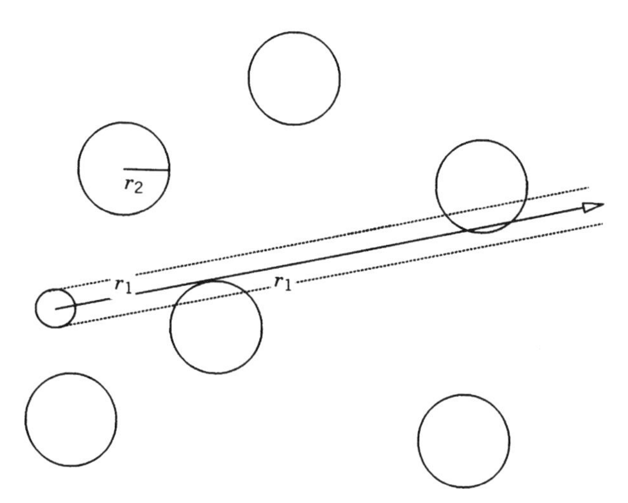

Figure 6.5. Collisions of a molecule of radius r_1 with molecules of radius r_2. The smaller molecule moves at constant velocity, sweeping out a cyclinder of radius r_1 and area πr_1^2, and the larger molecules are stationary. A collision will occur if the center of a stationary molecule lies within a cyclinder of radius $r_1 + r_2$ whose axis is the path of the center of the moving molecule.

are ignored, so our calculation will not be valid when the gas is dense and collisions are very frequent.)

Now one has to take into account the fact that the gas molecules are moving as well as the first molecule. One should use for v the average relative velocity of the first molecule and its collision partner. To derive the probability distribution for relative velocities, first multiply the probability that particle 1 has velocity components v_{x1}, v_{y1}, and v_{z1} [see (6.19)] by the probability that particle 2 has velocity components v_{x2}, v_{y2}, and v_{z2} [(6.19) again]. In the exponential factor of this joint probability distribution, make the substitution

$$m_1 v_1^2 + m_2 v_2^2 = \mu v^2 + (m_1 + m_2)V^2$$

with $\mu = m_1 m_2/(m_1 + m_2)$ being the reduced mass, v the relative velocity, and V the velocity of the center of mass of the two particles. Then integrate the joint probability distribution over all values of V_x, V_y, and V_z to get the probability distribution for relative velocity:

$$\begin{aligned}\mathscr{P}(v_x v_y v_z)\,dv_x\,dv_y\,dv_z &= \left(\frac{\mu}{2\pi kT}\right)^{3/2} \exp\left\{\frac{-\mu\left(v_x^2 + v_y^2 + v_z^2\right)}{2kT}\right\} dv_x\,dv_y\,dv_z \\ &= \left(\frac{\mu}{2\pi kT}\right)^{3/2} \exp\left\{\frac{-\mu v^2}{2kT}\right\} v^2\,dv \sin\theta\,d\theta\,d\phi\end{aligned}$$

We require $\langle v \rangle^*$, the average value of v for colliding molecules, which means that only positive values of v_x, v_y, and v_z should be considered. Thus we average v over only one octant of space, by dividing by 1/8 and integrating over θ from 0 to $\pi/2$ and over ϕ from 0 to $\pi/2$.

$$\begin{aligned}\langle v \rangle^* &= 8\left(\frac{\mu}{2\pi kT}\right)^{3/2} \int_0^\infty v^2\,dv \int_0^{\frac{1}{2}\pi} \sin\theta\,d\theta \int_0^{\frac{1}{2}\pi} d\phi \exp\left\{\frac{-\mu v^2}{2kT}\right\} v \\ &= \sqrt{\frac{8kT}{\pi\mu}}\end{aligned} \tag{6.29}$$

From the expression for C', with $\langle v \rangle^*$ replacing v, we obtain the number of collisions molecule 1 makes per unit time with the gas molecules.

$$R_c = \frac{N}{V}\sqrt{\frac{8kT}{\pi\mu}}\,\pi(r_1 + r_2)^2 \tag{6.30}$$

This is the collision rate of one molecule with mass m_1 with target molecules of mass m_2; N/V is the number density of target molecules.

The quantity $\pi(r_1 + r_2)^2$ in (6.30) is the collision cross-section. It is the area of a circle of radius $r_1 + r_2$, which is the target area that molecule 1, considered as a point particle, has to hit for there to be a collision. If m_1 is much smaller than m_2, the reduced mass μ is almost equal to m_1. Then R_c is the same as the rate of collision of molecules of mass m_1 with a gas of stationary molecules of number density N/V (6.28) with the mean speed (6.21) substituted for v. The mass of the heavy molecules is not relevant, because these molecules are stationary compared to the light ones.

If there are N_1 molecules of type 1 in the volume V as well as N_2 molecules of type 2, the number of collisions per second between all the molecules of type 1 and all the molecules of type 2 is R_c multiplied by N_1, that is,

$$Z_{12}V = \frac{N_1 N_2}{V}\sqrt{\frac{8kT}{\pi\mu}}\,\pi(r_1 + r_2)^2 \tag{6.31}$$

We have introduced a conventional abbreviation Z_{12}, for the collision number, equal to the number of collisions between type 1 and type 2 molecules per unit volume per unit time. Note that Z_{12} is proportional to the product of the number densities of molecules of the two species, to average relative speed, and to the collision cross-section. If some fixed fraction of collisions between a type 1 and a type 2 molecule result in a chemical reaction, the rate of the reaction (number of molecules of product formed per unit volume per unit time) will be proportional to Z_{12}. According to (6.31), Z_{12} is proportional to the product of the concentrations of the two species, so the reaction will follow a second-order rate law.

Equation (6.31) may be applied to collisions in a pure gas, containing only one type of molecule, with mass m. The reduced mass μ equals $m/2$ and the collision number from (6.31) must be divided by 2 because each collision would otherwise be counted twice, since the "moving" molecules and the "target" molecules are identical. Then

$$Z_{11} = \frac{N_1^2}{2V^2}\sqrt{\frac{16kT}{\pi m}}\,\pi(2r_1)^2 \tag{6.32}$$

is the number of collisions per unit volume per unit time. The number of collisions undergone by a single molecule per second is twice Z_{11} (since each collision involves two molecules) divided by N_1/V, that is,

$$z = \frac{N_1}{V}\sqrt{\frac{16kT}{\pi m}}\,\pi(2r_1)^2 \tag{6.33}$$

Note that z is higher by a factor of 1.4 than the number of collisions undergone by a molecule of mass m moving in a gas of stationary molecules.

In this model for a gas, a molecule travels in a straight line at constant velocity until it collides with another molecule or with the wall. The collision, which is almost instantaneous, changes the direction of motion of the particle and, in the case of collision with another molecule, its speed as well. A useful parameter for describing such behavior is the mean free path, the average distance a molecule travels (with constant velocity) from one collision to the next. The mean free path λ is equal to the average speed of a molecule [equation (6.21)] multiplied by the average time between collisions, which is the reciprocal of the collision rate z [equation (6.33)],

$$\lambda = \frac{c}{z} = \sqrt{\frac{8kT}{\pi m}} \frac{V}{N_1 \pi (2r_1)^2} \sqrt{\frac{\pi m}{16kT}} = \frac{V}{\sqrt{2} N_1 \pi (2r_1)^2} \tag{6.34}$$

The mean free path depends on the size and number density of molecules. At fixed number density, it does not depend on temperature, because both the mean speed and the collision rate are proportional to $T^{1/2}$.

The mean free path for a molecule of species 1 in a mixed gas is given by

$$1/\lambda_1 = \sum_j \left(\frac{N_j}{V}\right) \pi (r_1 + r_j)^2 \sqrt{1 + \frac{m_1}{m_j}} \tag{6.35}$$

where m_j is the mass of a molecule of species j. This result is obtained by dividing the average speed of a molecule of species 1 by the number of collisions such a molecule makes per unit time. If m_1 is much less than any of the other masses, and if N_1/V is much less than N_j/V for $j \neq 1$, (6.35) simplifies to

$$\frac{1}{\lambda_1} = \sum_j \left(\frac{N_j}{V}\right) \pi (r_1 + r_j)^2$$

This is applicable to electrons moving in a gas of atoms or molecules.

To get an idea of the mean free path for a typical gas, note that in an ideal gas at 1 atm pressure and $T = 300$ K, the number of molecules per unit volume is 2.45×10^{19} cm^{-3}. For air, almost all the molecules are O_2 and N_2, which are about the same mass and size (hard-sphere radii about 1.9×10^{-8} cm), so (6.34) gives $\lambda_1 = 6.36 \times 10^{-6}$ cm. The mean free path, while small on a macroscopic scale, is hundreds of times greater than the diameter of a molecule. On a molecular scale, the path of a molecule is mostly a straight line, justifying our neglecting bends in the cyclinder in Figure 6.5. The lower the pressure, the lower the density and the larger the mean free path. For pressures below 10^{-5} atm, λ is measured in centimeters and may become comparable to the dimensions of laboratory apparatus. In this case, gas molecules collide more frequently with the container walls than each other. In intersellar space, there is approximately 1 molecule per cubic centimeter, so, with a hard-sphere radius of 2×10^{-8} cm, λ is calculated as 1.4×10^9 km.

It must be emphasized that the mean free path is an average quantity. If one could monitor a particular molecule, one would find that the time between successive collisions with other molecules varies randomly, as does the distance traveled between successive collisions. From the point of view of the molecule, the collisions are random events. This means that the probability that a molecule moving with speed u suffers a collision during the time it moves through a distance δs depends only on δs, and not on the molecule's past history.

If δs is small, the probability that a molecule moving with speed u undergoes a collision during the time it moves through δs will be proportional to δs. Write this probability as $p_u\, \delta s$, so the probability that no collision occurs is $1 - p_u\, \delta s$. Let $P_u(s)$ be the probability that the molecule moves a distance s without suffering a collision. For the molecule to move a distance $s + \delta s$ without collision, it must (1) move through s and (2) not have a collision during δs, so $P_u(s + \delta s) = P_u(s)[1 - p_u\, \delta s]$ or

$$P_u(s) + \frac{dP_u}{ds}\, \delta s = P_u(s) - P_u(s) p_u\, \delta s$$

This means $d \ln P_u)/ds = -p_u$ and, since $P_u(0) = 1$, $P_u(s) = \exp(-p_u s)$.

The probability that, after moving through a distance s unmolested, the molecule undergoes a collision during the next distance δs is $p_u \delta s P_u(s)$, so the average distance moved by a molecule of speed u between collisions (mean free path) is

$$p_u \int_0^\infty ds\, P_u(s) s = p_u^{-1}$$

Thus p_u is the reciprocal of λ_u, the mean free path for a molecule with speed u, and

$$P_u(s) = e^{-p_u s} = e^{-s/\lambda_u} \tag{6.36}$$

$P_u(s)$ was defined as the probability that an observed particle in a gas with speed u will move through a distance s without experiencing a collision with another molecule. Because the laws of mechanics are symmetrical in time, it is also the probability that an observed molecule has moved a distance s in a straight line before being observed. Also, $P_u(s)$ is the probability that, after undergoing a collision, a molecule of speed u moves through a distance s without undergoing another collision.

The probability that a collision occurs within the distance ds after the molecule has moved through s is $P_u(s)$ multiplied by ds/λ_u. Thus, $P_u(s)/\lambda_u$

is the probability distribution function for free path lengths. It is normalized to unity:

$$\int_0^\infty ds\, \frac{P_u(s)}{\lambda_u} = 1$$

The probability that the distance traveled between collisions is less than the mean free path is

$$\int_0^{\lambda_u} ds\, \lambda_u^{-1}\, e^{-s/\lambda_u} = 1 - \frac{1}{e} = 0.632$$

or somewhat more than $\frac{1}{2}$. The mean-square deviation from the mean of the free-path length is

$$\left\langle (s - \langle s \rangle)^2 \right\rangle = \langle s^2 \rangle - \langle s \rangle^2 = \int_0^\infty ds\, s^2\, \lambda_u^{-1} P_u(s) - \lambda_u^2 = \lambda_u^2$$

This is quite a broad distribution, since the root-mean-square deviation from the mean is equal to the mean value.

6.4. TRANSPORT PROPERTIES

The mean free path of equations (6.34) and (6.35) is an average of λ_u, obtained by substituting the average relative speed for u. It is useful in constructing a simple model for transport properties. Transport properties we will consider are heat conduction (transport of energy), diffusion (transport of matter), and viscosity (transport of momentum). In each case, we are concerned with a gas which is not at thermal equilibrium, because some intensive property (energy density, mass density, momentum density) varies from one point to another. This variation produces a flow or transport (of energy, of matter, of momentum) in the direction of decreasing value of the intensive property.

If the value of the intensive property varies linearly with position, so that its gradient is constant, and if the gradient is not too large, the rate of flow is found experimentally to be proportional to the gradient. Thus the rate of heat flow (conduction) is proportional to the temperature gradient, the rate of mass flow (diffusion) is proportional to the concentration gradient, and the rate of momentum flow (viscosity) is proportional to the gradient of momentum density. The simple model to be discussed will explain why this is so and provide values for the constant of proportionality (thermal conductivity, diffusion constant, viscosity coefficient).

We consider first diffusion, the transport of matter in response to a concentration gradient. Imagine a gas between the planes at $x = 0$ and

$x = s$, with temperature T everywhere but with the concentration of molecules varying linearly with x. The concentration at a height x, $n(x)$, is equal to $n(0) + x(dn/dx)$, with the gradient, dn/dx, being constant. Considering an area A perpendicular to the x-axis, and at a height x_A, we calculate the number of molecules crossing A per unit time. Polar coordinates θ and ϕ are used to specify the direction of motion of a molecule, with the x-direction being the polar axis. It is assumed that all molecules have the same speed, c, and the same mean free path, λ.

The molecules that move at angles θ and ϕ, and pass upward through the area A, have been moving in straight lines, on average, for a distance λ. Their last collisions occurred when they were located in an area A at height $x = x_A - \lambda \cos\theta$ (see Figure 6.6). Since their speed in the x-direction is $c\cos\theta$, the number of such molecules passing through A per unit time is

$$n(x)(Ac\cos\theta)\left(\sin\theta\, d\theta\,\frac{d\phi}{4\pi}\right)$$

The last factor is the probability that a molecule's direction is between θ and $\theta + d\theta$, ϕ and $\phi + d\phi$. Substituting $x = x_A - \lambda\cos\theta$, and noting that θ must be between 0 and $\frac{1}{2}\pi$, we have the rate at which molecules move upward though A:

$$R_u = \frac{Ac}{4\pi}\int_0^{\pi/2} d\theta\cos\theta\sin\theta\int_0^{2\pi} d\phi\left\{n(0) + (x_A - \lambda\cos\theta)\frac{dn}{dx}\right\} \tag{6.37}$$

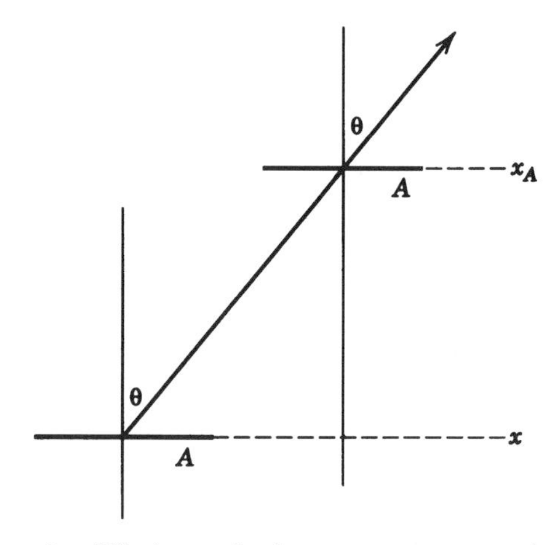

Figure 6.6. Geometry for diffusion and other transport properties. Molecules moving at an angle θ relative to the x-axis will pass through the area A located at height x_A, provided they are located in an area A at the height $x = x_A - \lambda\cos\theta$, where λ is the mean free path.

Molecules crossing A from above may be treated similarly. These have $\frac{1}{2}\pi \leq \theta \leq \pi$, and underwent their last collisions at a height $x_A - \lambda \cos \theta$ (note that $\cos \theta$ is negative for these molecules). The flux of downward-moving molecules through A is

$$R_d = \frac{Ac}{4\pi} \int_{\pi/2}^{\pi} d\theta \cos \theta \sin \theta \int_0^{2\pi} d\phi \left\{ n(0) + (x_A - \lambda \cos \theta) \frac{dn}{dx} \right\}$$

The net rate at which molecules pass through A is $R_u + R_d$ and the flux per unit area is

$$\frac{R_u + R_d}{A} = \frac{c}{2} \int_0^{\pi} d\theta \cos \theta \sin \theta \left\{ n(0) + (x_A - \lambda \cos \theta) \frac{dn}{dx} \right\}$$

$$= \frac{c}{2} \frac{dn}{dx} \left(\frac{-2\lambda}{3} \right)$$

This is proportional to the concentration gradient (but in the opposite direction).

A problem is that, if n is a function of x, so is the mean free path λ. We consider, therefore, a binary gas mixture of constant total number density, in which the mole fraction of the minor component, $n_a/(n_a + n_b)$, varies linearly with x. Since $n_a + n_b$ is constant, λ, the mean free path for molecules of component a, is independent of x. Now the number of a molecules crossing A from below per unit time is

$$R_{au} = \frac{Ac}{4\pi} \int_0^{\pi/2} d\theta \cos \theta \sin \theta \int_0^{2\pi} d\phi \left\{ n_a(0) + \frac{dn_a}{dx} (x_A - \lambda \cos \theta) \right\}$$

The number of a molecules crossing A from above per unit time, R_{ad}, is given by the same expression, except that $\frac{1}{2}\pi \leq \theta \leq \pi$. The net flux of a molecules through unit area is then

$$\frac{R_{au} + R_{ad}}{A} = \frac{c}{2} \int_0^{\pi} d\theta \cos \theta \sin \theta \left\{ n_a(0) + \frac{dn_a}{dx} (x_A - \lambda \cos \theta) \right\}$$

$$= \frac{-c\lambda}{3} \frac{dn_a}{dx} \tag{6.38}$$

The flux of a particles is thus proportional to the concentration gradient and opposite in direction, in accordance with the law of diffusion. Fick's first law of diffusion is normally written

$$j_{xa} = -D_{ab} \frac{dn_a}{dx} \tag{6.39}$$

with j_{xa} being the flux of a particles in the x-direction and D_{ab} the diffusion coefficient for a particles in a mixture of a and b particles. According to (6.38), D_{ab} is $c\lambda/3$.

A more correct treatment, taking into account the details of collisions, gives the same formula, but with a different numerical coefficient:

$$D_{ab} = \frac{3\pi\sqrt{2}}{64} c\lambda = 0.208 c\lambda \tag{6.40}$$

According to equation (6.35), the mean free path of a molecules in an a–b mixture is

$$\lambda = \left(n_a \pi (2r_a)^2 \sqrt{1+1} + n_b \pi (r_a + r_b)^2 \sqrt{1 + \frac{m_a}{m_b}} \right)^{-1}$$

If $n_a \ll n_b$, D_{ab} (referred to as the "tracer diffusion coefficient") is inversely proportional to the density of molecules b, as well as to $(T/m_a)^{1/2}$ because of the factor of c. If molecules a and b are of similar mass and size, (6.40) becomes

$$D_{ab} = \frac{3\pi\sqrt{2}}{64} \sqrt{\frac{8kT}{\pi m}} \frac{1}{4\pi r^2 \sqrt{2}\,(n_a + n_b)} = \frac{3\pi}{16\sqrt{2}} \sqrt{\frac{kT}{\pi m}} \frac{1}{4\pi r^2 n}$$

where n is the total number density of molecules, replacing $n_a + n_b$. If molecules a and b are identical, D_{ab} may be referred to as the "self-diffusion coefficient." The self-diffusion coefficient is inversely proportional to the molecular collision cross-section, $4\pi r^2$ or $\pi(2r)^2$ and to the density at a given T. Substituting the ideal-gas law, $n = P/kT$, one sees that, at fixed T, the self-diffusion coefficient is inversely proportional to the pressure, and at fixed P, proportional to $T^{3/2}$.

The "diffusion equation" is obtained by combining (6.39) with the conservation of the number of molecules. The latter means that the number of molecules in any region of space changes only because molecules enter and leave it by crossing the region's boundaries. Consider the box between x and $x + dx$, with unit cross-sectional area. Its volume is dx, so the number of molecules within it at time t is $n(x,t)\,dx$. The number of molecules entering through the bottom per unit time is $j_x(x)$, the number entering through the top (downward) is $-j_x(x + dx)$, and, if n depends only on x, there is no net flux of molccules through the sides. Therefore the conservation law reads

$$\frac{\partial n(x,t)}{\partial t} dx = j_x(x) - j_x(x + dx) = j_x(x) - \left[j_x(x) + \frac{\partial j_x}{\partial x} dx \right]$$

Using (6.39),

$$\frac{\partial n(x,t)}{\partial t} = D\frac{\partial^2 n(x,t)}{\partial x^2} \tag{6.41}$$

which is the "diffusion equation" in one dimension for a one-component system. Solutions to the diffusion equation are discussed in Section 8.4 [see equation (8.56) and following].

The next transport property to be considered is viscosity, which is related to the transport of momentum. Imagine gas to be present between two parallel plates, located at $x = 0$ and $x = s$, with the lower plate stationary and the other moving at a velocity U in the y-direction (see Figure 6.7). Gas molecules colliding with the upper plate are assumed to take on an average y-velocity U, and gas molecules colliding with the lower plate take on an average y velocity 0. When a steady state is attained, the average y-velocity of gas molecules will vary linearly with x:

$$v_y = U\frac{x}{s} \tag{6.42}$$

To maintain the steady state, work must be done, since molecules arriving at the upper plate from below have average y-velocity less than U, and have to be accelerated. This means that a force (force = rate of change of momen-

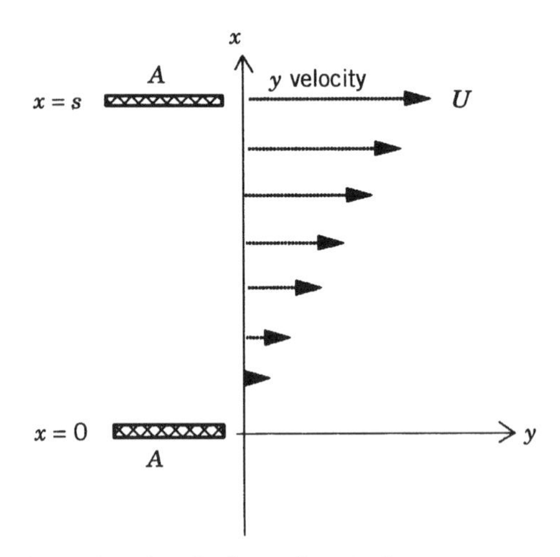

Figure 6.7. Effect of gas viscosity. A plate of area A at $x = s$ moves at a y velocity U while a plate of area A at $x = 0$ is stationary. At steady state, the y velocity of gas molecules varies linearly with height x. To maintain the velocity gradient dv_y/dx, a force must be applied to the upper plate, proportional to the viscosity coefficient η and to dv_y/dx.

tum) must be exerted continually to keep the upper plate moving at velocity U. By Newton's third law, this force is exerted by the molecules on the plate.

The treatment to follow, analogous to the treatment given of diffusion, will show that the force required to maintain the steady state is proportional to A, the area of the plates, and to U/s. The coefficient of proportionality between the force and AU/s is the coefficient of viscosity. Consider an area A at height x_A. Molecules crossing A from below, and moving at an angle θ relative to the x-direction ($0 \leq \theta \leq \frac{1}{2}\pi$), had their last collisions at a height $x_A - \lambda \cos\theta$, so their average y-momentum is $mU(x_A - \lambda\cos\theta)/s$. The same is true for molecules crossing A from above, except that $\frac{1}{2}\pi \leq \theta \leq \pi$. The net rate of transport of momentum is [see (6.38)]

$$T_p = \frac{ncA}{2}\int_0^{\pi} d\theta \cos\theta \sin\theta \frac{mU}{s}(x_A - \lambda\cos\theta) = \frac{ncA}{2}\frac{mU}{s}\left(\frac{-2\lambda}{3}\right)$$

Since the rate of change of momentum is equal to the force, the matter above an area A exerts a force T_p on the matter below the area A. This holds at each height x.

The force which must be exerted on the plate of area A to maintain the steady state is thus T_p. The coefficient of viscosity is

$$\eta = \frac{T_p}{-AU/s} = \frac{ncm\lambda}{3} \tag{6.43}$$

As for the other transport properties, the numerical coefficient is not correct, but the rest is. A more detailed theory replaces $1/3$ by $5\pi/32 = 0.491$.

Equation (6.43) makes several remarkable predictions. Since λ is inversely proportional to n, and $c^2 = 8kT/(\pi m)$, it implies that the coefficient of viscosity is independent of the gas density. Furthermore, being proportional to $\sqrt{T}$ and molecular parameters, η increases with temperature, contradicting common experience with liquids, whose viscosity decreases with temperature. The independence of η on gas density was first derived theoretically and then shown to hold experimentally by Maxwell. Of course, it cannot be true for all densities: if n approaches 0, there will be no gas and no viscosity.

The demonstration of the preceding paragraphs, since it invoked the notion of mean free path, is invalid unless gas molecules collide more frequently with each other than with the upper and lower plates. For this to hold, λ must be much less than s, the distance between the plates; since λ is inversely proportional to n, this puts a lower limit on n. There is also an upper limit on n. We consider only two-particle collisions, and neglect the kinks in a molecule's path. This is justified when λ is much larger than molecular diameters, so it fails for very high densities. The combined condition, $s \gg \lambda \gg d$, still leaves a large range of densities over which (6.43) is correct, because container sizes are many orders of magnitude larger than molecular sizes.

A third transport property to be considered is thermal conductivity. Imagine a gas between a pair of parallel plates perpendicular to the x-axis. One plate, at $x = 0$, is maintained at temperature T_1, and the other plate, at $x = s$, is maintained at temperature T_2 ($T_2 > T_1$). A steady state will be reached in which the temperature varies through the gas according to

$$T(x) = \frac{T_2 - T_1}{s} x$$

which corresponds to a constant temperature gradient $dT/dx = (T_2 - T_1)/s$. The basic idea is this: if the temperature is higher at point a than at point b, molecules moving from a to b carry more energy than molecules moving from b to a, resulting in a net flow of energy or heat from a to b. The energy flow will be shown to be proportional to dT/dx, as is found experimentally.

As for the other transport properties, it is assumed that all molecules move the same distance λ between collisions (note that the mean free path is independent of temperature), and have the same speed, c. The direction of motion of molecules is random. It is also assumed that the molecules at each height x are in thermal equilibrium at $T(x)$, so that a molecule undergoing a collision at x takes on the average energy corresponding to molecules at $T(x)$. (This may be the weakest of the assumptions.) Each such molecule carries an energy

$$\varepsilon[T(x)] = \varepsilon[T_1] + \frac{d\varepsilon}{dT}\frac{dT}{dx}x = \varepsilon[T_1] + c_v \frac{T_2 - T_1}{s} x \tag{6.44}$$

where c_v is a molecular heat capacity.

Consider an area A at $x = x_A$. Molecules that undergo collisions at a height x ($x < x_A$), move at angles θ and ϕ, and pass upward through the area A before undergoing another collision, are located in an area A at height x (see Figure 6.6). Since their speed in the z-direction is $c\cos\theta$, the number of such molecules passing through A per unit time is

$$(nA)(c\cos\theta)\left(\frac{\sin\theta\, d\theta\, d\phi}{4\pi}\right)$$

where n is the number density and the last factor is the probability that a molecule's direction is between θ and $\theta + d\theta$, ϕ and $\phi + d\phi$. Since a molecule passing through A has its last collision a distance λ from A, the average height at which it has its last collision is $x = x_A - \lambda\cos\theta$. Using this value of x in (6.44), and noting that θ must be between 0 and $\frac{1}{2}\pi$, we have the rate of energy transport through A by upward-moving molecules:

$$S_u = \frac{nAc}{4\pi}\int_0^{\pi/2} d\theta \cos\theta \sin\theta \int_0^{2\pi} d\phi \left\{ \varepsilon[T_1] + c_v \frac{T_2 - T_1}{s}(x_A - \lambda\cos\theta) \right\}$$

Molecules crossing A from above have $\frac{1}{2}\pi \leq \theta \leq \pi$, and underwent their last collisions at a height $x_A - \lambda \cos\theta$ ($\cos\theta$ is negative for these molecules). The rate of energy transport through A by downward-moving molecules is

$$S_d = \frac{nAc}{4\pi}\int_{\pi/2}^{\pi} d\theta \cos\theta \sin\theta \int_0^{2\pi} d\phi \left\{ \varepsilon[T_1] + c_v \frac{T_2 - T_1}{s}(x_A - \lambda\cos\theta) \right\}$$

The total energy transport rate per unit area or "heat flux" is therefore

$$\begin{aligned} \frac{S_u + S_d}{A} &= \frac{nc}{2}\int_0^{\pi} d\theta \cos\theta \sin\theta \left\{ \varepsilon[T_1] + c_v \frac{T_2 - T_1}{s}(x_A - \lambda\cos\theta) \right\} \\ &= \frac{nc}{2} c_v \frac{T_2 - T_1}{s}\left(\frac{-2\lambda}{3}\right) \end{aligned} \tag{6.45}$$

This is proportional to the temperature gradient (but in the opposite direction), in accord with the law of thermal conduction.

One normally writes the heat flux as $-\kappa(dT/dx)$, defining the coefficient of thermal conductivity κ. According to (6.45),

$$\kappa = \tfrac{1}{3} nc\lambda c_v \tag{6.46}$$

Mainly because of the assumption that the average energy of a molecule is determined by the place at which it collides with another molecule, the factor of $1/3$ is far from correct. A more correct theory, which requires detailed consideration of intermolecular collisions, replaces the $1/3$ by $25\pi/64 = 1.227$. The rest of (6.46) is correct.

The coefficient of thermal conductivity is proportional to the heat capacity of the gas molecules (which depends on their internal degrees of freedom), to the gas density, to the average speed, and to the mean free path. However, the mean free path is itself inversely proportional to the density, so the density dependence is illusory. If equation (6.34) is used for the mean free path, (6.46) becomes

$$\kappa = \frac{1}{3\sqrt{2}} \frac{cc_v}{\pi(2r_1^2)} = \frac{1}{3\sqrt{2}}\sqrt{\frac{8kT}{\pi m}} \frac{c_v}{\pi(2r_1^2)}$$

for a pure gas. Thus the coefficient of thermal conductivity is proportional to $(T/m)^{1/2}$, and is independent of the pressure. The pressure independence comes about because the density and the mean free path vary oppositely with pressure. The higher the pressure, the more molecules are available to carry the energy, but the shorter the distance each molecule moves between collisions.

6.5. DIPOLES IN A FIELD

Another application of classical statistical mechanics is to a rotating linear molecule. The coordinates defining its state are the polar and azimuthal angles θ and ϕ. The energy is kinetic energy only, and equal to

$$E = \frac{I}{2}\left[\left(\frac{d\theta}{dt}\right)^2 + \left(\frac{d\phi}{dt}\right)^2 \sin^2\theta\right]$$

where I is the moment of inertia. For a diatomic molecule AB, I is the reduced mass, $m_A m_B/(m_A + m_B)$, multiplied by the square of the internuclear distance. The momenta corresponding to the coordinates θ and ϕ are

$$p_\theta = I\frac{d\theta}{dt} \quad \text{and} \quad p_\phi = I\frac{d\phi}{dt}\sin^2\theta$$

In terms of these, the classical mechanical Hamiltonian is

$$H_0 = \frac{1}{2I}\left[p_\theta^2 + \frac{p_\phi^2}{\sin^2\theta}\right]$$

The expectation value of any function of the angular coordinates and angular momenta, say $F(\theta, \phi, p_\theta, p_\phi)$ is calculated as

$$\langle F\rangle = \frac{\int \sin\theta\, d\theta\, d\phi\, dp_\theta\, dp_\phi\, F e^{-H_0/kT}}{\int \sin\theta\, d\theta\, d\phi\, dp_\theta\, dp_\phi\, e^{-H_0/kT}} \tag{6.47}$$

where θ is integrated from 0 to π, ϕ is integrated from 0 to 2π, and the momenta are integrated from $-\infty$ to $+\infty$.

If the Hamiltonian H_0 governs the motion, all orientations are equally probable because H_0 is independent of the angles. The average value of $\cos\theta$, where θ is the angle between the molecular axis and the z or polar axis, is easily calculated to be 0, meaning it is equally likely for the molecule to point in the positive or in the negative z-direction. The average value of the cosine of the angle between the molecular axis and *any* space-fixed axis is zero.

If the rotating body has a dipole moment, addition of an external electric or magnetic field will favor orientations along the field direction. The energy associated with the interaction of a magnetic dipole μ and a magnetic field B pointing in the z-direction is $-\mu B\cos\theta$. For an electric dipole in an electric field E, B is replaced by E. If $-\mu B \cos\theta$ is added to the Hamiltonian, the average value of $\cos\theta$, the component of the dipole in the field direction, is

$$\langle \cos\theta \rangle = \frac{\int_0^\pi \cos\theta \, e^{-\beta\mu B\cos\theta} \sin\theta \, d\theta}{\int_0^\pi e^{-\beta\mu B\cos\theta} \sin\theta \, d\theta} \tag{6.48}$$

The denominator is

$$D = \frac{e^{\beta\mu B} - e^{-\beta\mu B}}{\beta\mu B}$$

and the numerator is $-[\partial D/\partial(\beta\mu B)]$ so that

$$\langle \cos\theta \rangle = \frac{e^{\beta\mu B} + e^{-\beta\mu B}}{e^{\beta\mu B} - e^{-\beta\mu B}} - \frac{1}{\beta\mu B} \tag{6.49}$$

An alternate way of obtaining (6.49) is to note that $\langle -\mu B \cos\theta \rangle$ is the average energy of a molecule, and is equal to $-(\partial \ln q/\partial\beta)$, where q is the partition function.

The first term in (6.49) is $\coth(\beta\mu B)$, with the hyperbolic cotangent (coth) equal to the hyperbolic cosine (cosh) divided by the hyperbolic sine (sinh), and

$$\sinh(x) \equiv \frac{[e^x - e^{-x}]}{2}, \qquad \cosh(x) \equiv \frac{[e^x + e^{-x}]}{2}$$

The function $\coth y - y^{-1}$ is called the Langevin function and denoted by $\mathscr{L}(y)$, so the average magnetic moment along the field direction is

$$\mu^B = \mu \mathscr{L}(\beta\mu B) \tag{6.50}$$

The Langevin function is plotted in Figure 6.8. For y large and positive, $\mathscr{L}(y)$ approaches 1; for y large and negative, $\mathscr{L}(y)$ approaches -1. Since $\langle \cos\theta \rangle = \mathscr{L}(\beta\mu B)$, this means that for large field or small T, $\langle \cos\theta \rangle = \pm 1$, corresponding to complete alignment. For small y, one can write the exponentials as power series in y, and

$$\mathscr{L}(y) = \frac{1 + y + \frac{1}{2}y^2 \cdots + 1 - y + \frac{1}{2}y^2 \cdots}{1 + y + \frac{1}{2}y^2 \cdots - 1 + y - \frac{1}{2}y^2 \cdots} - \frac{1}{y}$$

$$= \frac{1 + \frac{1}{2}y^2 \cdots}{y + \frac{1}{6}y^3 \cdots} - \frac{1}{y} \cong \frac{y}{3} \tag{6.51}$$

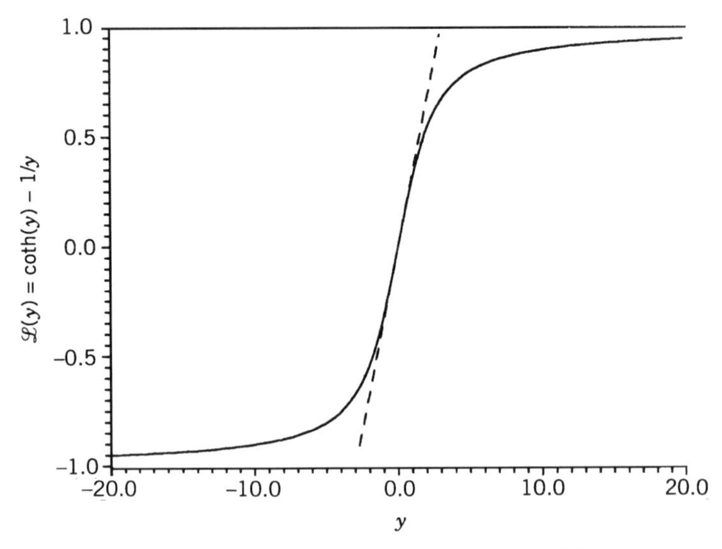

Figure 6.8. Langevin function $\mathscr{L}(y)$ for $-20 < y < 20$. $\mathscr{L}(y) = 0$ for $y = 0$ and approaches ± 1 for $y \to \pm\infty$. For small y, $\mathscr{L}(y) \cong y/3$; this approximation is plotted as the dashed line.

This linear approximation is shown as a dashed line in Figure 6.8. For small field or high T, $\langle \cos\theta \rangle = \beta\mu B/3$ and the average dipole moment in the field direction is $\mu^2 B/3kT$. Since the average dipole moment in any direction is zero with no field present, one can say that the field has induced a dipole moment in the field direction.

The average magnetic moment along the field direction for a single molecule, $\mu\mathscr{L}(\beta\mu B)$ or $\mu^2 B/3kT$ for small B, may be compared with the result of the quantum mechanical calculation in Section 3.5. In equations (3.34) to (3.36), the effect of a magnetic field on spin-J particles was calculated. The component of the magnetic moment along the field axis was $Mg\mu_B$ $(-J \le M \le J)$ with μ_B the Bohr magneton, and g the magnetogyric ratio. Since the magnitude of the square of the spin is $J(J+1)$, the magnitude of the magnetic moment is $\mu = g\mu_B\sqrt{J(J+1)}$. The average value of $\cos\theta$ for a spin J in the state with quantum number M is $M/\sqrt{J(J+1)}$, which can become unity only when $M = J$ and J is very large. Since the interaction energy with the field is $-Mg\mu_B B$, a positive field B gives states of higher M lower energies and hence larger populations. Considering all M, the average magnetic moment along the field direction for a molecule was found to be

$$\langle \mu^B \rangle = g\mu_B J B_J(y) \tag{6.52}$$

where B_J, the Brillouin function, was defined in (3.35) as

$$B_J(y) = \frac{2J+1}{2J}\coth\left(\frac{(2J+1)y}{2J}\right) - \frac{1}{2J}\coth\frac{y}{2J}$$

and $y = \beta g\mu_B B$.

Since $\coth y \to \pm 1$ for large $|y|$, $\langle \mu^B \rangle \to \pm g\mu_B J$ for large B (or small T). This is equal to $\pm\mu$, the value calculated from the Langevin function, for large J, $\sqrt{J(J+1)} \cong J$. For large J there are many quantum states ($2J+1$ values of M) with energies close together so that the partition function, a sum over M, may be approximated by an integral. To establish the equivalence between (6.52) and the Langevin result for small B (or large T), we use the approximation [see (6.51)]

$$\coth x = x^{-1}\left(1 + \tfrac{1}{3}x^2 + \cdots\right)$$

so that (6.52) becomes

$$\begin{aligned}\langle \mu^B \rangle &= g\,\mu_B J B_J(y)\\ &= g\mu_B J\left(y^{-1}\left[1 + \frac{1}{3}\left(\frac{(2J+1)y}{2}\right)^2\right] - y^{-1}\left[1 + \frac{1}{3}\left(\frac{y}{2}\right)^2\right]\right)\\ &= \frac{g\mu_B y}{3}\left[\left(J+\tfrac{1}{2}\right)^2 - \left(\tfrac{1}{2}\right)^2\right] = \frac{(g\mu_B)^2 \beta B(J^2+J)}{3}\end{aligned}$$

This is the classical mechanical result, $\mu^2 B/3kT$. For small B or large T, the energy levels are close together, so classical mechanics should give the same result as quantum mechanics. It can in fact be shown that (6.52) becomes equal to $\mu\mathscr{L}(\beta\mu B)$ for all fields for very large J (Problem 6.26), because there are many energy levels close together.

Using the classical mechanical (Langevin) theory, one can calculate the magnetization I of a gas of molecules with magnetic dipole moments. With n molecules per unit volume,

$$I = n\mu\mathscr{L}(\beta\mu B) \cong \frac{n(\beta\mu^2 B)}{3}$$

Practically, the approximation is valid except at very low temperatures. The magnetic susceptibility is defined as the ratio of the induced magnetization to the magnetic field. From I, one calculates the paramagnetic susceptibility

$$\chi_p = \frac{I}{B} = \frac{n\mu^2}{3kT} \tag{6.53}$$

The $1/T$ dependence of the paramagnetic susceptibility is Curie's law. The magnetic susceptibility includes the diamagnetic susceptibility as well as χ_p, but the paramagnetic term is normally much larger than the diamagnetic.

The magnetic field B in the definition of χ_p is the applied or external field. It may not be the same as the magnetic field that acts to orient a magnetic moment (and appears as the argument of the Langevin function). The orienting field, referred to as the local field, includes the field of nearby moments as well as the external field. For many applications, such as dilute gases, the former is small enough to neglect and the external field and the local field are identical. If this is not the case, one requires a relation between the two.

The field of nearby moments is expected to be proportional to the magnetization, so the local field is $B + C_w I$, with C_w a constant. The magnetization, in turn, is proportional to the local field and inversely proportional to the temperature:

$$I = \frac{n\mu^2}{3kT}(B + C_w I)$$

which is solved to give

$$I = \frac{n\mu^2 B/3k}{T - C_w n\mu^2/3k}$$

The proportionality of the magnetic susceptibility χ_p to $(T - \theta)^{-1}$, where θ is a constant, is known as the Curie–Weiss law. The reasoning which leads to the Curie–Weiss law may be compared to the derivation of the Clausius–Mossotti formula for the dielectric constant [see (6.58)–(6.60)].

The Langevin formula may be used to calculate the magnetic susceptibility of rare earth and other ions in the solid state. The magnetic susceptibility is given by (6.53) with

$$\mu = g\mu_B\sqrt{J(J+1)}$$

where J is the total electronic angular momentum of the ion. In this formula, μ_B is the Bohr magneton and g the Landé g-factor

$$g = 1 + \frac{J(J+1) + S(S+1) - L(L+1)}{2J(J+1)} \tag{6.54}$$

For the rare-earth ions from La^{3+} to Lu^{3+}, the number of 4f electrons varies from 0 to 14, the maximum possible. Using Hund's rules, one can calculate the values of L and S (total orbital and total spin angular momenta) for each ion, and then the value of J for the lowest-energy state. From the measured magnetic susceptibility one can calculate

$$p = (\mu_B)^{-1}\sqrt{\frac{3\chi_p kT}{n}}$$

which should be equal to $g\sqrt{J(J+1)}$. Agreement with calculated $g\sqrt{J(J+1)}$ is good for almost all the ions.

Applying the Langevin formula to an electric dipole, such as that carried by a rotating molecule, in an electric field E, we calculate the dipole polarizability for a single molecule as the ratio of the induced moment to the field: $\alpha_d = \mu^2/3kT$. For molecules, the total polarizability includes, in addition to α_d, the electronic polarizability α_e and the atomic polarizability α_a, corresponding to the dipole moments induced by distortion of the electron cloud and by displacement of atoms. If there are n molecules per unit volume, the electric polarization, or dipole moment per unit volume, is

$$\mathbf{P} = n(\alpha_e + \alpha_a + \alpha_d)\mathbf{E} = n\alpha\mathbf{E}$$

For the remainder of this chapter, we will use electrostatic units: the electrostatic unit of charge is the statcoulomb or esu with 1 C = 3×10^9 esu, and the electrostatic unit of potential is the statvolt with 1 statvolt = 300 V. The dielectric constant of vacuum is unity, so the Poisson equation is $\nabla \cdot \mathbf{E} = 4\pi\rho$.

6.6. DIELECTRIC CONSTANT

In the presence of an electric polarization, the Poisson equation is

$$\nabla \cdot \mathbf{E} = 4\pi\rho - 4\pi\nabla \cdot \mathbf{P} \tag{6.55}$$

where ρ is the charge density. This prompts the definition of the electric displacement $\mathbf{D}$ as $\mathbf{E} + 4\pi\mathbf{P}$, so that $\nabla \cdot \mathbf{D} = 4\pi\rho$. For an isotropic medium in an electric field, P and D will be in the direction of E. Then the static dielectric constant ε_s (for an electric field varying slowly, or not at all, in time) is defined as the ratio of D to E. For the gas of rotating dipolar molecules,

$$\varepsilon_s = 1 + \frac{4\pi P}{E} = 1 + 4\pi n(\alpha_e + \alpha_a + \alpha_d) \tag{6.56}$$

where α_e and α_a are temperature independent and $\alpha_d = \mu^2/3kT$. We refer to a medium with ε_s different from 1 as a dielectric. For dilute gases of molecules with permanent dipoles, the dielectric constant and its temperature dependence are given correctly by (6.56).

For more dense gases, and certainly for liquids, it is necessary to take into account the difference between the external electric field E and the local field, which is the field that acts to polarize a molecule. The local field is the external field plus the field of nearby molecules, which are also polarized. There are a number of methods to calculate the difference between the local field and the external field. The simplest discussed here is associated with the name of Lorentz.

Consider a point inside an isotropic dielectric in a uniform electric field E. Imagine a sphere S centered at this point, macroscopically small, but of radius large enough for the sphere to contain a large number of molecules, and for the molecules outside S to be considered as forming a continuum. Within S, individual molecules are considered, but they are assumed to be packed in a cubic lattice. The local field at the origin includes the external applied field E, the field due to molecules outside the sphere, and the field due to molecules inside the sphere. All fields are in the z-direction.

Outside the sphere, the polarization is P (uniform, in the direction of E). It is shown in electrostatics that, if one has a polarization P in some region of space, it is equivalent, as far as the electric field it produces outside that region, to a volume charge density $-\nabla \cdot \mathbf{P}$ [see (6.55)] and a surface charge density $\mathbf{P} \cdot \mathbf{n}$ on the surface bounding the region, where $\mathbf{n}$ is the outward normal on this surface. The polarization being uniform, $\nabla \cdot \mathbf{P}$ vanishes. One must now calculate the field at the center of S due to the charge density on the surface of the region, which is just the surface of S.

Suppose that S has radius r and surface area $4\pi r^2$. Consider the part of the surface of S between the circles formed by the intersection of S with the cones of angles θ and $\theta + d\theta$, as shown in Figure 6.9. On this part of the surface, which has area $2\pi r^2(\sin\theta)\,d\theta$, $\mathbf{P} \cdot \mathbf{n} = -P\cos\theta$ since P is in the z-direction. The electric field at the center of S due to this surface charge is in the z-direction, by symmetry, and has magnitude

$$2\pi r^2(\sin\theta)\,d\theta(-P\cos\theta)(-\cos\theta/r^2)$$

Integrating over θ, we obtain

$$2\pi \int_0^\pi P\cos^2\theta \sin\theta\,d\theta = \frac{4\pi P}{3} \tag{6.57}$$

for the field due to molecules outside S.

To calculate the field at the center due to the induced dipoles within S, one sums over individual molecules. If the mth dipole is located at (x_m, y_m, z_m), the field produced at the origin by these dipoles is

$$\mu_{\text{in}} \sum_m \frac{2z_m^2 - x_m^2 - y_m^2}{\left(x_m^2 + y_m^2 + z_m^2\right)^{5/2}} = \mu_{\text{in}} \sum_m \frac{3z_m^2 - r_m^2}{r_m^5}$$

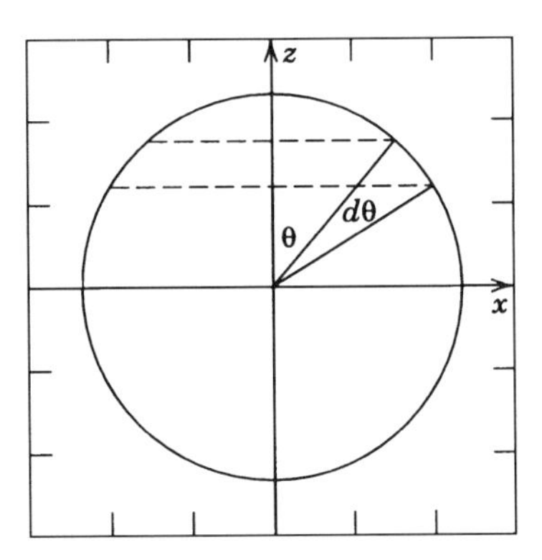

Figure 6.9. Geometry for calculation of local field. The polarization outside the sphere is equivalent to a volume charge density, which vanishes for a uniform field, and a surface charge density equal to the normal component of the polarization. On the part of the sphere's surface formed by cones of angles θ and $\theta + d\theta$, which has area $2\pi r^2 \sin\theta\, d\theta$, the normal component of the polarization is $-P\cos\theta$. Fields are in the z-direction.

where μ_{in} is the induced dipole, the same for all m. For a cubic arrangement of the molecules, or a random arrangement, $\sum_m x_m^2 = \sum_m y_m^2 = \sum_m z_m^2$ so that the sum vanishes. The random arrangement of dipoles would be appropriate for a dilute fluid.

Now the local field is $E_{\text{loc}} = E + 4\pi P/3$. The polarization, assuming n molecules with polarizability α per unit volume, is $n\alpha E_{\text{loc}}$:

$$P = n\alpha\left(E + \frac{4\pi P}{3}\right)$$

Solving for P, one obtains for the static dielectric constant

$$\varepsilon_s = \frac{D}{E} = 1 + \frac{4\pi P}{E} = 1 + \frac{4\pi n\alpha}{1 - (4\pi n\alpha)/3} \tag{6.58}$$

which is often written as

$$\frac{\varepsilon_s - 1}{\varepsilon_s + 2} = \frac{4\pi n\alpha}{3} \tag{6.59}$$

and referred to as the Clausius–Mossotti formula.

If the electric field varies very rapidly in time, only the electrons will be able to follow the changes, and only the electronic polarizability will be effective. The dielectric constant will be the high-frequency dielectric constant ε_∞ and the Clausius–Mossotti formula will be

$$\frac{\varepsilon_\infty - 1}{\varepsilon_\infty + 2} = \frac{4\pi n \alpha_e}{3} \tag{6.60}$$

This obtains for the electric field of light. In general, the index of refraction n_0 is equal to the square root of the product of the high-frequency dielectric constant and the magnetic permeability. The latter being close to unity for most dielectrics, $\varepsilon_\infty = n_0^2$ and

$$\frac{n_0^2 - 1}{n_0^2 + 2} = \frac{4\pi n \alpha_e}{3}$$

The static dielectric constant for a molecular fluid is typically several times the high-frequency dielectric constant because α_d is much larger than α_e and α_a.

Neglecting α_e and α_a and using the Langevin formula for α_d, the dielectric constant formula (6.58) becomes

$$\varepsilon_s = 1 + \frac{4\pi n \mu^2/3kT}{1 - (4\pi n \mu^2)/9kT} = \frac{1 + (8\pi n \mu^2)/3kT}{1 - (4\pi n \mu^2)/9kT}$$

This suggests that the static dielectric constant can become infinite at some low temperature, and negative for temperatures below that. Of course this "polarization catastrophe" does not take place. The problem is that the approximation of the local field as $E + 4\pi P/3$ is not sufficiently accurate for polar fluids and solids, because it assumes all the molecules have the same dipole moment. A more accurate model for a polarizable molecule in a dielectric, which takes into account correlations between the dipole moments of nearby molecules, is required. We now sketch such a model, developed by Onsager.

Each molecule is assumed to have an isotropic polarizability and to occupy a sphere of radius a, from which other molecules are excluded. The dipole moment of the molecule, $\mathbf{m}$, is at the center of the sphere. The matter outside the sphere is treated as a continuum with dielectric constant ε_s, with the value of ε_s to be determined. Within the sphere, the static dielectric constant is equal to unity, since the molecular moment is considered explicitly. One must calculate the electrostatic potential ϕ, assuming that $\mathbf{m}$, the external field, and the polarization are in the z-direction. Since there are no free charges, ϕ must satisfy the Laplace equation, $\nabla^2\phi = 0$, with boundary conditions: ϕ is to be finite everywhere, and ϕ and the normal component of the electric displacement $\mathbf{D}$ must be continuous on the surface of the sphere.

Also, ϕ must approach $-Er\cos\theta$ (corresponding to a constant electric field E in the z-direction) for very large r, and ϕ must approach the electrostatic potential of the dipole **m** for very small r.

Denoting the electrostatic potential inside and outside the molecular sphere by ϕ_i and ϕ_0, the continuity conditions are: $\phi_i = \phi_0$ at $r = a$ and

$$\left(\frac{\partial \phi_i}{\partial r}\right)_{r=a} = \varepsilon_s\left(\frac{\partial \phi_0}{\partial r}\right)_{r=1}$$

If there is an electric field **E** in the z-direction in the dielectric, but no dipole moment **m** at the origin, the Poisson equation and boundary conditions are satisfied by

$$\phi_o^E = -Ez\left[1 - \frac{a^3(1-\varepsilon_s)}{r^3(2\varepsilon_s+1)}\right]$$

and

$$\phi_i^E = -Ez\,\frac{3\varepsilon_s}{2\varepsilon_s+1}$$

with $z = r\cos\theta$. If there is a dipole moment **m** at the origin, but no electric field **E**, the Poisson equation and boundary conditions are satisfied by

$$\phi_o^m = \frac{(1-a^3g)mz}{r^3}$$

and

$$\phi_i^m = \frac{mz}{r^3} - gmz$$

with

$$g = \frac{2(\varepsilon_s - 1)}{a^3(2\varepsilon_s+1)} \tag{6.61}$$

Because of the linearity of the equations of electrostatics, the electrostatic potential in the presence of both an electric field and a dipole is the sum of ϕ^E and ϕ^m. Inside the cavity, the field in the z-direction is

$$-\frac{d(\phi_i^E + \phi_i^m)}{dz} = E\,\frac{3\varepsilon_s}{2\varepsilon_s+1} + gm + \frac{(3z^2-r^2)m}{r^5} \tag{6.62}$$

The last term in (6.62) is just the field of the dipole **m** located at the origin, which does not act on **m** itself; the other two must be the external field plus the field due to the polarized molecules in the surroundings of the central molecule. Thus

$$\mathbf{E}_{\text{loc}} = \mathbf{E}\frac{3\varepsilon_s}{2\varepsilon_s + 1} + g\mathbf{m}$$

is the local field.

The moment $\mathbf{m}$ consists of the permanent moment $\boldsymbol{\mu}$, oriented by the local field $\mathbf{E}_{\text{loc}}$, and the induced moment $\alpha'\mathbf{E}_{\text{loc}}$, with α' including the atomic and electronic parts of the polarizability [see (6.56)].

$$\mathbf{m} = \boldsymbol{\mu} + \alpha'\left(\mathbf{E}\frac{3\varepsilon_s}{2\varepsilon_s + 1} + g\mathbf{m}\right)$$

This may be solved for $\mathbf{m}$ to give

$$\mathbf{m} = (1 - \alpha' g)^{-1}\left(\boldsymbol{\mu} + \frac{3\varepsilon_s\alpha'}{2\varepsilon_s + 1}\mathbf{E}\right) \tag{6.63}$$

Substituting (6.63) in the expression for the local field, which is responsible for orienting the permanent moment, one obtains

$$\mathbf{E}_{\text{loc}} = \mathbf{E}\frac{3\varepsilon_s}{2\varepsilon_s + 1} + g(1 - \alpha' g)^{-1}\left(\boldsymbol{\mu} + \frac{3\varepsilon_s\alpha'}{2\varepsilon_s + 1}\mathbf{E}\right)$$

According to the Langevin theory [equations (6.48) to (6.51)],

$$\frac{\langle \mu_z \rangle}{\mu} = \frac{\int \cos\theta \exp(-\mu \cdot E_{\text{local}}/kT)\sin\theta\, d\theta}{\int \exp(-\mu \cdot E_{\text{local}}/kT)\sin\theta\, d\theta}$$

In this equation,

$$\boldsymbol{\mu} \cdot \mathbf{E}_{\text{loc}} = \boldsymbol{\mu} \cdot \mathbf{E}\frac{3\varepsilon_s}{2\varepsilon_s + 1}\left[1 + \alpha' g(1 - \alpha' g)^{-1}\right] + g(1 - \alpha' g)^{-1}\boldsymbol{\mu} \cdot \boldsymbol{\mu}$$

and the last term produces the same θ-independent factor in numerator and denominator. Therefore

$$\frac{\langle \mu_z \rangle}{\mu} = \frac{\int \cos\theta \exp(-\mu F\cos\theta/kT)\sin\theta\, d\theta}{\int \exp(-\mu F\cos\theta/kT)\sin\theta\, d\theta} = \mathscr{L}(\mu F/kT) \tag{6.64}$$

where

$$\mathbf{F} = \mathbf{E}\,\frac{3\varepsilon_s}{2\varepsilon_s + 1}\,\frac{1}{1 - \alpha' g}$$

[see equations (6.48) to (6.50)].

One can now calculate, using (6.63) and (6.64), the average value of the component of **m** (the dipole at the center of the sphere) along the field direction

$$\langle m_z \rangle = (1 - \alpha' g)^{-1}\left(\mu \mathscr{L}(\mu F/kT) + \frac{3\varepsilon_s \alpha'}{2\varepsilon_s + 1} E \right) \qquad (6.65)$$

The number of molecules per unit volume is $n = (4\pi a^3/3)^{-1}$, the reciprocal of the volume per molecule. Multiplying (6.65) by n yields the polarization P and P is related to the dielectric constant by (6.56). Thus

$$P = \frac{3}{4\pi a^3(1 - \alpha' g)}\left(\mu \mathscr{L}(\mu F/kT) + \frac{3\varepsilon_s \alpha'}{2\varepsilon_s + 1} E \right) = \frac{\varepsilon_s - 1}{4\pi} E$$

Except at very low temperatures, $\mathscr{L}(\mu F/kT)$ may be replaced by $\mu F/3kT$. Then one has, after some algebra,

$$\varepsilon_s - 1 = \frac{9\varepsilon_s}{a^3(1 - \alpha' g)(2\varepsilon_s + 1)}\left(\frac{\mu^2}{3kT}\,\frac{1}{1 - \alpha' g} + \alpha' \right) \qquad (6.66)$$

This equation for the dielectric constant may be put into a more convenient form by introducing ε_∞, the high-frequency dielectric constant [see (6.30)], and performing some algebraic manipulations.

If the electric field varies very rapidly, so that the permanent dipole cannot follow it, one must remove the term in μ and replace ε_s by ε_∞ in (6.66) and in g. The result is

$$\varepsilon_\infty - 1 = \frac{9\alpha' \varepsilon_\infty}{a^3(2\varepsilon_\infty + 1) - 2\alpha'(\varepsilon_\infty - 1)}$$

which rearranges to

$$\alpha' = \frac{a^3(\varepsilon_\infty - 1)}{(\varepsilon_\infty + 2)} \qquad (6.67)$$

This is exactly the Clausius–Mossotti formula (6.59), since $n = 3/(4\pi a^3)$. Substituting (6.67) for α' and using (6.61) for g in (6.66), one gets

$$\frac{a^3}{9\varepsilon_s(\varepsilon_\infty + 2)}(6\varepsilon_s^2 + 3\varepsilon_s - 6\varepsilon_s\varepsilon_\infty - 3\varepsilon_\infty) = \frac{\mu^2}{3kT}\frac{1}{1 - \alpha' g}$$

and finally

$$\frac{\mu^2}{3kTa^3} = \frac{4\pi n\mu^2}{9kT} = \frac{(\varepsilon_s - \varepsilon_\infty)(\varepsilon_\infty + 2\varepsilon_s)}{\varepsilon_s(\varepsilon_\infty + 2)^2} \tag{6.68}$$

This is Onsager's formula, relating the static dielectric constant, the high-frequency dielectric constant, the molecular dipole moment, and the volume of a molecule.

Unlike (6.59), Onsager's formula does not lead to the "polarization catastrophe." The static dielectric constant does not become infinite or negative because correlations between orientations of nearby molecules are taken into account. Onsager's formula gives reasonable results for many liquids. For example, acetone has a density of 0.792 g/cm^3 and a molecular dipole moment of 2.9 Debyes. The left-hand side of (6.68) is therefore 2.33 at $T = 300$ K. High-frequency dielectric constants for organic molecules tend to be about 2. Then (6.68) gives $\varepsilon_s = 19.7$, close to the actual value of about 22.

For water, the left-hand side of (6.68) is 3.77 at $T = 300$ K if the dipole moment of an isolated water molecule, 1.82 Debyes, is used for μ. With $\varepsilon_\infty = 2$, 3, and 4, (6.68) yields, respectively, $\varepsilon_s = 31.2$, 48.7, and 70.0. The measured value is higher, about 80. The problem is that water is a highly associated liquid; it is not reasonable to treat individual water molecules as independently orienting units (see Problem 6.34). Also, hydrogen bonding was not considered, so quantitative agreement should not be expected. Still, Onsager's theory is a vast improvement on theories such as Clausius–Mossotti (aside from being a tour de force of electrostatics and algebra).

PROBLEMS

6.1. Suppose the coordinate x enters the Hamiltonian only in the term cx^2, where c is a constant (a quadratic degree of freedom). Write an expression for the average value of cx^2 and show that this is equal to $(2\beta)^{-1}$

6.2. Calculate the average values of $x, p_x, dx/dt$, and $d(p_x^2/dt)$ for a one-dimensional harmonic oscillator, (Use 6.5 for the last two quantities.) Calculate the average values of these four quantities for a free particle.

6.3. For a gas subjected to a gravitational field in the z-direction, a potential energy term $\sum_i m_i g_z$ is added to the Hamiltonian of (6.17). Evaluate the partition function [equation (6.17)] in this case, and show that the Maxwell–Boltzmann distribution for momenta is still valid, although the particle density varies with position.

6.4. Calculate the spread or mean-square fluctuation in the kinetic energy, $\langle \mathcal{J}^2 \rangle - \langle \mathcal{J} \rangle^2$, for a single molecule in an ideal gas at equilibrium, with

$$\mathcal{J} = \frac{p_x^2 + p_y^2 + p_z^2}{2m}$$

Show that this result is consistent with the general theorem about the mean-square fluctuation in the energy proved in Section 2.4,

$$\langle E^2 \rangle - E^2 = kT^2 \left(\frac{\partial E}{\partial T} \right)_V = kT^2 C_V$$

6.5. Calculate the spread in the total kinetic energy for N molecules in an ideal gas at equilibrium at temperature T. The operator for the total kinetic energy is

$$\mathcal{J} = \sum_i \frac{p_{xi}^2 + p_{yi}^2 + p_{zi}^2}{2m}$$

Show that this result is consistent with the general theorem cited in the preceding problem.

6.6. The escape velocity for objects leaving earth (the minimum upward velocity required for an object to overcome the earth's gravitational field) is 11 km/sec. What fraction of air molecules at 300 K (assume these have a molecular mass of 30 g/mol) have their z-component of velocity greater than or equal to 11 km/sec? What would this fraction be at 1000 K?

6.7. A gas is at equilibrium at a temperature T and a pressure P in a container with a small opening in the wall, through which molecules escape. Outside this opening, where there is vacuum, is a series of collimating slits that stop all molecules except those moving in a direction perpendicular to the container wall. Show that the average velocity of these molecules is $(2\pi m/kT)^{1/2}$ where m is the mass of a molecule.

6.8. The hard-sphere diameter of a hydrogen molecule is about 2.5×10^{-8} cm and that of an atom of xenon about 5×10^{-8} cm. Calculate the number of collisions per second of one hydrogen molecule in a gas of xenon atoms at 1 atm pressure and a temperature of 300 K. What would this number be if the xenon atoms were stationary?

6.9. Hydrogen gas at $P = 1$ atm and $T = 300$ K is contained in a volume of 10 L. How many molecules, and how many moles of gas, strike an area of 1 cm^2 on the wall of the container per second? (*Answer:* For H_2 at 1 atm and 300 K, 1.1×10^{24} molecules or 1.8 moles.)

6.10. Gas effusion through a hole may be used to separate the components of a gaseous mixture, since lighter molecules leak through a hole faster than heavier molecules. Suppose one has a 4 : 1 mixture (mole ratio) of N_2 and O_2 in a container. What is the N_2-to-O_2 ratio in the gas that leaks out through a hole in the container wall?

6.11. As discussed in Section 5.5, radiation in an enclosed space may be considered to consist of photons, moving in all directions at the speed of light, $c = 2.997 \times 10^8$ m/sec. The energy density was shown to be aT^4 where a, the Stefan–Boltzmann constant, is

$$\frac{8\pi^5 k^4 T^4}{15h^3c^3} = 7.57 \times 10^{-15} \text{ ergs cm}^{-3} \text{ K}^{-4}$$

Considering the photons as gas molecules moving at speed c, show that the rate of emission of thermal radiation from a hole of area S in the container wall is equal to $\frac{1}{4}caT^4S$. The number of photons having frequency ν and x-component of velocity $c \cos\theta$ which will collide with area S in the time δt is

$$S\,\delta t\,\frac{8\pi}{c^3}\,\frac{\nu^2}{e^{h\nu/kT} - 1}\,c\cos\theta\,\tfrac{1}{2}\sin\theta$$

[see (5.79) for the distribution over frequency]. See equation (6.24); possible values of θ are between 0 and $\frac{1}{2}\pi$.

6.12. By multiplying the distribution function [equation (6.19)] for the velocity of a particle of mass m_1 in a gas by the corresponding distribution function for a particle of mass m_2, obtain the joint probability distribution for the velocities of two particles. Make the substitution

$$m_1 v_1^2 + m_2 v_2^2 = \mu v^2 + (m_1 + m_2)V^2$$

where v is the relative velocity, μ is the reduced mass, and V is the velocity of the center of mass, so that $V_x = (m_1 v_{x1} + m_2 v_{x2})/(m_1 + m_2)$. Then integrate over all values of V_x, V_y, and V_z to obtain the distribution function for relative velocity.

6.13. A tank of volume V contains gas of molecules of mass M at pressure P_i and temperature T. Surrounding the tank is a gas of the same molecules at pressure P_s and temperature T, with $P_s > P_i$. A small hole, of area A, develops in the wall of the tank, such that every gas particle striking the hole from either the inside or the outside passes through. Derive an expression for P_i as a function of time.

6.14. Derive (6.35) for the mean free path of a molecule of mass m_1 moving in a gas which is a mixture of molecules of different species.

6.15. Let λ_u be the mean free path for molecules moving at speed u. What fraction of the free paths are greater than $2\lambda_u$?

6.16. Show that $D_{12} = D_{21}$, where D_{ij} is the diffusion coefficient for molecules of type i moving through a gas of molecules of type j; the number density of the gas is the same in both cases.

6.17. Values for the coefficient of thermal conductivity κ for air, hydrogen, and helium at 0°C and a pressure below 1 atm (according to gas kinetic theory, κ does not vary with pressure) are, respectively, 0.014, 0.100, and 0.081 J K^{-1} m^{-1}. What are the hard-sphere radii for these molecules? Use (6.46) with the factor $1/3$ replaced by $25\pi/64$.

6.18. One method of measuring thermal conductivity of gases, called the "plate method," is as follows: the gas is introduced into the space between two concentric cylindrical surfaces of radii r_1 and r_2, and of average area A, so that the gas occupies a volume of $\pi(r_1^2 - r_2^2)A$. The inner cylinder, which may be solid, is brought to a temperature T_1 and the outer cylinder is maintained at a temperature T_2 throughout the experiment ($T_2 < T_1$). Since the gas conducts heat from higher to lower temperature, T, the temperature of the inner cylinder, decreases with time. If C_i, the heat capacity of the inner cylinder, is independent of temperature, the change in its energy during time dt is $C_i(dT/dt)\,dt$. This is equal to the heat conducted by the gas during dt, plus the heat conducted by the solid parts of the apparatus. The latter may be written as $R_a(T - T_2)\,dt$, with the value of R_a being characteristic of the apparatus. Using the law of thermal conductivity for a gas, derive an equation for dT/dt and show that

$$C_i \ln \frac{T_1 - T_2}{T - T_2} = \left[R_a + \frac{\kappa A}{(r_1 - r_2)}\right] t$$

To obtain κ, the logarithm of the temperature difference is plotted vs. time, R_a having previously been determined by performing the experiment with no gas between the cylinders.

6.19. The viscosity of Cl_2 gas at 100°C is $\eta = 1.7 \times 10^{-5}$ poise (1 poise $= 10^{-1}$ J m^{-3} sec). What is the mean free path? What is the hard-sphere collision cross-section? What is the average time between collisions for a Cl_2 molecule at 100°C and a pressure of 1 atm?

6.20. The theory of the viscosity coefficient for a gas, and the predicition that it is independent of the number density n at constant T, is valid as long as the dimension of the container of the gas is much larger than the mean free path, and the mean free path is much larger than the molecular diameter. Suppose the container dimension is 1 cm and the molecular diameter is 4×10^{-8} cm, and suppose that "much larger than" means "ten times as big as." What is the range of n over which the theory is valid? At $T = 300$ K, to what range of pressure does this correspond?

6.21. What is the most probable kinetic energy of a molecule in a gas at equilibrium at temperature T? (Assume a Maxwell–Boltzmann distribution of velocities.)

6.22. A thin-walled container of volume V is held at constant temperature T and contains a gas at equilibrium at pressure P. Unfortunately, there is a small hole in the wall, of area A, through which the gas escapes (slowly enough that the equilibrium is not disturbed). In terms of V, A, and the mean speed of the gas molecules c, how long will it take for the pressure to decrease to $P/2$?

6.23. For Ar at 0°C and 1 atm, the coefficient of viscosity η is 2.10×10^{-5} kg m^{-1} sec^{-1}, the diffusion coefficient D is 0.156×10^{-4} m^2 sec^{-1}, and the coefficient of thermal conductivity κ is 1.65×10^{-2} J m^{-1} sec^{-1} K^{-1}. Calculate the hard-sphere atomic radius r_1 of Ar from each of the three coefficients, using (a) the simplest formulas [with factors of 1/3, like (6.43)] and (b) the formulas with the more correct numerical factors [like (6.40)]. Six values for r_1 should result; the three from the more correct theory should be fairly close to each other. (*Answer:* from η, 1.51, and 1.83A; from D, 2.62 and 2.06A; from κ, 3.50 and 1.82A.)

6.24. Since the formulas for all three transport coefficients involve the same kinetic parameters, the kinetic theory predicts a number of relationships between them. Show that, in the simplest theory, $m\kappa/c_v\eta = 1$. What does the more exact theory, in which the factors of 1/3 are replaced by more correct numerical factors, predict for $m\kappa/c_v\eta$? Another relationship involves D and η (and the density). Find this relationship in the simplest theory and in the more exact theory.

6.25. **(a)** Assuming that the Hamiltonian for a rigid rotor (H_0 in Section 6.5) involves p_θ and p_ϕ, but not θ and ϕ, calculate the average value of $\cos\theta$. Calculate the average value of $\sin\theta\cos\phi$, which is the cosine of the angle between the rotor and the x-axis.

(b) Now suppose a term $A\cos\theta$ is added to the Hamiltonian, representing the effect of a field along the z-axis. Calculate the average value of $\cos\theta$ and the average value of $\sin\theta\cos\phi$, assuming that $\beta A\cos\theta$ is small.

6.26. If a rotating magnetic moment μ in a magnetic field B is treated by classical mechanics, the average magnetic moment along the field direction is found to be $\mu\mathscr{L}(\beta\mu B)$, where $\mathscr{L}(x) = \coth x - x^{-1}$ [equation (6.50)]. The quantum mechanical treatment of the magnetic moment associated with angular momentum J gives the average magnetic moment along the field direction as $\sqrt{J/(J+1)}\,\mu B_J(y)$ where $y = \beta\mu B/\sqrt{J(J+1)}$ and

$$B_J(y) = J^{-1}\left[\left(J + \tfrac{1}{2}\right)\coth\left[\left(J + \tfrac{1}{2}\right)y\right] - \tfrac{1}{2}\coth\left(\tfrac{1}{2}y\right)\right]$$

[equation (6.52)]. Show that the quantum mechanical result becomes equal to the classical mechanical result for $J \to \infty$.

6.27. The magnetic susceptibilities of salts of the rare earth ions have been measured (with difficulty because of problems of purification), and the quantity $p = \mu_B\sqrt{3\chi_p kT}$ calculated from the susceptibilities (for La^{3+}, which is diamagnetic, $\chi_p = 0$). Theoretically, p should be equal to $g\sqrt{J(J+1)}$ where g is the Landé g-factor of (6.54). Some values of p are listed below. Calculate p for each ion and compare with the measured value.

ion	ground state	p
La^{4+}	$^1S_0$	0
Ce^{3+}, Pr^{4+}	$^2F_{5/2}$	2.4
Pr^{3+}	$^3H_4$	3.6
Nd^{3+}	$^4I_{9/2}$	3.6
Sm^{3+}	$^6H_{5/2}$	1.5
Eu^{3+}, Sm^{2+}	$^7F_0$	3.6
Gd^{3+}, Eu^{2+}	$^8S_{7/2}$	8.0
Dy^{3+}	$^6H_{15/2}$	10.6
Ho^{3+}	$^5I_8$	10.4
Yb^{3+}	$^2F_{7/2}$	4.5

6.28. In an n-type semiconductor (Section 5.4), N_d donor atoms, each of which has an electronic energy level ε_d, are present in addition to the atoms of the host. Because of interelectronic repulsion, each donor-atom state may be occupied by no more than one electron. A magnetic field B splits ε_d into two energy levels, $\varepsilon_d - gM\mu_B B$, where the g-factor is 2, $M = \frac{1}{2}$ or $-\frac{1}{2}$, and μ_B is the Bohr magneton. If there are N_+ electrons in donor-atom states with spin up $(M = \frac{1}{2})$ and N_- electrons in donor-atom states with spin down ($M = -\frac{1}{2}$), the canonical partition function is

$$Q = \frac{N_d!\, e^{-\beta E(N_+, N_-)}}{N_+!\, N_-!\,(N_d - N_+ - N_-)!}$$

where $E(N_+, N_-) = N_+(\varepsilon_d - \mu_B B) + N_-(\varepsilon_d + \mu_B B)$.

(a) Using the fact that $(\partial A/\partial N_+) = (\partial A/\partial N_-) = \mu$ (A is the Helmholtz free energy), find N_+ and N_- in terms of N_d, β, μ, ε_d, and B.

(b) Calculate the total magnetic moment I as $N_+(-\mu_B) + N_-(\mu_B)$ and show it is equal to

$$N_d\,\mu_B\,\frac{2e^{\beta(\mu-\varepsilon_d)}\sinh(\mu_B B\beta)}{1 + 2e^{\beta(\mu-\varepsilon_d)}\cosh(\mu_B B\beta)}$$

(c) Calculate the magnetic susceptibility, $\chi = I/B$, and show it is equal to

$$\frac{N_d}{\frac{1}{2}e^{-\beta(\mu-\varepsilon_d)} + 1}\,\mu_B^2\,\beta$$

for small fields ($\mu_B B\beta \ll 1$). Note that the first factor is the number of electrons in donor levels ($N_+ + N_-$) and the second factor is the Langevin result $\mu^2/3kT$, since the square of the magnetic moment is $g^2 S(S+1)$ and $S = \frac{1}{2}$.

6.29. Low-frequency dielectric constants have been measured for a number of halomethanes at varying temperatures and gas densities. Extrapolated to low density, the values determined for $A = \mathcal{N}_A n^{-1}(\varepsilon_s - 1)/(\varepsilon_s + 2)$ are given below ($\mathcal{N}_A$ is Avogadro's number). For the first five molecules, use the Clausius–Mossotti formulas (6.59) and (6.60) to find the permanent electric dipole moment in Debyes (1 D = 10^{-18} esu cm = 3.34×10^{-30} C m). Also find values for $\alpha_a + \alpha_e$ in cm^3/mol. (*Hint:* plot A vs. T^{-1}.)

CH_3F		CH_3Cl		CHF_3		$CClF_3$	
$T(K)$	A (cm^3/mol)	$T(K)$	A (cm^3/mol)	$T(K)$	A (cm^3/mol)	$T(K)$	A (cm^3/mol)
323.2	72.13	303.2	84.17	303.2	63.62	323.2	19.167
369.5	64.03	323.2	79.55	323.2	60.35	369.5	18.566
416.5	57.64	369.5	71.04	369.5	53.87		
		404.8	65.87	416.5	48.81		

$CHClF_2$		CCl_2F_2		CCl_3F	
$T(K)$	A (cm^3/mol)	$T(K)$	A (cm^3/mol)	$T(K)$	A (cm^3/mol)
303.2	56.88	323.2	24.90	369.5	27.42
323.1	54.16				
373.4	48.72				
425.0	44.50				

(*Answer:* Dipole moments are: 1.853, 1.904, 1.649, 0.504, 1.467 D.)

6.30. For CCl_2F_2 and CCl_3F, there is data (Problem 6.29) at only one temperature. To get dipole moments, one must guess a value for $\alpha_a + \alpha_e$. Assuming $\alpha_a + \alpha_e$ is the average of the values found for CHF_3, $CClF_3$, and $CClF_2$, 2.95 cm^3/mol, derive the dipole moments of Cl_2F_2 and CCl_3F.

6.31. If the electronic and atomic polarizabilities $\alpha_a + \alpha_e$ are neglected, the Clausius–Mossotti formula becomes: $\varepsilon_s = (1 + 2A)/(1 - A)$ with $A = 4\pi n \mu^2/3kT$. This leads to the "polarization catastrophe," ε_s becoming infinite and negative. For a gas of molecules with $\mu = 2$ D $= 2 \times 10^{-18}$ esu cm, below what temperature would the catastrophe occur ar $P = 1$ atm? Above what pressure would it occur at $T = 300$ K?

6.32. For NH_3, $\alpha_a + \alpha_e$ is about 2.4×10^{-24} cm^3 and μ is about 1.5 D. Calculate the importance of the Lorentz field at 300 K, $P = 1$ atm, by evaluating ε_s according to (6.56), which neglects the Lorentz field, and according to (6.59), the Clausius–Mossotti formula.

6.33. For high values of the electric field E, the dipole moment induced in a rotating dipolar molecule is not proportional to the field, so that the dielectric constant becomes a function of E. This "dielectric saturation" can be seen by continuing the expansion (6.51) of the Langevin function past the first term; one obtains $\mathscr{L}(y) = \frac{1}{3}y - \frac{1}{45}y^3 + \cdots$. Here, $y = \mu E/kT$ and the induced moment is $\mu\mathscr{L}(y)$. According to the discussion preceding (6.56), the polarization is

$$P = n(\alpha_e + \alpha_a)E_{\text{loc}} + n\mu\mathscr{L}(\mu E_{\text{loc}}/kT)$$

with $E_{\text{loc}} = E + 4\pi P/3$.

(a) Show that the dielectric constant, defined as $\varepsilon = D/E = 1 + 4\pi P/E$, is given by

$$\varepsilon = \frac{1 + (8\pi n\alpha)/3 - (4\pi n\mu^4 E^2)/45(kT)^3}{1 - 4\pi n\alpha/3}$$

where $\alpha = \alpha_e + \alpha_a + \mu^2/3kT$.

(b) For NH_3, $\mu = 1.5$ D and $\alpha_e + \alpha_a = 2.3 \times 10^{-24}$ cm^3. What is ε at $P = 1$ atm, $T = 300$ K, for $E = 0$ and $E = 10^7$ volts/cm (3.33×10^4 esu/cm^2)? What is the difference in ε for $P = 10$ atm for $E = 0$ and $E = 10^7$ volts/cm? (*Answer:* 1.00630 and 1.00684 for 1 atm.)

6.34. Because water is a strongly associated liquid, the Onsager equation (6.68) cannot reproduce the dielectric constant unless unreasonable values are used for certain parameters.

(a) What value must be used for ε_∞ if one wants to get the correct dielectric constant at 300 K, $\varepsilon_s = 78$, with the molecular dipole moment of water, 1.8 D?

(b) To take association into account, water is sometimes considered to consist of tetramers; the tetramers, rather than single water molecules, orient in an electric field. Using $\varepsilon_\infty = 2$ and $T = 300$ K, calculate what the dipole moment of a water tetramer must be if (6.68) is to predict $\varepsilon_s = 78$. (*Answer:* 5.8 D.)

(c) SO_2 is not likely to be much associated in the liquid. Using the molecular dipole moment of 1.67 D, $\varepsilon_\infty = 2$, and $n = 0.8$ g/cm^3, predict ε_s from the Onsager formula (6.68). The actual value is 12.35.

CHAPTER 7

FLUIDS

In this chapter, we consider imperfect gases and liquids, using classical statistical mechanics. With no quantum statistical effects, the source of the imperfection is the interaction between the molecules. The Hamiltonian, which represents the energy, has a potential energy, as well as a kinetic energy, component. The potential energy depends on the intermolecular distances and, if the molecules are not spherically symmetrical, on the molecular orientations as well. We are interested in its effect on the fluid's properties.

7.1. VIRIAL EXPANSION

First, we consider a pure gas, consisting of N identical and indistinguishable molecules in a volume V. If there are no external fields, the gas will be homogeneous and isotropic. Let the interaction potential energy between the N molecules be $U_N(\{\mathbf{r}_i\})$, a function of the coordinates of all the molecules. By writing it in this form, we imply that it is independent of the momenta. Then the Hamiltonian is

$$H(\mathbf{p}, \mathbf{q}) = \sum_{i=1}^{N} \frac{p_{xi}^2 + p_{yi}^2 + p_{zi}^2}{2m} + U_N \tag{7.1}$$

where U_N depends on $\{\mathbf{r}_i\}$. Initially, we neglect internal motions of the molecules, except insofar as the internal structure determines the intermolecular interactions. The Hamiltonian is a sum of two terms, one involving the momenta only and the other involving the coordinates only.

The partition function is [see equation (6.12)]

$$Q = \frac{1}{N!\,h^{3N}} \prod_j \left\{ \int_V dx_j\, dy_j\, dz_j \int_{-\infty}^{\infty} dp_{xj} \int_{-\infty}^{\infty} dp_{yj} \int_{-\infty}^{\infty} dp_{zj} \right\}$$
$$\times \prod_i \left[e^{-(p_{xi}^2 + p_{yi}^2 + p_{zi}^2)/2mkT} \right] e^{-U_N/kT}$$

and can be written as a product of integrals over momenta and an integral over coordinates. Furthermore, the momentum integrals are the same ones that arise for the ideal gas, so that

$$Q = \frac{(2\pi mkT)^{3N/2}}{N!\,h^{3N}} \prod_i \left[\int dx_i \int dy_i \int dz_i \right] e^{-U_N/kT}$$

This is conveniently written as

$$Q = Q^{\text{ideal}} \frac{Z}{V^N} \tag{7.2}$$

where

$$Q^{\text{ideal}} = \frac{V^N}{N!} \left(\frac{2\pi mkT}{h^2} \right)^{3N/2} = \frac{V^N}{N!} \Lambda^{-3N}$$

and Z is called the configuration integral:

$$Z = \prod_i \left[\int dx_i \int dy_i \int dz_i \right] e^{-U_N/kT} \tag{7.3}$$

We have used Λ to represent the de Broglie wave length, $\Lambda = h/\sqrt{2\pi mkT}$. Because Q is written as a product, the Helmholtz free energy A is a sum of terms,

$$A = A^{\text{ideal}} - kT \ln\left(\frac{Z}{V^N} \right) \tag{7.4}$$

Other thermodynamic properties can likewise be written as sums of an ideal-gas contribution and a correction for nonideality. If U_N is much smaller than kT, $\exp(-U_N/kT) \simeq 1$ and the correction vanishes because the configurational integral Z becomes equal to V^N and $\ln(Z/V^N)$ becomes equal to 0.

Sometimes the nonideality correction is referred to as an "excess" property: (7.4) would be written as

$$A = A^{\text{ideal}} + A^{\text{exc}}; \qquad -\beta A^{\text{exc}} = \ln\left(\frac{Z}{V^N}\right) \tag{7.5}$$

The inclusion of internal motions (rotation, vibration, electronic, and others) would multiply Q by a factor of $(q^{\text{int}})^N$, where q^{int} is the internal-motion partition function, discussed in Sections 4.2 and 4.3. The internal motions give an additive contribution to $\ln Q$. It is also easy to describe a mixed gas, containing molecules of several species. If m_a is the mass of species a, the de Broglie wavelength for this species is

$$\Lambda_a = h/\sqrt{2\pi m_a kT}$$

Let there be N_a molecules of species a, with $\Sigma_a N_a = N$. The partition function is still a product of Q^{ideal} and Z/V^N, but now

$$Q^{\text{ideal}} = V^N \prod_a \frac{\Lambda_a^{-3N_a}}{N_a!} \left(q_a^{\text{int}}\right)^{N_a} \tag{7.6}$$

and

$$Z = \prod_a \prod_i \left[\int dx_i^a \int dy_i^a \int dz_i^a\right] e^{-U_N/kT} \tag{7.7}$$

U_N depends on all coordinates of all particles of all species, and Z involves integration over all these coordinates. Thermodynamic properties are again sums of "ideal" and "excess" contributions.

It is easy to see where corrections to ideal-gas behavior come from. With $\ln Q = \ln Q^{\text{ideal}} + \ln(Z/V^N)$,

$$\begin{aligned} P &= kT\left(\frac{\partial \ln Q^{\text{ideal}}}{\partial V}\right)_T + kT\left(\frac{\partial \ln(Z/V^N)}{\partial V}\right)_T \\ &= \frac{kT}{V}\sum_a N_a + kT\left(\frac{\partial \ln(Z/V^N)}{\partial V}\right)_T \end{aligned} \tag{7.8}$$

and

$$\begin{aligned} E &= kT^2\left(\frac{\partial \ln Q^{\text{ideal}}}{\partial T}\right)_V + kT^2\left(\frac{\partial \ln Z}{\partial T}\right)_V \\ &= \frac{3kT}{2}\sum_a N_a + \frac{\Pi_a \Pi_i[\int dx_i^a \int dy_i^a \int dz_i^a] U_N e^{-U_N/kT}}{Z} \end{aligned} \tag{7.9}$$

Obviously, the second term in (7.9) is the weighted average of the potential energy U_N, the exponential $\exp(-U_N/kT)$ being the weighting factor. This means that $\exp(-U_N/kT)/Z$ gives the probability of finding molecule 1 at position $(x_1, y_1, z_1), \ldots$, molecule N at position (x_N, y_N, z_N). If the intermolecular forces are repulsive, U_N and the potential energy are positive; attractive intermolecular forces correspond to a negative potential energy. Similarly, the second term in (7.8) is positive or negative according to whether the forces are repulsive or attractive. This is seen by noting that Z/V^N is simply $C = \exp(-U_N/kT)$ averaged over the positions of all the molecules. Increasing V means increasing all coordinates and hence increasing the average distance between molecules, which decreases the size of U_N. Repulsive forces have $C < 1$, so increasing V will bring C closer to unity, and

$$\left(\frac{\partial \ln(Z/V^N)}{\partial V}\right)_T > 0.$$

Attractive forces have $C > 1$, so increasing V will decrease C, making the above quantity negative.

Equation (7.8) may be rewritten simply as

$$P = kT\left(\frac{\partial \ln(Z)}{\partial V}\right)_T \tag{7.10}$$

since $N = \Sigma_a N_a$. To simplify the notation, we write

$$Z = \int_V d^{3N}q\, e^{-\beta U_N(q)}$$

instead of (7.7). The derivative with respect to volume is conveniently carried out by using scaled coordinates, that is, writing each q_i $(i = 1 \cdots 3N)$ as Ls_i where $V = L^3$, so that (7.10) becomes

$$\begin{aligned}
P &= \frac{kT}{Z}\frac{\partial}{\partial V}\left[\int_V d^{3N}q\, e^{-\beta U_N(\{q\})}\right] \\
&= \frac{kT}{Z}\frac{\partial L}{\partial V}\frac{\partial}{\partial L}\left[\int L^{3N} d^{3N}s\, e^{-\beta U_N(\{Ls\})}\right] \\
&= \frac{kT}{Z}\frac{1}{3L^2}\Bigg[3NL^{3N-1}\int d^{3N}s\, e^{-\beta U_N(\{Ls\})} \\
&\qquad + L^{3N}\int d^{3N}s\, e^{-\beta U_N(\{Ls\})}\left\{-\beta\sum_j \frac{\partial U_N}{\partial q_j}s_j\right\}\Bigg]
\end{aligned}$$

The integrations over s in the last two expressions extend over the unit cube, that is, each s_j varies from 0 to 1. The sum over j arises because U_N depends on L through all of the coordinates. Now, remembering that $q_i = Ls_i$,

$$P = \frac{kT}{Z}\left[NL^{-3}Z + \tfrac{1}{3}L^{-2}\int d^{3N}q\, e^{-\beta U_N(\{Ls\})}\left\{-\beta \sum_j s_j \frac{\partial U_N}{\partial q_j}\right\}\right]$$

$$= \frac{NkT}{V} - \frac{1}{3VZ}\int_V d^{3N}q\, e^{-\beta U_N(\{q\})} \sum_j q_j \frac{\partial U_N}{\partial q_j} \qquad (7.11)$$

In equation (7.11), the pressure is given as the ideal-gas pressure plus a correction term from the potential energy of intermolecular interaction.

Equation (7.11) is often referred to as the 'virial equation,' since the second term is the average value over coordinates of

$$\sum_{j=1}^{3N} q_j \frac{\partial U_N}{\partial q_j}$$

which is called the virial. The average is calculated with $\exp(-\beta U_N)$ as the weighting factor. The virial has already appeared in Chapter 6; according to equation (6.16),

$$-3NkT = \left\langle \sum_{i=1}^{3N} q_i \frac{dp_i}{dt} \right\rangle = -\left\langle \sum_{i=1}^{3N} q_i \frac{\partial H}{\partial q_i} \right\rangle \qquad (7.12)$$

We have used Hamiltonian's equations of motion to replace dp_i/dt by $-(\partial H/\partial q_i)$; since the kinetic energy part of H depends on momenta and not positions, H may be replaced by the potential energy. However, the potential energy is not just U_N, the potential energy of intermolecular interactions, for the following reason: in deriving (7.12), it was assumed that the molecules were confined to the volume V. Therefore the potential energy in (7.12) includes, in addition to U_N, the repulsive interactions of molecules with the walls of the container, which are responsible for confining the molecules.

If we denote the wall-molecule potential by W_N, the virial is

$$\sum_{i=1}^{3N} q_i \frac{\partial U_N}{\partial q_i} + \sum_{i=1}^{N}\left(x_i \frac{\partial W_N}{\partial x_i} + y_i \frac{\partial W_N}{\partial y_i} + z_i \frac{\partial W_N}{\partial z_i}\right)$$

For simplicity, suppose the volume is a cube of side length L ($V = L^3$). Now $(\partial W_N/\partial x_i)$ is the force exerted on molecule i by the walls perpendicular to the x-direction. It is nonzero only when x_i is near L or near 0. The average value of $x_i(\partial W_N/\partial x_i)$ is therefore L times the average force exerted by the wall at L on molecule i. There is no contribution to $x_i(\partial W_N/\partial x_i)$ from the force exerted by the other wall on molecule i, since $x_i = 0$. Thus (7.12) reads

$$-3NkT = -\left\langle \sum_{i=1}^{3N} q_i \frac{\partial U_N}{\partial q_i} \right\rangle - \left\langle \sum_{i=1}^{N} \left[LF_{xi} + Lf_{yi} + LF_{zi} \right] \right\rangle \tag{7.13}$$

The average force exerted by molecule i on the wall perpendicular to the x-axis, F_{xi}, is the same for each molecule; summed over molecules, the average force is the pressure times the area of the wall, L^2. [Here, the average is over positions; in deriving (6.26), we averaged over time. The equivalence between space average and time average has come up before.] For the isotropic gas in a cubic box, the average of $y_i(\partial W_N/\partial y_i)$ is the same as the average of $x_i(\partial W_N/\partial x_i)$, and the same for the term in z_i. The pressure on the walls perpendicular to the y- and z-axes is the same as the pressure on the wall perpendicular to the x-axis. Equation (7.13) has now been transformed into

$$-\left\langle \sum_{i=1}^{3N} q_i \frac{\partial U_N}{\partial q_i} \right\rangle - 3L(PL^2) = -3NkT$$

The volume V is L^3 and the average value of the virial of U_N is

$$\frac{\int_V d^{3N}q\, e^{-\beta U_N} \sum_{i=1}^{3N} q_i(\partial U_N/\partial q_i)}{\int_V d^{3N}q\, e^{-\beta U_N}} = \frac{1}{Z} \int_V d^{3N}q\, e^{-\beta U_N} \sum_i q_i \frac{\partial U_N}{\partial q_i}$$

so that we rediscover (7.11).

Equation (7.11) states that the pressure is equal to NkT/V minus the average of the virial of intermolecular forces. If βU_N is small, one may expect this term (see section 7.2) to be proportional to $(N/V)^2$. In general, it can be written as a power series in the number density $\rho = N/V$. Therefore (7.11) rearranges to

$$\frac{P}{kT} = \rho + \sum_{k=2}^{\infty} B_k \rho^k \tag{7.14}$$

Equation (7.14) is called a virial expansion. The values of the $\{B_k\}$, called virial coefficients, reflect the intermolecular forces. A virial series is a completely general equation of state for any fluid, but, the higher the density, the more terms in the series are required to give the pressure accurately.

The next task is to find the virial coefficients in terms of the intermolecular forces. For this purpose, it is natural to use the grand partition function Ξ, since $\ln \Xi = PV/kT$. Another reason for using the grand partition function is that, as just seen, taking the walls of the container into account is somewhat tricky, and it is simpler to deal with a gas in a volume V without walls. This would be an open system, in which the number of molecules N is not fixed; it is described by the grand canonical ensemble.

For the grand canonical ensemble, as discussed in Section 2.5,

$$e^{PV/kT} = \Xi = \sum_{N=0}^{\infty} e^{N\beta\mu/N_A} Q_N$$

where N_A is Avogadro's number. For the one-component fluid, according to (7.2),

$$Q_N = \frac{Z_N}{N!}\Lambda^{-3N}$$

(the subscript N is affixed to the configurational integral to indicate that it is for N particles). From its definition as a sum over states for 0 particles, $Q_0 = 1$. Combining the above two equations,

$$\frac{PV}{kT} = \ln\left\{1 + \sum_{N=1}^{\infty} \frac{Z_N}{N!}\left(\frac{e^{\beta\mu/N_A}}{\Lambda^3}\right)^N\right\} \tag{7.15}$$

This will be the basis for the virial expansion.

The quantity $\Lambda^{-3} e^{\beta\mu/N_A}$ will be abbreviated as z. Note that, for $z \to 0$, (7.15) becomes $PV/kT = zZ_1$. Since $Z_1 = V$ because $U_1 = 0$ (there are no potential energy terms involving just one particle), $P/kT \to z$ as $z \to 0$. Since P/kT must approach ρ $(= \langle N \rangle / V)$ for $\rho \to 0$, z approaches ρ as $\rho \to 0$. In general, one may use the expansion

$$\ln(1 + x) = x - \frac{x^2}{2} + \frac{x^3}{3} - \cdots$$

so that (7.15) becomes

$$\begin{aligned} \frac{P}{kT} &= \frac{1}{V}\left(\sum_{N=1}^{\infty} \frac{Z_N}{N!} z^N\right) - \frac{1}{2V}\left(\sum_{N=1}^{\infty} \frac{Z_N}{N!} z^N\right)^2 + \cdots \\ &= \frac{zZ_1}{V} + \frac{z^2}{V}\left(\frac{Z_2}{2} - \frac{Z_1^2}{2}\right) + \frac{z^3}{V}\left(\frac{Z_3}{3!} - \frac{Z_1 Z_2}{2!} + \frac{Z_1^3}{3}\right) + \cdots \end{aligned} \tag{7.16}$$

This is an expansion of P/kT in powers of z. We are seeking an expansion of P/kT in powers of ρ, like (7.14). To get this, we note that, according to equations (2.42) to (2.45),

$$\langle N \rangle = \frac{N_A}{\beta}\left(\frac{\partial \ln \Xi}{\partial \mu}\right)_{\beta, V} = \left(\frac{\partial \ln \Xi}{\partial \ln z}\right)_{\beta, V}$$

Since $\ln \Xi = \beta PV$ and $\langle N \rangle$ is the average number of particles in the volume V,

$$\frac{\langle N \rangle}{V} = \rho = z\left(\frac{\partial(\beta P)}{\partial z}\right)_{\beta,V} \tag{7.17}$$

With (7.16) for βP, (7.17) gives ρ as a power series in z. If this can be transformed to give z as a power series in ρ, substitution of that series into (7.16) will give the virial series.

Transformations of this kind were carried out in Section 5.3, in connection with Fermi statistics, so the procedure is only summarized here. Substitution of equation (7.16) in (7.17), with V substituted for Z_1, gives

$$\rho = z + z^2(Z_2V^{-1} - V) + z^3\left(\tfrac{1}{2}Z_3V^{-1} - \tfrac{3}{2}Z_2 + V^2\right) + \cdots \tag{7.18}$$

Let $z = c_1\rho + c_2\rho^2 + c_3\rho^3 \cdots$. Then substitution of this makes (7.18) into an identity. The coefficient of ρ on the right side must be 1, and the coefficients of higher powers of ρ must be zero. This allows one to find the $\{c_j\}$ in terms of the coefficients in (7.18). The result is

$$z = \rho + \rho^2(V - Z_2V^{-1}) + \rho^3\left(2Z_2^2V^{-2} - \tfrac{5}{2}Z_2 + V^2 - \tfrac{1}{2}Z_3V^{-1}\right) + \cdots$$

Substitution of this, in turn, into (7.16) gives the virial expansion (7.14), with explicit expressions for the coefficients.

The second virial coefficient, the coefficient of ρ^2 in the virial expansion of P, is

$$B_2 = \left(\frac{V}{2} - \frac{Z_2}{2V}\right) \tag{7.19}$$

and the coefficient of ρ^3 is

$$B_3 = \left(\frac{Z_2^2}{V^2} - Z_2 - \frac{Z_3}{3V} + \frac{V^2}{3}\right) \tag{7.20}$$

B_4 is even more complicated, and involves Z_4, which is an integral over the coordinates of four molecules. One may hope that, if the density ρ is not too high, the first two or three terms of the virial expansion give the pressure accurately enough. B_2 may be rewritten as

$$\frac{1}{2V}\int dx_1\, dy_1\, dz_1\, dx_2\, dy_2\, dz_2[1 - e^{-\beta U_2}]$$

For large interparticle distance r_{12}, U_2 will approach 0 and the integrand, in square brackets, will vanish, so the integral is finite even if the range of particle coordinates is infinite. Something similar can be shown for B_3 and the higher virial coefficients.

7.2. PAIRWISE INTERACTION

Not much has been said about the potential U_N, which gives the intermolecular interactions in the fluid. For simplicity, the internal structure of the molecules has been ignored, and only $3N$ coordinates (giving the positions of the molecules) have been considered. This means that U_N is assumed to depend only on the positions of the N molecules and not on their orientations, which would require consideration of additional coordinates of molecules. This simplifying assumption will be maintained, so the "molecules" being discussed are in fact atoms. Now an additional simplification will be introduced.

In general, one could have

$$U_N = s + \sum_i t(\mathbf{r}_i) + \sum_{ij} u(\mathbf{r}_i, \mathbf{r}_j) + \sum_{ijk} v(\mathbf{r}_i, \mathbf{r}_j, \mathbf{r}_k) + \cdots$$

The first term simply changes the zero of energy, and the t-terms correspond to external fields, which we assume are not present, so s and t may be dropped. The u-terms involve pairs of particles, the v-terms triplets, and so on. The existence of the terms after u is referred to as nonadditivity; they make the energy of interaction between particles i and j depend on the position of a third particle in addition to r_{ij}. It is probable that U_N for a real system does include such terms, but we will hope they are not very important and limit U_N to pair interactions. Also, U_N will be assumed to depend only on the distances between particles, not on the direction of interparticle vectors, so we write

$$U_N = \sum_{i<j=1}^{N} u(r_{ij}) \tag{7.21}$$

The sum runs over all pairs of particles (a particle does not interact with itself); r_{ij} is the distance between particle i and particle j. The particles all being equivalent, the same function, u, is used for each pair.

The interatomic pair potential $u(r)$ is expected to be repulsive at short distances, due to overlap of electron clouds and internuclear repulsion, and attractive at large distances, due to van der Waals or dispersion forces. The familiar form of a typical potential is shown in Figure 7.1, along with the force, $-du/dr$. A short-range repulsion with a long-range attraction necessarily leads to a minimum in the energy.

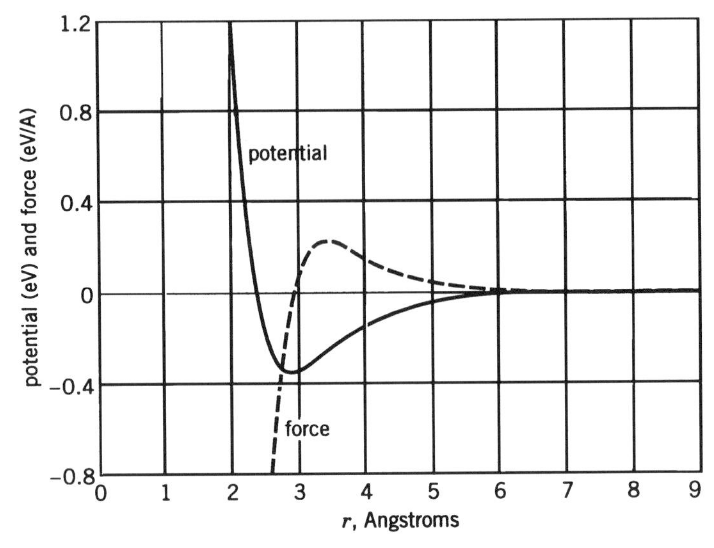

Figure 7.1. Morse potential and force. The form of the potential u is given in equation (7.22). The parameter D, which is the dissociation energy or depth of the well, is here equal to 0.3429, and r_0, the equilibrium internuclear distance or position of the well, is 2.866 A. The force is $-du/dr$.

A great variety of functional forms, empirically and theoretically based, have been used to represent interatomic potentials. These include the Buckingham, Lennard–Jones or 6–12, Morse, and many others. The potential in Figure 7.1 is a Morse potential

$$u(r) = D[e^{-2\alpha(r-r_0)} - 2\,e^{-\alpha(r-r_0)}] \tag{7.22}$$

parametrized to represent the interaction between two Cu atoms. In this case, D (which can be shown to be the dissociation energy) is 0.3429 eV, r_0 (the equilibrium internuclear distance) is 2.866 A, and α (related to the force constant) is 1.3588 A^{-1}. The Lennard–Jones potential, which in shape is similar to the Morse, is

$$u(r) = D\left[\left(\frac{r}{\sigma}\right)^{-12} - \left(\frac{r}{\sigma}\right)^{-6}\right] \tag{7.23}$$

The distance of closest approach of two molecules interacting according to (7.23) is σ when their energy is low ($u(\sigma) = 0$). An algebraically simpler function which mimics the behavior shown in Figure 7.1 is the square well:

$$u(r) = \infty, \quad r < r_1; \qquad u(r) = -D, \quad r_1 \leq r \leq r_2; \qquad u(r) = 0, \quad r > r_2 \tag{7.24}$$

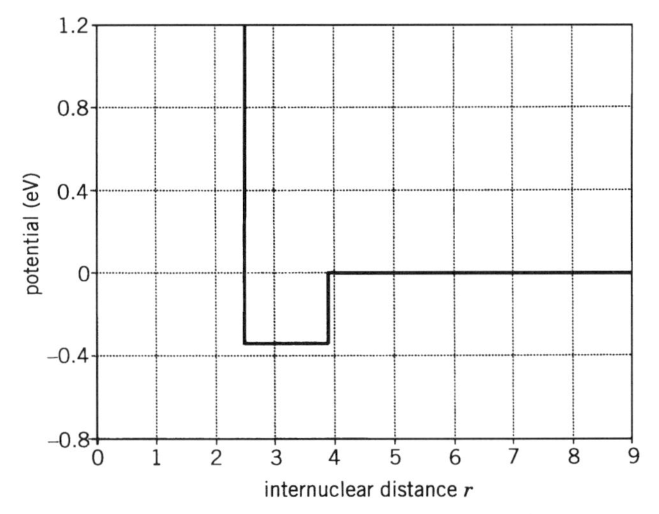

Figure 7.2. Square-well potential. This is algebraically convenient because it consists of vertical and horizontal lines, but physically unreasonable because of the discontinuities in slope.

The square well potential is shown in Figure 7.2; since it is a collection of horizontal and vertical lines it is easy to deal with in integrals. The hard-sphere potential is (7.24) with $D = 0$; there is no attractive part, and r_1 is the hard-sphere diameter, twice the hard-sphere radius. Often, r_1 is identified with σ of the Lennard–Jones and related potentials.

Assuming that U_N involves only pair interactions, the partition function Q_N for N particles is [see (7.2) and following]

$$Q_N = \frac{V^N}{N!} \Lambda^{-3N} \frac{Z_N}{V^N}$$

with

$$Z_N = \prod_i \left\{ \int dx_i \int dy_i \int dz_i \right\} \exp\left[-\beta \sum_{i<j=1}^{N} u(r_{ij}) \right]$$

The energy is

$$E = \frac{3NkT}{2} + \frac{\Pi_i \{ \int dx_i \int dy_i \int dz_i \} \Sigma_{i<j=1}^{N} u(r_{ij}) \exp\left[-\beta \, \Sigma_{i<j=1}^{N} u(r_{ij}) \right]}{Z_N} \tag{7.25}$$

Although U_N is a sum over particle pairs, Z_N and E are much more complicated, because U_N appears in the exponential. The pressure is

$$P = \frac{NkT}{V} - \frac{1}{3VZ_N} \prod_i \left\{ \int dx_i \int dy_i \int dz_i \right\} \exp\left[-\beta \sum_{i<j=1}^{N} u(r_{ij}) \right]$$
$$\times \sum_{k=1}^{N} \left(x_k \frac{\partial}{\partial x_k} + y_k \frac{\partial}{\partial y_k} + z_k \frac{\partial}{\partial z_k} \right) \sum_{i<j=1}^{N} u(r_{ij}) \tag{7.26}$$

The second term, which involves the virial, can be simplified.

Note that $u(r_{ij})$ depends only on the distance between particles i and j, where

$$r_{ij}^2 = (x_i - x_j)^2 + (y_i - y_j)^2 + (z_i - z_j)^2$$

Thus the derivative of $u(r_{ij})$ with respect to x_k is zero unless $k = i$ or $k = j$, and the sums over k, i, and j reduce to

$$\sum_{i<j=1}^{N} \left(x_i \frac{\partial}{\partial x_i} + y_i \frac{\partial}{\partial y_i} + z_i \frac{\partial}{\partial z_i} + x_j \frac{\partial}{\partial x_j} + y_j \frac{\partial}{\partial y_j} + z_j \frac{\partial}{\partial z_j} \right) u(r_{ij})$$
$$= \sum_{i<j=1}^{N} \left(x_i \frac{x_i - x_j}{r_{ij}} + y_i \frac{y_i - y_j}{r_{ij}} + z_i \frac{z_i - z_j}{r_{ij}} \right.$$
$$\left. + x_j \frac{x_j - x_i}{r_{ij}} + y_j \frac{y_j - y_i}{r_{ij}} + z_j \frac{z_j - z_i}{r_{ij}} \right) \frac{du(r_{ij})}{dr_{ij}}$$
$$= \sum_{i<j=1}^{N} \left(\frac{(x_i - x_j)^2 + (y_i - y_j)^2 + (z_i - z_j^2)}{r_{ij}} \right) \frac{du(r_{ij})}{dr_{ij}}$$

Then the pressure is

$$P = \frac{NkT}{V} - \frac{1}{3VZ_N} \prod_i \left\{ \int dx_i \int dy_i \int dz_i \right\}$$
$$\times \exp\left[-\beta \sum_{i<j=1}^{N} u(r_{ij}) \right] \sum_{i<j=1}^{N} r_{ij} \frac{du(r_{ij})}{dr_{ij}} \tag{7.27}$$

For an attractive interaction, du/dr_{ij} is positive, the virial is positive, the integral in (7.27) is positive, and the interaction between molecules decreases the pressure. If u is repulsive, du/dr_{ij} is negative, the integral is negative, and the interaction between molecules increases the pressure. In general (see

Figure 7.1) u will have attractive and repulsive parts; whether the integral is positive or negative depends on which dominates.

The virial coefficients simplify when the interaction between molecules is a sum of pair interactions. The second virial coefficient, since it involves U_2, will always include just the pair interaction, even if three-body or more complicated terms are present. According to (7.19),

$$B_2 = \frac{V}{2} - \frac{Z_2}{2V} = \frac{1}{2V} \int_V d^3q_1 \int_V d^3q_2 [1 - e^{-\beta u(r_{12})}]$$

$$= \frac{1}{2V} \int d^3q_1 \int d^3q_{12} [1 - e^{-\beta u(r_{12})}]$$

In the last expression, we have changed variables from the six coordinates of particles 1 and 2 to the three coordinates of particle 1 plus the three relative coordinates: $x_{12} = x_2 - x_1$, and so on. Since the integrand depends only on the relative coordinates, one can perform the integration over $\mathbf{q}_1$ to get V. Then one can use spherical coordinates for $\mathbf{q}_{12}$ and integrate over angles, obtaining

$$B_2 = 2\pi \int_0^\infty r^2\, dr [1 - e^{-\beta u(r)}] \tag{7.28}$$

The third virial coefficient [see (7.20)] is

$$B_3 = Z_2 \left(\frac{Z_2}{V^2} - 1 \right) + \left(\frac{V^3 - Z_3}{3V} \right) = \frac{Z_2}{V} (-2B_2)$$

$$+ \frac{1}{3V} \int d^3q_1 \int d^3q_2 \int d^3q_3 (1 - e^{-\beta [u(r_{12}) + u(r_{13}) + u(r_{23})]})$$

In the integral, we change variables to $\mathbf{q}_1$, $\mathbf{q}_{12}$, and $\mathbf{q}_{13}$; integration over $\mathbf{q}_1$ can then be performed, giving a factor of V. The integrand depends only on $\mathbf{q}_{12}$ and $\mathbf{q}_{13}$ because r_{23} can be written in terms of $\mathbf{q}_{12}$ and $\mathbf{q}_{13}$: particles 1, 2, and 3 define a triangle with sides r_{23}, r_{13}, and r_{12}, and the law of cosines can be used to calculate r_{23}:

$$r_{23}^2 = r_{12}^2 + r_{13}^2 - 2r_{12}r_{13} \cos\theta = r_{12}^2 + r_{13}^2 - 2\mathbf{r}_{12} \cdot \mathbf{r}_{13}$$

with the last term the scalar product of the vectors $\mathbf{r}_{12}$ and $\mathbf{r}_{13}$. Even if there is no three-body term in U_3, one cannot separate the contributions of the three two-body potentials in B_3. This is because the two-body interactions $u(r_{12})$ and $u(r_{13})$ affect the probabilities of finding molecules 2 and 3 at various distances from molecule 1, and therefore affect the average over r_{23}. These probabilities will be discussed more in Section 7.3.

The second virial coefficient for the Lennard–Jones potential (7.23), calculated according to the method of Problem 7.13, is shown as a function of the reduced temperature, $t = kT/D$, in Figure 7.3. For low temperatures ($t < 4$), B_2 is negative, as attractive forces dominate; for higher temperatures, repulsive forces are more important, and B_2 becomes positive. Also, B_2 passes through a maximum, which has been observed experimentally. This is because, at very high temperatures, the molecules have very high kinetic energy, and can approach each other more and more closely. The decrease in the distance of closest approach is equivalent to a softening of the repulsive potential.

If the interatomic potential is not very large, and if only low densities are considered, the equation of state obtained by keeping terms through ρ^2 in

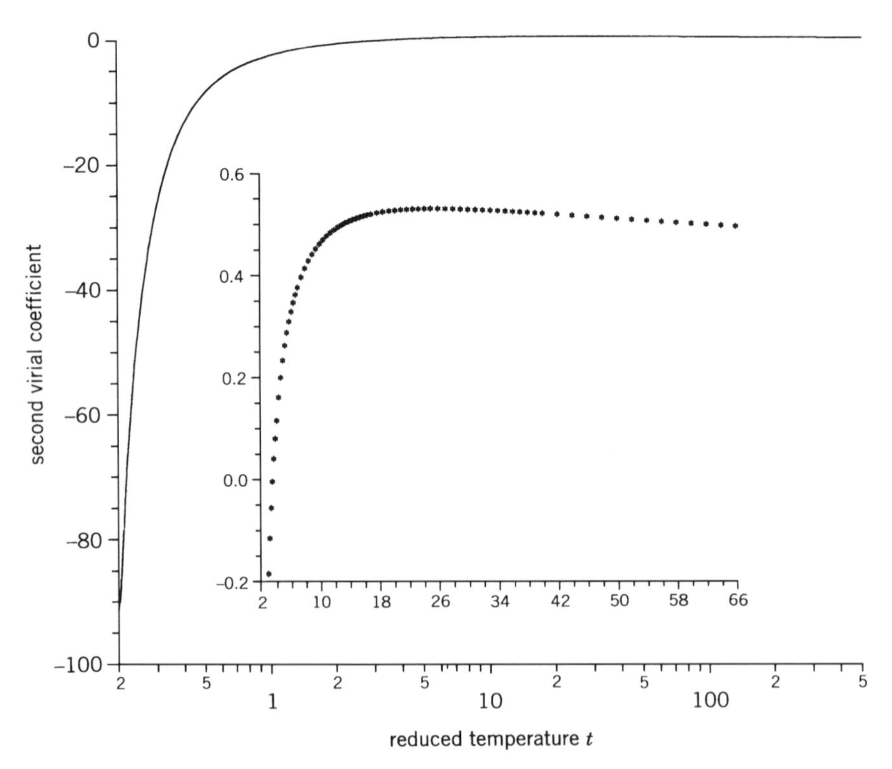

Figure 7.3. Calculated second virial coefficient for the Lennard–Jones potential, equation (7.23). The reduced temperature t is kT/D. See Problem 7.13 for the method of calculation. The second virial coefficient is negative for low temperatures, positive for high temperatures, and goes through a maximum. The scale of t in the inset is chosen to show the maximum more clearly.

the virial series may be sufficiently accurate. Then

$$P = \frac{NkT}{V} + \frac{kTN^2}{2V^2} 4\pi \int_0^\infty r^2\, dr[1 - e^{-\beta u(r)}]$$

A better equation of state can be obtained by keeping terms through ρ^3. In principle, one needs $u(r)$ to get B_2 and, if $u(r)$ is known, one can do the more complicated integral to get B_3. However, B_2 is given exactly by (7.28), whereas our formula for B_3 involves assuming the three-body interaction is zero. This equation can be simplified if $\beta u(r)$ is small enough so that $\exp[-\beta u(r)] \simeq 1 - \beta u(r)$. Then

$$P = \frac{NkT}{V} - \frac{N^2 kT}{2V^2} 4\pi \int_0^\infty r^2\, dr\, \beta u(r) \tag{7.29}$$

This corresponds to neglecting the effect of the interatomic potential u on the distribution of particles around a given particle and calculating the average interatomic potential energy assuming all the particles are randomly and independently distributed in space.

Equation (7.29) may also be obtained directly from (7.26). If the exponential in (7.26) is approximated by unity, the second term in P becomes a sum of $6N(N-1)/2$ terms ($N(N-1)/2$ is the number of pairs of particles). Integrating each by parts, and dropping boundary terms like $x_i u(r_{ij})$ because $u(r_{ij}) = 0$ for large r_{ij},

$$P = \frac{NkT}{V} - \frac{1}{3VZ_N} \sum_{i<j=1}^{N} \prod_k \left\{ \int dx_k \int dy_k \int dz_k \right\} 3u(r_{ij})$$

All the terms in the sum are identical, and there are $N(N-1)/2$ of them. For each choice of i and j, the integrations may be carried out over all variables except i and j, giving $N-2$ factors of V. If the exponential in the configurational integral Z_N is also approximated by unity, Z_N becomes V^N. Therefore

$$P = \frac{NkT}{V} - \frac{1}{V^{N+1}} \frac{N(N-1)}{2} V^{N-2} \int d^3q_i \int d^3q_j\, u(r_{ij})$$

When the relative coordinate $\mathbf{q}_{ij}$ is introduced and the integration over $\mathbf{q}_i$ performed,

$$P = \frac{NkT}{V} - \frac{N(N-1)}{2} V^{-2} \int d^3q_{ij}\, u(r_{ij})$$

For N large, this is identical to (7.29). The approximation of $\exp[-\beta U]$ by unity, which neglects the influence of the interatomic potential on the distribution of particles, may also be made in the energy expression (7.25), which becomes

$$
\begin{aligned}
E &= \frac{3NkT}{2} + \frac{\Pi_i[\int dx_i \int dy_i \int dz_i]U_N}{Z_N} \\
&= \tfrac{3}{2}NkT + \frac{N(N-1)}{2V} \int dq_{ij}\, u(r_{ij}) \\
&= \tfrac{3}{2}NkT + \frac{N(N-1)}{2V} 4\pi \int_0^\infty r^2\, dr\, u(r) \qquad (7.30)
\end{aligned}
$$

Note that the second term is simply the average value of u, assuming that all values of r_{ij} are equally probable.

If u includes a hard-sphere potential [equation (7.24)] the preceding formulas are clearly incorrect. When u becomes infinite for some range of r, the integrals in (7.29) and (7.30) do not converge. In this case, one can never say that u is small compared to kT. Potentials including a hard-sphere potential require special consideration.

Consider first a purely hard-sphere potential [(7.24) with $D = 0$]; the effect of an added attractive potential will be calculated subsequently. It is easy to calculate B_2 analytically. For $r \leq r_1$, u is infinite and the exponential in (7.28) vanishes; for $r > r_1$, u vanishes, the exponential is unity, and the square bracket vanishes.

$$B_2 = 2\pi \int_0^{r_1} r^2[1]\, dr = \frac{2\pi r_1^3}{3}$$

This is half the volume of a sphere of radius r_1. Each atom excludes other atoms from a sphere of radius r_1, twice the radius of a single atom. The volume of a single atom is

$$v = \frac{4\pi}{3}\left(\frac{r_1}{2}\right)^3 = \frac{\pi}{6} r_1^3$$

so $B_2 = 4v$. Also, B_3 can be found exactly, although it is harder to calculate than B_2; the result is $10v^2$. Successive virial coefficients are obtained from computer simulations of hard-sphere fluids or from approximate theories; B_4 is a little less than $20v^3$, and B_5 a little less than $30v^4$. The virial series for a hard-sphere fluid therefore starts with

$$\frac{P}{kT} = \rho + 4v\rho^2 + 10v^2\rho^3 + 20v^3\rho^4 + \cdots \qquad (7.31)$$

That all of the virial coefficients are temperature independent reflects the infinite strength of the hard-sphere potential.

Something like this equation of state can be obtained from a simple argument. We have noted that each atom excludes other atoms from a volume $8v$. This "excluded volume" corresponds to a pair of atoms; the excluded volume per atom is half of this. For any atom in V, the $N-1$ other atoms reduce the available volume by $(N-1)(4v)$. Since there are no forces between atoms not in contact, it is reasonable to assume that the atoms are distributed randomly in the available volume. The equation of state is then the ideal-gas equation of state with the volume reduced:

$$\frac{P}{kT} = \frac{N}{V - N(4v)} = \frac{\rho}{1 - \rho(4v)}$$
$$= \rho + 4v\rho^2 + 16v^2\rho^3 + 64v^3\rho^4 + \cdots \tag{7.32}$$

The difference between $N-1$ and N has been ignored (N is large) and the power series for $(1-x)^{-1}$ has been used. The fact that this differs from (7.31), in B_3 and higher virial coefficients, shows that the atoms are not really distributed randomly. Detailed calculations (see later discussion) show that the exclusion of atoms from a sphere of radius r_1 centered at a given atom leads to a buildup in the density of atoms for r slightly greater than r_1.

The reasoning that led to (7.32) can be extended to derive the equation of state for N atoms interacting with a potential u that has an attractive part in addition to the hard-sphere potential. [The result is (7.36).] It is necessary to assume that the attractive potential can be considered small compared to kT. Suppose $u(r) = u^{hs}(r) + u^a(r)$ (hard-sphere and attractive parts) with $u^a \ll kT$. Correspondingly, $U_N = U^{hs} + U^a$, each part being a sum of $N(N-1)/2$ pair potentials. Equation (7.10) is rewritten as

$$\frac{P}{kT} = \frac{\partial \ln Z^{hs}}{\partial V} + \frac{\partial \ln(Z/Z^{hs})}{\partial V} \tag{7.33}$$

where Z^{hs} is the configurational integral for vanishing U^a. If $U^a = 0$, the equation of state is (7.32), so

$$\frac{N}{V - 4vN} = \left(\frac{\partial \ln Z^{hs}}{\partial V}\right)$$

which implies that

$$Z^{hs} = (V - 4vN)^N$$

In the second term in (7.33), $\exp(-\beta U^a)$ may be approximated as $1 - \beta U^a$ because βU^a is small, so

$$\ln \frac{Z}{Z^{hs}} = \ln\left(\frac{\int d^{3N}q \exp(-\beta U^{hs})\exp(-\beta U^a)}{Z^{hs}} \right)$$

$$\simeq \ln\left(1 - \frac{\int d^{3N}q \exp(-\beta U^{hs})(\beta U^a)}{Z^{hs}} \right)$$

Using the approximation $\ln(1 - x) \simeq -x$ (for x small compared to 1), and writing U^a as a sum of pair interactions,

$$\ln \frac{Z}{Z^h \mathbf{s}} \simeq -\beta \sum_{i<j} \frac{\int d^{3N}q \exp(-\beta U^{hs}) u^a(r_{ij})}{Z^{hs}} \tag{7.34}$$

This represents an average of the attractive potential, with $\exp(-\beta U^{hs})/Z^{hs}$ being the weighting function.

Consistently with writing Z^{hs} as $(V - 4vN)^N$, we consider that, in (7.34), integration over the coordinates of any particle other than i or j gives a factor of $V - 4vN$ [we know that $\exp(-\beta u^{hs}) = 1$ wherever it is not zero]. Then

$$\ln \frac{Z}{Z^{hs}} \simeq -\beta \frac{N(N-1)}{2} \frac{\int d^3q_1\, d^3q_2 \exp[-\beta u^{hs}(r_{12})] u^a(r_{12})}{(V - 4vN)^2}$$

As done previously, we put $N - 1 \simeq N$, substitute the relative coordinate $\mathbf{q}_{12}$ for $\mathbf{q}_2$, and integrate over $\mathbf{q}_1$ and over angles in $\mathbf{q}_{12}$. Inserting the result into (7.33),

$$P = kT \frac{N}{V - 4vN} + kT \frac{\partial}{\partial V}\left(\frac{(-\beta N^2/2)4\pi \int_{r_1}^{\infty} dr\, r^2 u^a(r)}{V - 4vN} \right)$$

$$= \frac{NkT}{V - 4vN} + \frac{2\pi N^2}{(V - 4vN)^2} \int_{r_1}^{\infty} dr\, r^2 u^a(r) \tag{7.35}$$

Since u^a is an attractive potential, the integral is negative; call it $-a/2\pi$.

If one agrees to treat only low densities, so that $4vN$ is much less than V, one can replace $V - 4vN$ by V in the second term. Then (7.35) becomes

$$P = \frac{NkT}{V - Nb} - \frac{aN^2}{V^2} \tag{7.36}$$

which is the van der Waals equation of state. The van der Waals constant b is four times the hard-sphere volume of a molecule. The size of b is a measure of the importance of the hard-sphere repulsion; larger b increases the pressure. The second term in (7.36), due to the attractive forces, decreases the pressure. Isotherms (P as a function of V/N at T constant) for the van der Waals equation are shown in Figure 7.4 for CO_2 at temperatures of 250, 270, 290, 310, and 330 K (bottom to top). The values of a (0.3637 Pa m^6 mol^{-2}) and b (4.27×10^{-5} m^3 mol^{-1}) were determined to give the best fit to measured isotherms for CO_2. In fact, the fit is good only for high temperatures (for which the isotherms of the van der Waals equation resemble the isotherms for an ideal gas), but the van der Waals equation is often used to discuss gas imperfections because of its simple algebraic form.

P must be a monotonic function of V, so the oscillations in the isotherms for lower temperatures are unphysical; they are interpreted as the best attempt of a continuous function to describe condensation. For a real gas which can condense, the isotherm should have a discontinuous slope, and have a horizontal segment, such as is shown in the figure on the isotherm for 270 K. The horizontal segment corresponds to the coexistence of two phases,

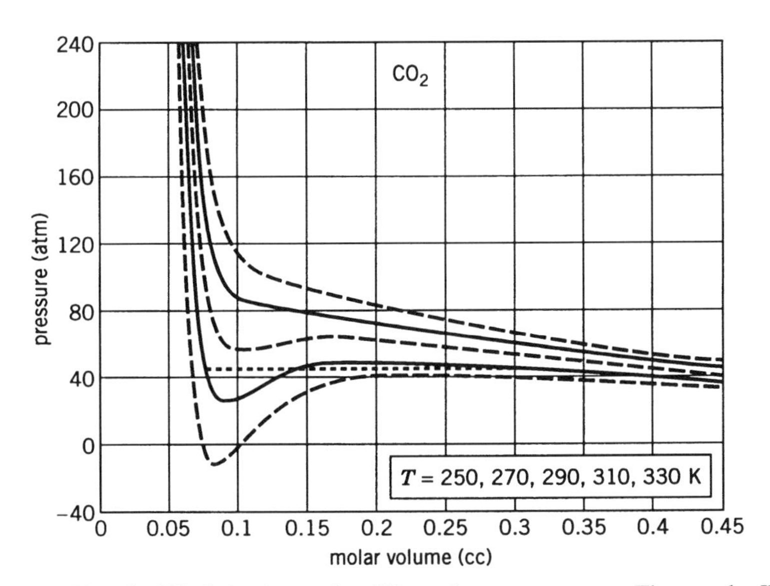

Figure 7.4. Van der Waals isotherms for CO_2 at five temperatures. The van der Waals equation of state is (7.36). The coefficients a and b were chosen to give the best fit to experimental isotherms. For the lower temperatures (three lowest curves) the isotherms are nonmonotonic, indicating condensation. The two higher temperatures are above the critical point, so the corresponding isotherms are monotonic. A real isotherm below the critical point would have a horizontal segment, as shown by the dashed line on the 270 K curve.

with different densities, at the same temperature and pressure. The highest temperature at which condensation can occur is the critical temperature. For the van der Waals equation, the critical temperature is the highest temperature for which the isotherm is nonmonotonic. The critical temperature, along with the critical pressure and critical molar volume, can be found by seeking the point for which $(\partial P/\partial V)_T$ and $(\partial^2 P/\partial V^2)_T$ both vanish.

7.3. CLUSTERS

Even with only pair interactions, getting higher virial coefficients than the second is difficult. The mathematics, to be given in part in this section, is tricky (it took a long time before a succession of workers found and eliminated errors), but it gives insights into the meaning of the partition function and the virial coefficients, and introduces diagrams, which are important in statistical mechanics. Some students may want to go directly to Section 7.4, where another approach is discussed.

The virial coefficient B_2 is simple to determine [equation (7.28)]. It involves the interaction of just two particles. However, it would be a mistake to conclude that B_3 includes interaction of three particles, and so on. In the configurational integral Z_N the exponential of U_N appears; since U_N is a sum of pair interactions, it is a product of exponentials, one for each pair. To derive the virial series, one writes Z_N in terms of the Mayer f-functions, defined by

$$e^{-\beta u(r_{ij})} \equiv f(r_{ij}) + 1 \tag{7.37}$$

For an ideal gas, $f = 0$, and $f(r) \to -\beta u(r)$ as $r \to \infty$. The f-function for the Morse potential of equation (7.22), with $T = 900$ K, is shown in Figure 7.5. Note that it has a pronounced maximum where the potential has its minimum.

With $\exp[-\beta u(r_{ij})]$ written in terms of $f(r_{ij})$, Z_N becomes

$$Z_N = \int \left[1 + \sum_{i<j} f(r_{ij}) + \sum_{i<j} f(r_{ij}) \sum_{k<l} f(r_{kl}) + \cdots \right] d^{3N}q \tag{7.38}$$

In the second sums, the *pair* (i, j) must be different from the *pair* (k, l). The "1" in Z_N integrates to V^N, the first sum to something multiplied by V^{N-1}, the next involves lower powers of V, and so on. Unfortunately, all the terms in the double sum, for instance, do not correspond to the same power of V, as will be seen below. It is extremely helpful to one's mental equilibrium to represent the terms in any sum by diagrams. Figure 7.6 is such a diagram: the particles are represented by circles and the f-functions by lines between circles (they are sometimes referred to as bonds).

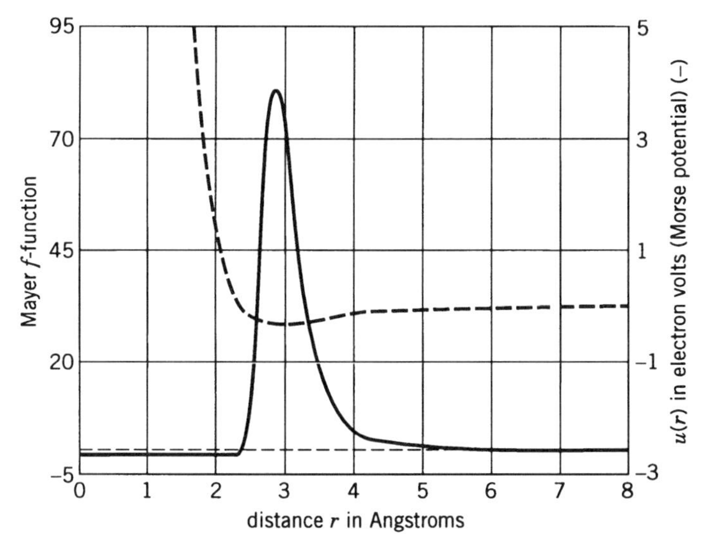

Figure 7.5. Mayer f-function for Morse potential (7.22), $T = 900$ K. The potential is shown as the dashed curve. The Mayer function, $f(r) = \exp[-\beta u(r)] - 1$ where u is the potential, has a maximum where u has its minimum.

In Figure 7.6, there are 21 numbered objects representing the molecules ($N = 21$) and 8 lines, so this represents one of the integrals from the ninth group of terms in (7.38). In particular, this is

$$f_{34}f_{56}f_{78}f_{7,15}f_{10,13}f_{12,17}f_{17,19}f_{17,21}$$

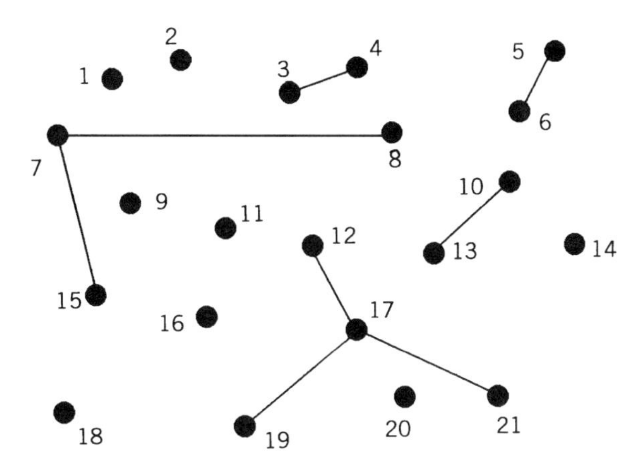

Figure 7.6. Diagram for $N = 21$ with five clusters. A cluster is a group of two or more molecules connected by lines representing Mayer f-functions.

to be integrated over the coordinates of all particles. There is no restriction on the number of lines that end on the same object (this corresponds to f_{ij} with the same i but different j), but there can be only one line between the same two objects (f_{ij} for the same i and j appears only once). For every diagram, there are many terms in Z_N, corresponding to renumberings of the particles involved in the bonds. For example, the diagram with two unconnected lines represents $N(N-1)(N-2)(N-3)/4!$ terms, each equal to

$$\int f(r_{12})f(r_{34})\,d^{3N}q = V^{N-4}\left[\int f(r_{12})\,\mathbf{dr}_1\,\mathbf{dr}_2\right]^2$$

The diagram with two connected lines represents $N(N-1)(N-2)/2$ terms, each equal to

$$\int f(r_{12})f(r_{23})\,d^{3N}q = V^{N-3}\int f(r_{12})f(r_{23})\,\mathbf{dr}_1\,\mathbf{dr}_2\,\mathbf{dr}_3$$

An important concern will be determining how many terms or renumberings correspond to a diagram.

In every diagram, there are groups of molecules connected by lines. These are referred to as "clusters." In the diagram of Figure 7.6, there are three two-molecule clusters, one three-molecule cluster, and one four-molecule cluster. A two-molecule cluster is two molecules connected by a line with neither molecule connected to any other molecules. Three-molecule clusters come in two varieties: there may be two lines (and hence two factors of f_{ij}: $f_{78}f_{7,15}$ in the figure), or three lines forming a triangle (corresponding to $f_{ij}f_{jk}f_{ik}$). Four-molecule clusters are even more varied; Figure 7.7 shows some of them. All arrangements of clusters of all sizes appear in the integrals that make up Z_N.

Letting ν_n represent the number of n-molecule clusters in a diagram, one has

$$\sum_{n=1}^{\infty} n\nu_n = N$$

The cluster integral b_n is defined as

$$b_n \equiv \frac{1}{n!\,V}\int\cdots\int \sum \prod f_{ij}\,\mathbf{dr}_1\cdots\mathbf{dr}_n \tag{7.39}$$

where "$\sum\prod$" means a sum over all products of f_{ij} consistent with a cluster of size n. The first few cluster integrals are: b_1 (no bonds) $= 1$

$$b_2 = \frac{1}{2V}\iint f_{12}\,\mathbf{dr}_1\,\mathbf{dr}_2$$

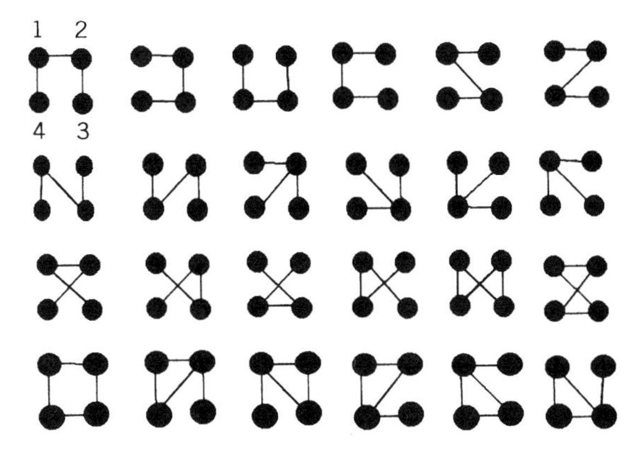

Figure 7.7. Clusters of four molecules. Not all the possible 4-molecule clusters are shown. In addition, each cluster can be assigned molecule numbers in many different ways.

and

$$b_3 = \frac{1}{6V} \iiint [f_{12}f_{13} + f_{12}f_{23} + f_{13}f_{23} + f_{12}f_{13}f_{23}]\, \mathbf{dr}_1\, \mathbf{dr}_2\, \mathbf{dr}_3$$
$$= \tfrac{1}{6}\left[(2b_2)^2 + (2b_2)^2 + (2b_2)^2 + \iint f_{12}f_{13}f_{23}\, \mathbf{dr}_{12}\, \mathbf{dr}_{13}\right]$$

Since one may always integrate over one of the n coordinates and use relative coordinates for the others, the value of a cluster integral is independent of V.

Each set of clusters (values of $\{\nu_n\}$ and a particular numbering of the atoms) contributes

$$\prod_{n \geq 1} (n!\, Vb_n)^{\nu_n}$$

to Z_N and the number of terms (numberings of the atoms) for a given set of ν_n is

$$\frac{N!}{\prod_{n \geq 1}\{\nu_n!\,(n!)^{\nu_n}\}}$$

(This last quantity is the number of ways of distributing N numbered objects into groups so that there are ν_n groups of n objects.) Therefore the canonical partition function is

$$Q_N = \frac{1}{N!\,\Lambda^{3N}} \sum_{\nu_1}^{(\Sigma n\nu_n = N)} \sum_{\nu_2} \cdots \left(\frac{N!\prod_{n \geq 1}(n!\, Vb_n)^{\nu_n}}{\prod_{n \geq 1}\{\nu_n!\,(n!)^{\nu_n}\}} \right) \tag{7.40}$$

There is a troublesome restriction on the summations, which ceases to make trouble when one goes to the grand partition function. The grand partition function is constructed by multiplying Q_N by $\exp(N\beta\mu)$ and summing over N. Since all values of N are considered, one can forget about the restriction on the summations.

With $z = e^{\beta\mu}/\Lambda^3$, the grand partition function is

$$\Xi = \sum_{\nu_1} \sum_{\nu_2} \cdots \prod_{n \geq 1} \frac{z^{n\nu_n}(Vb_n)^{\nu_n}}{\nu_n!} = \prod_{n \geq 1} \sum_{\nu_n} \frac{z^{n\nu_n}(Vb_n)^{\nu_n}}{\nu_n!}$$

With no restriction on the $\{\nu_n\}$, the sum of products can be written as a product of sums. Now the sum over ν_n is the power series for an exponential, so

$$\Xi = \prod_{n \geq 1} e^{Vb_n z^n}$$

Since $\ln \Xi = \beta PV$,

$$\beta P = V^{-1} \ln\left\{ \prod_{n \geq 1} e^{Vb_n z^n} \right\} = \sum_{n \geq 1} b_n z^n \tag{7.41}$$

To transform this into the virial series, we require z as a power series in the density ρ. Since $\langle N \rangle = \partial(\ln \Xi)/\partial\mu = \beta z\, \partial(\ln \Xi)/\partial z$,

$$\rho = \frac{\langle N \rangle}{V} = \beta \sum_{n \geq 1} nb_n z^n$$

from which the required power series for z may be found. Substituting this series in (7.41) gives the virial series for the pressure, with virial coefficients given in terms of the cluster integrals of (7.39):

$$\beta P = \rho - b_2\rho^2 - 2(b_3 - 2b_2^2)\rho^3 - 3(b_4 - 6b_2 b_3 + \tfrac{20}{3}b_2^3)\rho^4 + \cdots \tag{7.42}$$

Remember that each cluster integral b_n may involve cluster integrals of lower n.

The coefficient of ρ^3 in (7.42) is, using the expressions for b_2 and b_3 given above,

$$\frac{-1}{3} \iint f_{12} f_{13} f_{23}\, \mathbf{dr}_{12}\, \mathbf{dr}_{13}$$

corresponding to the fully connected cluster (represented as a triangle) only.

The coefficient of ρ^4 becomes, after a significant amount of algebra,

$$\frac{-1}{8}\iiint[f_{12}f_{23}f_{34}f_{41}f_{13}f_{24}+6f_{12}f_{23}f_{34}f_{41}f_{13}+3f_{12}f_{23}f_{34}f_{41}]\,\mathbf{dr}_{12}\,\mathbf{dr}_{13}\,\mathbf{dr}_{14}$$

The only clusters which remain are the "irreducible" clusters, shown for $n = 4$ in Figure 7.8. An irreducible cluster of order n is represented by a diagram in which n vertices are connected by bonds such that there are at least two independent nonintersecting paths from any vertex to any other (intersections mean meetings at vertices). The irreducible cluster integral β_n is defined like (7.39), except that it is the integral of a sum of products of f_{ij} corresponding to irreducible clusters only:

$$\beta_n \equiv \frac{1}{n!\,V}\int\cdots\int\sum(\text{irr})\prod f_{ij}\,\mathbf{dr}_1\cdots\mathbf{dr}_n$$

It can be shown in general by rather elegant mathematics that only the irreducible cluster integrals enter the virial expansion, and that

$$\beta P = \rho\left[1-\sum_{k>0}\frac{k}{k+1}\beta_k\,\rho^k\right] \tag{7.43}$$

In β_k one may integrate over one (and only one) of the coordinates explicitly to give a factor of V, so that β_k is independent of volume.

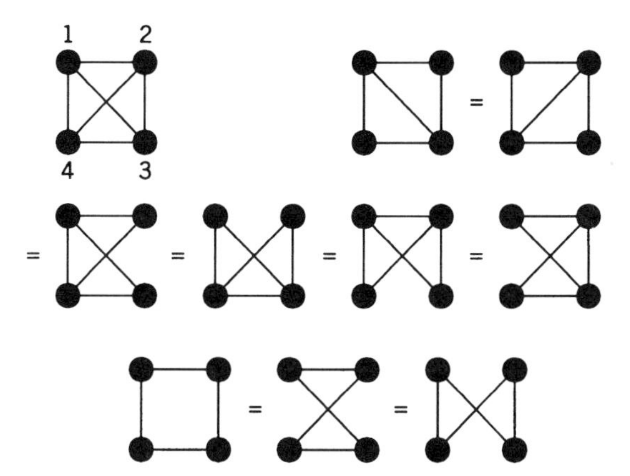

Figure 7.8. Irreducible clusters for four molecules. The second six are topologically equivalent, leading to the same terms in the expansion of the partition function. The same is true for the last three clusters.

Now that the partition function has been represented by diagrams, diagrams can be used to represent any property of the system. For instance,

$$-\left(\frac{\partial \ln \Xi}{\partial \beta}\right)_{z,V} = E - \frac{3N}{2\beta} = E^{\text{exc}}$$

[This can be derived from (2.43), in which $\gamma = \beta\mu$, rather than z, is held constant for the derivative; $\ln(z) = \gamma - \Lambda^3$.] Using (7.41) for $\ln \Xi = \beta PV$, the excess energy (which is the interaction energy between atoms) is

$$E^{\text{exc}} = -\frac{\partial}{\partial \beta}\left[V \sum_{n\geq 1} b_n z^n\right]$$

$$= \sum_{n\geq 1} \frac{z^n}{n!} \int \cdots \int \sum \frac{-\partial}{\partial \beta}\left(\prod f_{ij}\right) \mathbf{dr}_1 \cdots \mathbf{dr}_n \qquad (7.44)$$

Taking the negative derivative of the product of f_{ij}'s with respect to β has the effect of replacing one of the f_{ij}'s by $u(r_{ij})\exp[-\beta u(r_{ij})]$. The diagrams representing the terms in (7.44) are thus the same as those representing the terms in $\ln \Xi$, except that one of the lines or bonds, which indicate factors of f_{ij}, is made heavy, which indicates a factor of $u(r_{ij})\exp[-\beta u(r_{ij})]$. This is shown in Figure 7.9.

Now for a system in which there are only pair interactions E^{exc} is equal to the number of molecule pairs multiplied by the average value of any one $u(r_{ij})$, since all pairs are equivalent. One could write this as

$$E^{\text{exc}} = \frac{N(N-1)}{2} \int u(r_{12})\, e^{-\beta u}(r_{12}) G(\mathbf{r}_1, \mathbf{r}_2)\, \mathbf{dr}_1\, \mathbf{dr}_2 \qquad (7.45)$$

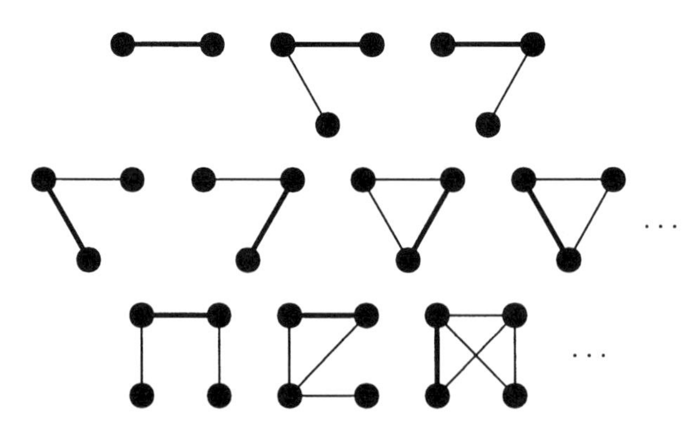

Figure 7.9. Diagrams representing terms in the expansion of the excess energy. Some of the diagrams for $n = 2$, 3, and 4 are shown. Each thin line represents a Mayer f-function, and each heavier connection represents $u(r_{ij})\exp[-\beta r_{ij}]$.

with G representing the result of the integrations over all other coordinates. Since this is an average value, G (which could be represented by the diagrams of Figure 7.9) is the probability of finding atoms 1 and 2 such that the distance between them is r_{12}. It contains the structural information about the system. Naturally, it is difficult to get at, but one must think about it and find out whatever one can about it. This is what we discuss next.

7.4. DISTRIBUTION FUNCTIONS

We have seen how thermodynamic properties of a liquid or gas can be found from the partition function and how the partition function is to be calculated from the interaction potential energy. The partition function contains all information about the distributions of the particles in momentum space as well as in coordinate space. The distribution in momentum is simply the Maxwell–Boltzmann distribution which is independent of the distribution of the particles in space. It holds at every point in space, regardless of the strength of the interparticle interaction, for liquids as well as gases.

This section deals with the space probability distributions for a many-particle system. To start, there is the N-particle probability distribution,

$$P(\mathbf{q}_1, \mathbf{q}_2, \ldots, \mathbf{q}_N)\, \mathbf{dq}_1 \mathbf{dq}_2 \cdots \mathbf{dq}_N = \frac{e^{-\beta U_N}}{Z_N} \mathbf{dq}_1 \mathbf{dq}_2 \cdots \mathbf{dq}_N \quad (7.46)$$

with

$$Z_N = \int d^{3N}q\, e^{-\beta U_N} = \int \mathbf{dq}_1 \mathbf{dq}_2 \cdots \mathbf{dq}_N\, e^{-\beta U_N}$$

being the configurational integral. Bold face is used to identify vector quantities: $\mathbf{q}_i$ is the set of three coordinates giving the position of particle i and $\mathbf{dq}_i$ is the three-dimensional infinitesimal volume element at q_i. As we noted after (7.9), (7.46) represents the probability of finding particle 1 in the volume $\mathbf{dq}_1$ at $\mathbf{q}_1$, particle 2 in the volume $\mathbf{dq}_2$ at $\mathbf{q}_2, \ldots,$ particle N in the volume $\mathbf{dq}_N$ at $\mathbf{q}_N$. P is normalized according to

$$\int d^{3N}q\, P(\mathbf{q}_1, \mathbf{q}_2, \ldots, \mathbf{q}_N) = 1$$

From $P(\mathbf{q}_1, \mathbf{q}_2, \ldots, \mathbf{q}_N)$ other probabilities of interest will be derived.

To get the probability of finding particle 1 in the volume $\mathbf{dq}_1$ at $\mathbf{q}_1$, regardless of the locations of the other particles, one would integrate $P(\mathbf{q}_1, \ldots, \mathbf{q}_N)$ over the coordinates of all the particles except particle 1. For the probability of finding particle 1 in the volume $\mathbf{dr}$ at $\mathbf{r}$ (some point in space), one would simply substitute $\mathbf{r}$ for $\mathbf{q}_1$ in the expression. To get the probability of finding particle 2 in the volume $\mathbf{dr}$ at $\mathbf{r}$ (that same point in space), one would put $\mathbf{r}$ for $\mathbf{q}_2$ and integrate $P(\mathbf{q}_1, \ldots, \mathbf{q}_N)$ over the coordinates of all particles except particle 2. The particles being equivalent, one would obtain the same value for the probability of finding particle 2 in the volume $\mathbf{dr}$ at $\mathbf{r}$ as for the probability of finding particle 1 in the volume $\mathbf{dr}$ at $\mathbf{r}$.

Now, what is the probability of finding *a* particle in the volume $\mathbf{dr}$ at $\mathbf{r}$? Since no one cares *which* particle, this probability would be N times the probability of finding any particular particle in the volume $\mathbf{dr}$ at $\mathbf{r}$, that is,

$$\rho^{(1)}(\mathbf{r})\,\mathbf{dr} = N\,\mathbf{dr}\int \mathbf{dq}_2\,\mathbf{dq}_3 \cdots \mathbf{dq}_N\, P(\mathbf{r}, \mathbf{q}_2, \ldots, \mathbf{q}_N)$$

$$= \frac{N}{Z_N}\,\mathbf{dr}\int \mathbf{dq}_2\,\mathbf{dq}_3 \cdots \mathbf{dq}_N \exp[-\beta U_N(\mathbf{r}, \mathbf{q}_2, \mathbf{q}_3, \ldots, \mathbf{q}_N)] \quad (7.47)$$

The function $\rho^{(1)}(\mathbf{r})$ will be referred to as the one-particle probability density. It is clear from (7.47) that, if $\rho^{(1)}(\mathbf{r})$ is integrated over $\mathbf{r}$, the integral over N particle coordinates will be just Z_N. Therefore

$$\int \rho^{(1)}(\mathbf{r})\,\mathbf{dr} = N \quad (7.48)$$

the total number of particles.

Similarly to the one-particle probability, the probability of finding particle 1 in the infinitesimal volume element $\mathbf{dr}$ at the position $\mathbf{r}$ and particle 2 in the infinitesimal volume element $\mathbf{dr}'$ at the position $\mathbf{r}'$, regardless of the positions of other particles, is given by

$$\int P(\mathbf{r}, \mathbf{r}', \mathbf{q}_3, \ldots, \mathbf{q}_N)\,\mathbf{dq}_3\,\mathbf{dq}_4 \cdots \mathbf{dq}_N$$

This probability must be the same regardless of which two particles are chosen, so the probability of finding a particle at $\mathbf{r}$ and another particle at $\mathbf{r}'$ is given by

$$\rho^{(2)}(\mathbf{r}, \mathbf{r}')\,\mathbf{dr}\,\mathbf{dr}' = \frac{N(N-1)}{Z_N}\,\mathbf{dr}\,\mathbf{dr}'$$

$$\times \int \mathbf{dq}_3 \cdots \mathbf{dq}_N \exp[-\beta U_N(\mathbf{r}, \mathbf{r}', \mathbf{q}_3, \ldots, \mathbf{q}_N)] \quad (7.49)$$

The factor of $N(N-1)$ is there because the first particle can be any of the N particles, and the second one can be any of the $N-1$ remaining particles, since it must be different from the first one. The two-particle probability density $\rho^{(2)}(\mathbf{r}, \mathbf{r}')$ is a function of two sets of particle coordinates.

From their definitions it follows that integrating $\rho^{(2)}(\mathbf{r}, \mathbf{r}')$ over $\mathbf{r}'$ will give $(N-1)\rho^{(1)}(\mathbf{r})$. Integrating $\rho^{(2)}(\mathbf{r}, \mathbf{r}')$ over $\mathbf{r}$ and $\mathbf{r}'$ will give $N(N-1)$. One can define three-particle, four-particle, etc. probability densities, but only the two introduced above concern us. Notice that one can always get a lower-order probability density from a higher one by integrating, but it is very hard to go the other way.

For reasons of algebraic convenience and (perhaps) clarity of interpretation, the probability densities are sometimes written in terms of the "density operator"

$$\boldsymbol{\rho}(\mathbf{r}) = \sum_{j=1}^{N} \delta(\mathbf{q}_j - \mathbf{r}) \tag{7.50}$$

(here, bold face $\boldsymbol{\rho}$ is to remind us that this is an operator). The quantity $\delta(\mathbf{q}_j - \mathbf{r})$ is a three-dimensional delta-function. The function $\delta(\mathbf{p})$ is supposed to be equal to zero everywhere except for $\mathbf{p} = 0$, and to become infinite at $\mathbf{p} = 0$ in such a way that

$$\int \mathbf{dp}\, \delta(\mathbf{p}) = 1$$

This means that

$$\int \mathbf{dq}_j\, F(\mathbf{q}_j)\, \delta(\mathbf{q}_j - \mathbf{r}) = F(\mathbf{r})$$

for any well-behaved function F. The delta-function, sine it is nonzero only at $\mathbf{q}_j - \mathbf{r} = 0$, picks out the value of F at $\mathbf{q}_j - \mathbf{r} = 0$. In terms of the density operator,

$$\begin{aligned} \rho^{(1)}(\mathbf{r}) &= \int \mathbf{dq}_1\, \mathbf{dq}_2\, \mathbf{dq}_3 \cdots \mathbf{dq}_N\, P(\mathbf{q}_1, \mathbf{q}_2, \ldots, \mathbf{q}_N)\boldsymbol{\rho}(\mathbf{r}) \\ &= Z_N^{-1} \int \mathbf{dq}_1\, \mathbf{dq}_2\, \mathbf{dq}_3 \cdots \mathbf{dq}_N\, \boldsymbol{\rho}(\mathbf{r}) \exp[-\beta U_N(\mathbf{r}, \mathbf{q}_2, \mathbf{q}_3, \ldots, \mathbf{q}_N)] \\ &= \langle\, \rho(\mathbf{r}) \rangle \end{aligned} \tag{7.51}$$

with the fences signifying an average value, taken with $\exp[-\beta U_N]$ as the weighting factor. In the same way, the potential energy in equation (7.9) is $\langle U_N \rangle$.

If there are no external fields, U_N depends only on interparticle coordinates. Instead of using the N sets of coordinates $\{\mathbf{q}_j\}$, one could use the N sets of coordinates $\mathbf{q}_1, \mathbf{q}_{21}, \mathbf{q}_{31}, \ldots, \mathbf{q}_{N1}$. Since any $\mathbf{q}_{jk}$ can be expressed in terms of the $\{\mathbf{q}_{i1}\}$, U_N depends only on the last $N - 1$ of these. Thus

$$\rho^{(1)}(\mathbf{r}) = N \frac{\int \exp[-\beta U_N(\mathbf{q}_{21}, \mathbf{q}_{31} \cdots \mathbf{q}_{N1})]\, \mathbf{dq}_{21} \cdots \mathbf{dq}_{N1}}{\int \exp[-\beta U_N(\mathbf{q}_{21}, \mathbf{q}_{31} \cdots \mathbf{q}_{N1})]\, \mathbf{dq}_1\, \mathbf{dq}_{21} \cdots \mathbf{dq}_{N1}}$$

The integration over $\mathbf{q}_1$ in the denominator produces a factor of V; what is left is identical to the numerator. Therefore, for no external fields, $\rho^{(1)}(\mathbf{r})$ is independent of position and equal to the density N/V.

In contrast, consider a system in which two fluid phases coexist. Let there be a gravitational field along the z-direction, and suppose the phase boundary is near the plane $z = 0$. Then the denser phase (liquid) is found for $z < 0$ and the less dense (gas) phase for $z > 0$. The one-particle density $\rho^{(1)}$ depends on z, perhaps in the manner shown in Figure 7.10. The two-particle density $\rho^{(2)}$ [equation (7.49)] depends on both z and z'; if either z or z' is greater than 0, $\rho^{(2)}$ will be much smaller than if both are in the liquid region, and, if both z and z' are greater than 0, $\rho^{(2)}$, will be even smaller. In addition, $\rho^{(2)}$ will depend on $\mathbf{r} - \mathbf{r}'$.

To separate out the dependence on the distance between $\mathbf{r}$ and $\mathbf{r}'$, it is usual to write the two-particle density as

$$\rho^{(2)}(\mathbf{r}, \mathbf{r}') = \rho^{(1)}(\mathbf{r})\,\rho^{(1)}(\mathbf{r}')\,g(\mathbf{r}, \mathbf{r}') \tag{7.52}$$

defining the "pair correlation function" g. If $\mathbf{r}$ and $\mathbf{r}'$ are far apart, the probability of finding one particle at $\mathbf{r}$ and another at $\mathbf{r}'$ will be the product of two independent probabilities (the densities at the two positions), since the particle at $\mathbf{r}$ cannot have any influence on what happens very far away at $\mathbf{r}'$; then g is 1. If $\mathbf{r}$ is close to $\mathbf{r}'$ the two-particle probability will not be a product of two independent factors. In particular, $\rho^{(2)}(\mathbf{r}, \mathbf{r}')$ must vanish for a real

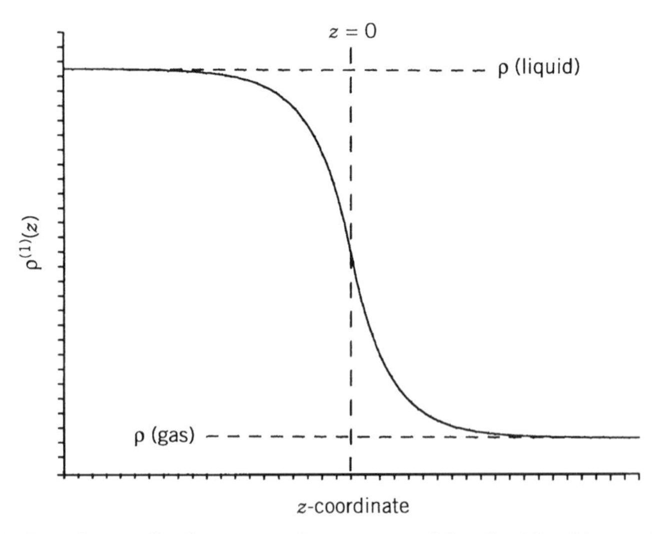

Figure 7.10. Density profile for a two-phase system. The liquid, with number density ρ(liquid), is found at z much less than 0 (to the left) and the gas, with number density ρ(gas), is found at z much greater than 0.

system when $\mathbf{r}$ is very close to $\mathbf{r}'$ because two particles cannot be in the same place. For an ideal gas (not a real system), there is absolutely no correlation between particles, so that g is a constant, independent of $\mathbf{r} - \mathbf{r}'$.

When there are no external fields and one has a homogeneous, one-phase system, (7.52) becomes

$$\rho^{(2)}(\mathbf{r}, \mathbf{r}') = (N/V)^2 g(\mathbf{r}, \mathbf{r}')$$

The pair correlation function cannot really depend on $\mathbf{r}$ and $\mathbf{r}'$ for a homogeneous system, but only on the relative coordinate $\mathbf{r} - \mathbf{r}'$. Furthermore, the system is isotropic: all directions are the same. Thus g depends only on the magnitude of $\mathbf{r} - \mathbf{r}'$. Being a function of only one variable, it is easy to visualize. For a hard-sphere potential, $g = 0$ for $r \leq \sigma$, the hard-sphere diameter, and g approaches $(N - 1)/N$ for large r. Since there are no forces between molecules not in contact, one might expect g to rise abruptly to $(N - 1)/N$ at $r = \sigma$ and remain constant for $r > \sigma$. It will be seen later that this is true only for a gas at low pressure, and we will try to understand why.

Now we specialize to a fluid for which U_N is a sum of pair interactions, as in equation (7.21). Each of the $N(N - 1)/2$ terms in the sum, $u(r_{ij})$, is a function of one interparticle distance. The energy, equation (7.9), is a sum of an ideal term, $3NkT/2$, and an "excess" term due to the potential energy U_N which may be written as

$$\begin{aligned} E^{\text{exc}} &= Z_N^{-1} \int d\mathbf{q}_1\, d\mathbf{q}_2 \cdots d\mathbf{q}_N \sum_{i<j}^{N} u(r_{ij})\, e^{-\beta U_N} \\ &= \frac{N(N-1)}{2Z_N} \int d\mathbf{q}_1\, d\mathbf{q}_2 \cdots d\mathbf{q}_N\, u(r_{12})\, e^{-\beta U_N} \end{aligned}$$

Performing the integrations over $\mathbf{q}_3 \cdots \mathbf{q}_N$ first, we recognize the two-particle density of (7.49). Therefore,

$$\begin{aligned} E^{\text{exc}} &= \frac{1}{2} \int d\mathbf{q}_1\, d\mathbf{q}_2\, u(r_{12}) \rho^{(2)}(\mathbf{q}_1, \mathbf{q}_2) \\ &= \frac{N^2}{2V^2} \int d\mathbf{q}_1\, d\mathbf{q}_{12}\, u(r_{12}) g(r_{12}) \\ &= \frac{2\pi N^2}{V} \int_0^\infty dr\, r^2 g(r) u(r) \end{aligned} \tag{7.53}$$

It is only necessary to know $g(r)$, which defines the correlations between the particles, to calculate the excess energy.

The virial equation of state can be treated similarly. With the present notation, (7.26) reads

$$P - \frac{NkT}{V} = \frac{-1}{3VZ_N} \int \mathbf{dq}_1 \, \mathbf{dq}_2 \cdots \mathbf{dq}_N \exp\left[-\beta \sum_{i<j=1}^{N} u(r_{ij}) \right]$$
$$\times \frac{N(N-1)}{2} r_{12} \frac{du(r_{12})}{dr_{12}}$$
$$= \frac{-1}{6V} \int \mathbf{dq}_1 \, \mathbf{dq}_2 \, \rho^{(2)}(\mathbf{q}_1, \mathbf{q}_2) r_{12} \frac{du(r_{12})}{dr_{12}}$$
$$= \frac{-N^2}{6V^2} \int \mathbf{dq}_{12} \, g(r_{12}) r_{12} \frac{du(r_{12})}{dr_{12}}$$

This is the contribution of the potential energy to the pressure. In terms of the density, $\rho = N/V$, the equation may be written as

$$\frac{\beta P}{\rho} = 1 - \frac{2\pi}{3} \beta \rho \int_0^\infty dr \, r^3 g(r) \frac{du}{dr} \tag{7.54}$$

The important role of $g(r)$ is again seen.

The formulas above are easily generalized to multicomponent fluids. Using s to index the different species, one has to consider a set of one-particle densities:

$$\rho_s^{(1)} = \langle \boldsymbol{\rho}_s(\mathbf{r}) \rangle = \left\langle \sum_{j=1}^{N_s} \delta(\mathbf{q}_{sj} - \mathbf{r}) \right\rangle \tag{7.55}$$

where $\mathbf{q}_{sj}$ is the set of coordinates for the jth particle (of N_s particles) of species s. For a multicomponent fluid, U_N involves the coordinates of all the particles of all the species, with N equal to the sum of $\{N_s\}$. The one-particle density $\rho_s^{(1)}$ is obtained by putting $\mathbf{r}$ for $\mathbf{q}_{s1}$, integrating $\exp[-\beta U_N]$ over the coordinates of all other particles, multiplying by N and dividing by Z_N, that is,

$$\rho_s^{(1)} = N_s \langle \delta(\mathbf{q}_{s1} - \mathbf{r}) \rangle$$

There is also a set of two-particle densities,

$$\rho_{st}^{(2)}(\mathbf{r}, \mathbf{r}') = \frac{N_s(N_t - \delta_{st})}{Z_N} \int \mathbf{dq}_1 \, \mathbf{dq}_{12} \cdots \delta(\mathbf{q}_{sj} - \mathbf{r}) \, \delta(\mathbf{q}_{tk} - \mathbf{r}') \, e^{-\beta U_N} \tag{7.56}$$

Here, $\delta_{st} = 0$ for $s \neq t$ and 1 for $s = t$.

If U_N includes only pair interactions, we write

$$U_N = \frac{1}{2}\sum_s \sum_t^{(t\neq s)} \sum_{j=1}^{N_s} \sum_{k=1}^{N_t} u_{st}(\mathbf{q}_{sj}, \mathbf{q}_{tk}) + \frac{1}{2}\sum_s \sum_{j=1}^{N_s} \sum_k^{(k\neq j)} u_{ss}(\mathbf{q}_{sj}, \mathbf{q}_{sk})$$

to take into account the fact that the pair interaction potential is different between particles of different species. Then the excess energy is, instead of (7.53) for a one-component fluid,

$$\begin{aligned}
E^{\text{exc}} &= \frac{1}{2}\sum_s \sum_t \int \mathbf{dr}\,\mathbf{dr}'\, \rho_{st}^{(2)}(\mathbf{r}, \mathbf{r}')u_{st}(\mathbf{r}, \mathbf{r}') \\
&= \frac{1}{2}\sum_s \sum_t \rho_s^{(1)}\rho_t^{(1)} \int \mathbf{dr}_1\,\mathbf{dr}_{12}\, g_{st}(\mathbf{r}_{12})u_{st}(\mathbf{r}_{12}) \\
&= \sum_s \sum_t \frac{N_s N_t}{2V^2} \int \mathbf{dr}_1\,\mathbf{dr}_{12}\, g_{st}(\mathbf{r}_{12})u_{st}(r_{12}) \\
&= \sum_s \sum_t \frac{4\pi N_s N_t}{2V} \int dr_{12}\, r_{12}^2\, g_{st}(r_{12})u_{st}(r_{12})
\end{aligned} \tag{7.57}$$

and the virial equation of state is

$$\begin{aligned}
P &= \sum_s \frac{N_s kT}{V} - \frac{1}{6V}\sum_s \sum_t \int \mathbf{dr}\,\mathbf{dr}'\, \rho_{st}^{(2)}(\mathbf{r}, \mathbf{r}')|\mathbf{r} - \mathbf{r}'| \frac{du_{st}}{d\,|\mathbf{r} - \mathbf{r}'|} \\
&= \sum_s \frac{N_s kT}{V} - \sum_s \sum_t \frac{N_s N_t}{6V^3} \int \mathbf{dr}_1\,\mathbf{dr}_{12}\, r_{12}\, g_{st}(r_{12}) \frac{du_{st}}{dr_{12}} \\
&= \sum_s \frac{N_s kT}{V} - \sum_s \sum_t \frac{2\pi N_s N_t}{3V^2} \int dr_{12}\, r_{12}^3\, g_{st}(r_{12}) \frac{du_{st}}{dr_{12}}
\end{aligned} \tag{7.58}$$

We return to a one-component fluid for consideration of the one-particle and two-particle densities in the grand canonical ensemble. The formulas we get are easily generalized to a multispecies system.

In the grand canonical ensemble for a one-component fluid, all properties are calculated in terms of the grand partition function of equation (2.41):

$$\Xi = \sum_{N=0}^{\infty} e^{\beta\mu N} Q(N, V, \beta) = \sum_{N=0}^{\infty} \frac{z^N}{N!} Z_N(V, \beta) \tag{7.59}$$

The N-particle canonical partition function $Q(N, V, \beta)$ has been written as $(N!)^{-1}\ \Lambda^{-3N} Z_N$ with Λ the de Broglie wavelength [equations (3.6) and

(3.7)], and the activity $z = \Lambda^{-1} e^{\beta\mu}$ has been used. The average number of particles is

$$\langle N \rangle = \Xi^{-1} \sum_{N=0}^{\infty} N \frac{z^N}{N!} Z_N(V, \beta) = z\left(\frac{\partial \ln \Xi}{\partial z}\right)_{V,\beta}$$

The density of a uniform system is $\langle N \rangle / V$, so that

$$\frac{\rho}{z} = \frac{1}{V} \frac{\partial \ln \Xi}{\partial z} \tag{7.60}$$

For $z \to 0$, $\Xi \to 1 + zZ_1 = 1 + zV$, so (7.60) becomes

$$\frac{\rho}{z} = \frac{1}{V} \frac{V}{1 + zV}$$

which approaches unity; thus $\rho/z \to 1$ as $z \to 0$.

Now consider the two-particle density. Since the number of particles is not fixed, the two-particle density gives the *average* number of pairs of particles such that one particle is in **dr** at **r** and the other in **dr**′ at **r**′:

$$\rho^{(2)}(\mathbf{r}, \mathbf{r}') = \Xi^{-1} \sum_{N=0}^{\infty} \frac{z^N}{N!} N(N-1) \int e^{-\beta U_N(\mathbf{r}, \mathbf{r}', \mathbf{q}_3 \cdots \mathbf{q}_N)} \, \mathbf{dq}_3 \cdots \mathbf{dq}_N \tag{7.61}$$

Note that the terms for $N = 0$ and $N = 1$ vanish. Integrating (7.61) over **r** and **r**′ gives

$$\begin{aligned}\iint \mathbf{dr} \, \mathbf{dr}' \, \rho^{(2)}(\mathbf{r}, \mathbf{r}') &= \Xi^{-1} \sum_{N=0}^{\infty} \frac{z^N}{N!} N(N-1) \\ &\quad \times \int e^{-\beta U_N(\mathbf{r}\mathbf{r}'\mathbf{q}_3 \cdots \mathbf{q}_N)} \, \mathbf{dr} \, \mathbf{dr}' \, \mathbf{dq}_3 \cdots \mathbf{dq}_N \\ &= \langle N(N-1) \rangle = \langle N^2 \rangle - \langle N \rangle \end{aligned} \tag{7.62}$$

For a homogeneous system, $\rho^{(2)}(\mathbf{r}, \mathbf{r}') = \rho^2 g(|\mathbf{r} - \mathbf{r}'|)$ with $\rho = \langle N \rangle / V$ so (7.61) may be rewritten as

$$\begin{aligned}\Xi g(|\mathbf{r} - \mathbf{r}'|) \left(\frac{\rho}{z}\right)^2 &= \sum_{N=2}^{\infty} \frac{z^{N-2}}{N!} N(N-1) \int e^{-\beta U_N(\mathbf{r}\mathbf{r}'\mathbf{q}_3 \cdots \mathbf{q}_N)} \, \mathbf{dq}_3 \, \mathbf{dq}_4 \cdots \mathbf{dq}_N \\ &= \sum_{M=0}^{\infty} \frac{z^M}{M!} \int e^{-\beta U_{M+2}(\mathbf{r}\mathbf{r}'\mathbf{q}_3 \cdots \mathbf{q}_N)} \, \mathbf{dq}_3 \, \mathbf{dq}_4 \cdots \mathbf{dq}_N \end{aligned} \tag{7.63}$$

with the new variable M used instead of $N-2$ [note that $N(N-1)/N! = 1/(N-2)!$]. In the limit of $z \to 0$ (or $\rho \to 0$), so $\rho/z \to 1$, (7.63) becomes

$$g(|\mathbf{r}-\mathbf{r}'|) = z^0\, e^{-\beta U_2(\mathbf{r},\mathbf{r}')} = e^{-\beta u(|\mathbf{r}-\mathbf{r}'|)} \tag{7.64}$$

since only one term remains of the sum. Thus, for small density, g is given by a Boltzmann factor with the pair interaction potential as potential energy. This is true even if there are many-body interactions.

One also has a normalization condition on g. From (7.62),

$$\iint \mathbf{dr}\,\mathbf{dr}'\left[\rho^{(2)}(\mathbf{r},\mathbf{r}') - \rho^{(1)}(\mathbf{r})\rho^{(1)}(\mathbf{r}')\right] = \langle N^2\rangle - \langle N\rangle - \langle N\rangle^2$$

which, for a homogeneous system, is

$$\frac{\langle N\rangle^2}{V^2} V \int \mathbf{d}(\mathbf{r}-\mathbf{r}')\left[g(|\mathbf{r}-\mathbf{r}'|) - 1\right] = \langle N^2\rangle - \langle N\rangle - \langle N\rangle^2$$

or

$$\frac{\langle N\rangle}{V} \int \mathbf{dr}_{12}\left[g(\mathbf{r}_{12}) - 1\right] = \frac{\langle N^2\rangle - \langle N\rangle^2}{\langle N\rangle} - 1 \tag{7.65}$$

The fraction on the right-hand side is equal to ρ/β times the compressibility

$$\kappa_T = -\left(\frac{\partial \ln V}{\partial P}\right)_{N,T}$$

[see equation (2.53)]. Therefore

$$\rho \int \mathbf{dr}_{12}\left[g(r_{12}) - 1\right] + 1 = \frac{\kappa_T \rho}{\beta} \tag{7.66}$$

Equation (7.66) is the compressibility equation of state.

Return to the ideal gas: the equation of state is simply $P = \rho kT$ and $g(r)$ is equal to 1 for all r. Both the equation of state and the constancy of g express the independence of the particles. Of course, since U_N is identically zero, there is no E^{exc} and the virial equation of state (7.54) is just $\beta P/\rho = 1$ again. The compressibility κ_T of an ideal gas is $1/P$ so (7.66), the compressibility equation of state, reads

$$\rho \int \mathbf{dr}_{12}\left[g(r_{12}) - 1\right] + 1 = 1 \tag{7.67}$$

which is consistent with $g = 1$. Now suppose there is a hard-sphere repulsion at distance σ, so $g = 0$ for $r < \sigma$. At low densities, (7.64) shows that g is a step function, equal to 0 for $r < \sigma$ and to 1 for $r \geq \sigma$. The behavior of g is more complicated at higher densities, as can be seen from the compressibility equation of state. Calculating the compressibility from (7.32) (which is correct through terms in ρ^2) and putting $g = 0$ for $r_{12} < \sigma$, the compressibility equation of state (7.66) becomes

$$-\frac{4\pi\rho\sigma^3}{3} + 4\pi\rho \int_\sigma^\infty dr_{12}\, r_{12}^2[g(r_{12}) - 1] + 1 = (1 - 4\rho v)^2$$

Since the atomic volume v is $\pi\sigma^3/6$,

$$4\pi\rho \int_\sigma^\infty dr_{12}\, r_{12}^2[g(r_{12}) - 1] - \frac{4\pi\rho\sigma^3}{3} = -4\rho\pi\frac{\sigma^3}{3} + 4\rho^2\pi^2\frac{\sigma^6}{9}$$

This shows that, on the average, $g(r)$ must exceed 1 for $r > \sigma$; apparently the exclusion of particles from a sphere around a given particle leads to a "pile-up" of particles for r greater than σ.

The virial equation of state and the compressibility equation of state provide connections between measurable properties and the pair correlation function $g(r)$. However, both involve integrals over $g(r)$, so one cannot obtain $g(r)$ itself from the compressibility or other bulk properties. It is possible to determine $g(r)$ experimentally from neutron or X-ray diffraction, or by computer simulation (see Section 7.5). One now has quite a bit of information about $g(r)$ for the hard-sphere as well as more complicated potentials. The general form of g is almost always the same. Apparently, the structure of the fluid is determined mostly by the short-range repulsive interactions, which can be approximated as hard-sphere repulsions; the long-range attractive interactions, while they contribute to average values of energy and other properties, have only a minor effect on structure. Because they are long-range, their contribution to the force between molecules (force = derivative of potential) is not large.

For low densities, one may invoke (7.64); the pair correlation function $g(r)$ is approximately $\exp[-\beta u(r)]$. Figure 7.11 shows this for the Morse potential (7.22) and T = 400 K, 500 K, 600 K, and 700 K (the temperatures used are high because the potential is a very strong one). Note that, the higher the temperature, the less important is the potential, and the more g approaches ideal-gas behavior (except for the exclusionary effect of the short-range repulsion). For any potential which includes a short-range repulsion and a long-range attraction, the form of $g(r)$ given by (7.64) is the same. Equation (7.64) can never yield a pair correlation function like that shown in Figure 7.12, which is what one deduces from experimental measurements or computer simulations for dense gases or liquids. In addition to the exclusion sphere produced by the short-range repulsion and the large peak just outside of the exclusion sphere, $g(r)$ shows a damped oscillation about $g(r) = 1$. This is true even for the hard-sphere fluid with no attractive forces.

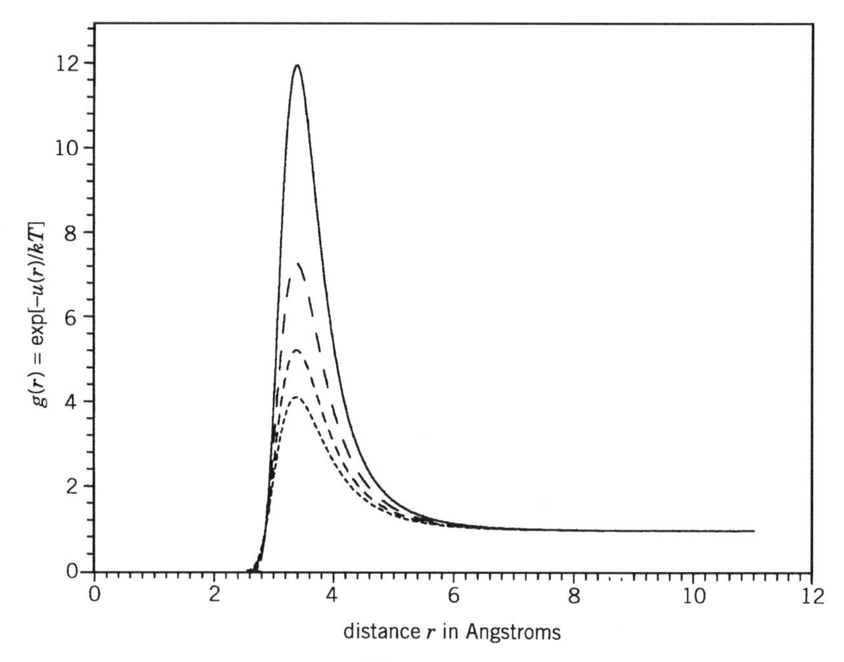

Figure 7.11. Correlation function $g(r)$ for a gas at various temperatures. For low densities, $g(r) = \exp[-u(r)/kT]$. The potential $u(r)$ is the Morse potential of (7.22) and Figure 7.1 and the temperatures are (top to bottom) 400 K, 500 K, 600 K, and 700 K.

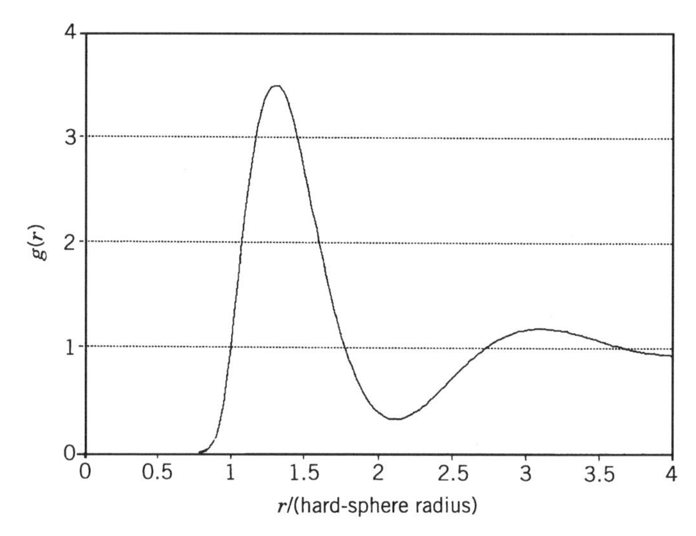

Figure 7.12. Correlation function $g(r)$ for a liquid. In addition to the exclusion sphere with the hard-sphere radius and the large maximum for slightly larger r, $g(r)$ has damped oscillations before approaching 1 for large r.

The oscillation is because, at higher densities, more than two atoms are found in close proximity; (7.64), since it is for low densities, describes the correlation between two atoms in the absence of others. The first peak in $g(r)$ is due to the repulsive potential, which, by expelling atoms from the exclusion sphere around any atom ($r < \sigma$), increases $g(r)$ just outside this sphere. The subsequent peaks and valleys are due to indirect correlations: if atoms at a distance σ from the central atom are probable, their exclusion spheres make it unlikely to find atoms at a distance near $3\sigma/2$ and more likely to find atoms at a distance near 2σ from the central atom. In the next section we give some idea of how these indirect correlations can be taken into account.

7.5. PROPERTIES OF CORRELATION FUNCTIONS

Thermodynamic and structural properties of a fluid depend on the pair correlation function $g(r)$, which gives the probability of finding a particle at a distance r from a particle at $r = 0$. If $g(r)$ is equal to unity for all r, there is no correlation; the presence of the particle at $r = 0$ does not affect the distribution of other particles. In a uniform fluid, the density (number of particles per unit volume) at any point fluctuates around its average value because the particles are always moving. The quantity $g(r) - 1$ measures the correlation between the density fluctuations. This correlation is responsible for the scattering of radiation, so $g(r)$ may be deduced from the scattering intensity as a function of angle, as shown below. Another source of information about $g(r)$ is computer simulation. By generating configurations for many particles interacting with a known force law, one obtains a detailed picture of a fluid on a molecular scale. The limitation of this method is the amount of computer time available. Finally, $g(r)$ may be calculated by making approximations in the exact formula or from approximate theories that derive differential equations satisfied by $g(r)$. Such theories are touched on later.

Diffraction of neutrons is due to nonuniformities in the number density of nuclei in the fluid; diffraction of X-rays is due to nonuniformities in the electron density. Both densities follow the number density of atoms. Incident radiation, represented as a plane wave, polarizes the scattering particles and causes them to emit spherical waves. The interference between these waves makes the intensity of radiation scattered by the system depend on the angle at which the radiation is scattered. It will now be shown that from this angle-dependence one can obtain the pair correlation function for a fluid. The particles scattered will be referred to as radiation, whether they are neutrons, electrons, or photons.

Suppose a beam (plane wave) of radiation enters a one-component fluid in a volume V. Its amplitude at position $\mathbf{r}$ is $A \exp[i\mathbf{k}_0 \cdot \mathbf{r}]$, where the direction of $\mathbf{k}_0$ is the direction of propagation and the magnitude of $\mathbf{k}_0$ is $k = 2\pi/\lambda$, λ

being the wavelength. Each scattering particle in the path of the radiation becomes the origin of an outgoing spherical wave with intensity proportional to that of the incident plane wave (attenuation of the beam is neglected). The constant of proportionality, c, depends on k, on the nature of the radiation, and on the nature of the particles. The scattered waves from all the particles add together, and the intensity of the resulting wave in the direction $\mathbf{k}$, at angle ϕ to $\mathbf{k}_0$, is to be found. (In reality, the ratio of the intensity scattered by each particle to the incident intensity depends on ϕ, but the dependence is small if one limits consideration to small angles, and it is neglected here.)

The wave in the direction $\mathbf{k}$ scattered from a particle at point $\mathbf{O}$ (see Figure 7.13) has amplitude

$$A \frac{c}{r} e^{i\mathbf{k}_0 \cdot \mathbf{O}} e^{i\mathbf{k} \cdot \mathbf{r}}$$

where r is the vector from $\mathbf{O}$ to the observation point. The factor of $1/r$ comes in because this is a spherical wave, for which the intensity (amplitude squared) integrated over all angles must be the same for all r. The wave in the direction $\mathbf{k}$ scattered from a particle at point $\mathbf{P}$ has amplitude

$$A \frac{c}{r} e^{i\mathbf{k}_0 \cdot \mathbf{P}} e^{i\mathbf{k} \cdot (\mathbf{r} - \mathbf{P} + \mathbf{O})} = A \frac{c}{r} e^{i\mathbf{k}_0 \cdot \mathbf{O}} e^{i\mathbf{k} \cdot \mathbf{r}} e^{i(-\mathbf{k} + \mathbf{k}_0) \cdot \mathbf{s}}$$

where $\mathbf{s} = \mathbf{P} - \mathbf{O}$. Since the observation point is far away from the volume V, the distance r from any point within the sample to the observation point is essentially the same. To get the amplitude of the total scattered wave, one

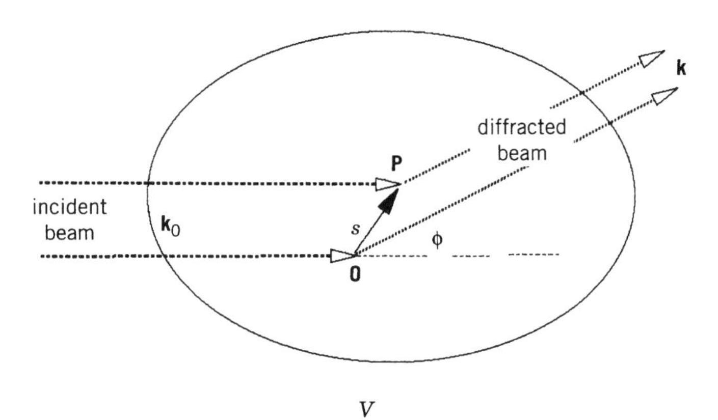

Figure 7.13. Diffraction from a fluid. The incident beam with wave vector $\mathbf{k}_0$ is scattered from particles in the volume V. Diffracted rays of wave vector $\mathbf{k}$, scattered from particles at points O and P, interfere, so the amplitude of the diffracted beam depends on the angle between $\mathbf{k}_0$ and $\mathbf{k}$.

multiplies the amplitude of the scattered wave from a particle at every location in the sample by the density of particles at that location, and integrates over the volume of the sample. The intensity is the amplitude multiplied by its complex conjugate.

Let the density of particles at point **P** be denoted by $\nu(\mathbf{P})$. The total scattered intensity is

$$A^2 \frac{c^2}{r^2}\left[\int_V d\mathbf{O}\, \nu(\mathbf{O})\, e^{i\mathbf{k}_0\cdot\mathbf{O}}\, e^{i\mathbf{k}\cdot\mathbf{r}}\right]^* \left[\int_V d\mathbf{P}\, \nu(\mathbf{P})\, e^{i\mathbf{k}_0\cdot\mathbf{O}}\, e^{i\mathbf{k}\cdot\mathbf{r}}\, e^{i(-\mathbf{k}+\mathbf{k}_0)\cdot\mathbf{s}}\right]$$

$$= A^2 \frac{c^2}{r^2}\left[\int_V d\mathbf{O} \int_V d\mathbf{P}\, \nu(\mathbf{O})\, \nu(\mathbf{P})\, e^{i(-\mathbf{k}+\mathbf{k}_0)\cdot\mathbf{s}}\right]$$

This should be averaged over time, since the particles are in constant motion. The quantity $d\mathbf{O}\, d\mathbf{P}\, \nu(\mathbf{O})\, \nu(\mathbf{P})$ is the number of pairs of particles such that one particle is in the volume $d\mathbf{O}$ at **O** and the other in the volume $d\mathbf{P}$ at **P**; averaging over time, this is just $d\mathbf{O}\, d\mathbf{P}\, \rho^{(2)}(\mathbf{O},\mathbf{P})$. Therefore the scattered intensity is

$$A^2 \frac{c^2}{r^2}\left[\int_V d\mathbf{O} \int_V d\mathbf{P}\, \rho^{(2)}(\mathbf{O},\mathbf{P})\, e^{i(-\mathbf{k}+\mathbf{k}_0)\cdot\mathbf{s}}\right]$$

$$= A^2 \frac{c^2}{r^2}\left[\int_V d\mathbf{O} \int_V d\mathbf{s}\, \rho^2 g(|\mathbf{s}|)\, e^{i(-\mathbf{k}+\mathbf{k}_0)\cdot\mathbf{s}}\right] \tag{7.68}$$

with ρ the particle density, N/V. Integration over **O** produces a factor of V (sample volume), and the remaining integral,

$$\int d\mathbf{s}\, g(s) e^{i(\mathbf{k}_0-\mathbf{k})\cdot\mathbf{s}} = \int s^2\, ds \sin\theta\, d\theta\, d\phi\, e^{i\,|\mathbf{k}_0-\mathbf{k}|\, s\cos\theta} \tag{7.69}$$

is a Fourier transform (θ is the angle between $\mathbf{k}_0 - \mathbf{k}$ and **s**). Thus the intensity of radiation scattered by the particles in a fluid is proportional to the Fourier transform of the pair correlation function g.

A few basic equations relating to Fourier transforms are given here. The Fourier transform of a function $f(\mathbf{r})$ is a function $\tilde{f}(k)$ where

$$\tilde{f}(\mathbf{k}) = \int d\mathbf{r}\, e^{i\mathbf{k}\cdot\mathbf{r}} f(\mathbf{r}) \tag{7.70}$$

The original function may be recovered from the transform function by inverting the Fourier transform:

$$f(\mathbf{r}) = (2\pi)^{-3} \int d\mathbf{k}\, e^{-i\mathbf{k}\cdot\mathbf{r}} \tilde{f}(\mathbf{k}) \tag{7.71}$$

To show how this works, substitute (7.70) for $\tilde{f}(\mathbf{k})$, so the right-hand side of (7.71) becomes

$$(2\pi)^{-3} \int \mathbf{dk}\, e^{-i\mathbf{k}\cdot\mathbf{r}} \int \mathbf{dr}'\, e^{i\mathbf{k}\cdot\mathbf{r}'} f(\mathbf{r}') = (2\pi)^{-3} \int \mathbf{dr}'\, f(\mathbf{r}') \int \mathbf{dk}\, e^{i\mathbf{k}\cdot(\mathbf{r}'-\mathbf{r})}$$

Because of interference between waves of different k, the integral

$$\int \mathbf{dk}\, e^{i\mathbf{k}\cdot(\mathbf{r}'-\mathbf{r})} = (2\pi)^3\, \delta(\mathbf{r}' - \mathbf{r}) \tag{7.72}$$

where $\delta(\mathbf{r}' - \mathbf{r})$ is the three-dimensional delta-function, equal to zero for $\mathbf{r}' - \mathbf{r} \neq 0$ and infinite for $\mathbf{r}' - \mathbf{r} = 0$. Using (7.72) makes (7.71) into an identity.

Returning to (7.68), we write the scattered intensity, using the notation just introduced, as

$$A^2 \frac{c^2}{r^2} \frac{N^2}{V} \left[\int_V ds [g(s) - 1]\, e^{i(\mathbf{k}_0 - \mathbf{k})\cdot\mathbf{s}} + \int_V ds\, e^{i(\mathbf{k}_0 - \mathbf{k})\cdot\mathbf{s}} \right]$$

$$= A^2 \frac{c^2}{r^2} \frac{N^2}{V} \tilde{h}(|\mathbf{k}_0 - \mathbf{k}|) + A^2 \frac{c^2}{r^2} \frac{N^2}{V} \int_V ds\, e^{i(\mathbf{k}_0 - \mathbf{k})\cdot\mathbf{s}}$$

where $h(s) = g(s) - 1$. Since $g(s) - 1 = 0$ when s exceeds a few molecular diameters, the first integral (extending over the volume V, which is very large on a molecular scale) may be considered to extend to infinity, so that it is in fact equal to the Fourier transform of h. The last integral vanishes when V is large, except for $\mathbf{k}_0 - \mathbf{k} \approx 0$ (the range of $|k_0 - k|$ for which the integral differs appreciably from zero actually varies inversely to V), and $\mathbf{k}_0 - \mathbf{k} = 0$ corresponds to a scattering angle ϕ of 0. This integral is thus a contribution to forward or zero-angle scattering; since the integral is just V when $\mathbf{k}_0 - \mathbf{k} = 0$, the contribution is proportional to N^2.

The scattering intensity for nonzero angles is proportional to $\tilde{h}$. To get $h(s)$ from the measured scattering intensity, it is only necessary to invert the Fourier transform, which can be done according to (7.71). Because g and h depend only on the magnitude of $\mathbf{s}$, their Fourier transforms depend only on the magnitude of $\mathbf{k}_0 - \mathbf{k}$, as shown by (7.69).

Another often-used route to information about $g(r)$ is computer simulation. Assuming a pair interaction potential $u(r)$, Newton's or Hamilton's equations of motion are integrated numerically for some convenient number of particles (limited by available computational resources and time). When the positions and momenta of the particles reach thermal equilibrium it is straightforward to find properties such as $g(r)$. Integrating the equations of motion is the hard part; even with today's computers, much ingenuity is necessary to integrate for a reasonable time, even for a few thousand particles. The method of integrating the equations of motion is referred to as molecular dynamics. It is useful in understanding nonequilibrium as well as equilibrium states of a system.

An alternative scheme, the Monte Carlo method, is for equilibrium states only. It is used to evaluate the configurational integral or any integral of the form

$$\mathscr{I}_F = \int F(\mathbf{q})\, e^{-\beta U_N}\, d^{3N}q \tag{7.73}$$

where $\mathbf{q}$ stands for the $3N$ coordinates of N particles. Suppose the $3N$-dimensional space, of total volume $\mathscr{V} = V^N$ with V the volume available to one particle, is divided into k volumes $\mathscr{V}/k$, and the value of $F \exp[-\beta U_N]$ is found at a point in the center of each such volume. Then $\mathscr{I}_F$ could be approximated as $\mathscr{V}/k$ times the sum of the values found, or $\mathscr{V}$ multiplied by the average value. Equivalently, one could choose k points at random in $\mathscr{V}$, evaluate $F \exp[-\beta U_N]$ at each, and approximate $\mathscr{I}_F$ by $\mathscr{V}$ multiplied by the average of these values. This approximation, like the previous one, would be very unreliable unless k was very large. For reliable sampling, one might require that at least two values for each coordinate be taken; with $N = 1000$ (a very small number compared to Avogadro's number); this would require $k = 2^{3000} = 10^{903}$ points!

Furthermore, most of the computation would be wasted, because $\exp[-\beta U_N]$ would be found to be very small for most of the points, and they would contribute nothing to the average. One requires a method of picking evaluation points for which $\exp[-\beta U_N]$ is larger, so that the number of points needed is small enough to be practical. The Monte Carlo method accomplishes this.

One starts with the particles in some reference configuration (set of $3N$ coordinates) and generates new evaluation points by a series of random "moves" of individual particles (changes of three of the $3N$ coordinates). Thus, the x coordinate of one particle is increased by $\alpha_1 \delta_1$, the y coordinate by $\alpha_2 \delta_2$, and the z coordinate by $\alpha_3 \delta_3$, where δ_1, δ_2, and δ_3 are random numbers between -1 and 1, and $\alpha_1 \alpha_2 \alpha_3 = V$. If the move takes the particle out of the volume V, the particle is imagined to reenter V from the face opposite to the face by which it left, moving into V by the same distance it went out of V. It is as if the volume V with its N particles is surrounded on all sides by identical copies of itself. After the move, the quantity F is evaluated at the new configuration and the result stored.

Then an additional calculation involving random numbers is made, to ensure that one goes to evaluation points for which $\exp[-\beta U_N]$ is large. Let ΔU_N be the change in the value of U_N in going from the old configuration to the new one. If $\Delta U_N < 0$, the new configuration becomes the starting point for a new move, in which another particle has its x, y, and z coordinates changed according to new random numbers δ_1, δ_2, and δ_3. If, however, U_N is increased by the move ($\Delta U_N > 0$), another "spin of the wheel" decides whether or not the move is accepted. If ξ, a random number between 0 and 1, exceeds $\exp[-\beta\Delta U_N]$, the move is not accepted and the particle is returned to its position before the move. Then the next particle move starts from the same configuration as the previous move and the value of F is counted twice.

The process of generating random numbers, moving individual particles, and finding new evaluation points for F is repeated until all the particles have been considered. Then one starts over with the first particle, and repeats the cycle as many times as is practical. The use of random number generators in making the moves and in deciding whether a move should be kept is why the model is named for the city of the famous gambling casino. The more moves made, the closer one approximates an equilibrium configuration, for which U_N is minimized. The average value of F approximates $\mathcal{I}_F/\mathcal{I}_1$.

The reason for this is seen by considering an ensemble of identical N-particle systems in volumes V, such that n_r is the number of systems in configuration r. A move is made in each system of the ensemble, producing a new distribution of systems over the possible configurations. The probability that a move takes a system from configuration s to configuration t is p_{st}. Although $p_{ts} = p_{st}$, the procedure for accepting or rejecting a move causes the number of systems in the ensemble changing their configurations from s to t to be greater than the number changing their configurations from t to s if $U_N(s) > U_N(t)$. The net rate of movement of systems in the ensemble from configuration s to configuration t is

$$n_s p_{st} - n_t p_{ts} = p_{st}\{n_s - n_t\, e^{-\beta[U_N(t) - U_N(s)]}\} \tag{7.74}$$

Eventually, the ensemble reaches a stationary equilibrium state for which there is no net movement between configurations; the $\{n_r\}$ no longer change. Then

$$n_s = n_t\, e^{-\beta[U_N(t) - U_N(s)]} = n_t\, e^{-\beta[\Delta U_N]} \tag{7.75}$$

Thus the evaluation points (configurations) generated by the Monte Carlo procedure correspond to an equilibrium (canonical) distribution, in which each configuration is weighted by $\exp[-\beta U_N]$. (For this to work, it must be possible to reach all configurations from the initial configuration.)

The number of moves one makes is limited by the computational capacity and time available. The more moves made, the more closely one can approximate the canonical distribution and the more closely $\mathcal{I}_F/\mathcal{I}_1$ is approximated. Larger values of N require more moves for comparable accuracy. It may be noted that the procedure is particularly simple for hard-sphere interactions. If a move brings a particle within the exclusion sphere of another particle, it must be rejected; if not, it is accepted.

Monte Carlo, molecular dynamics, and experiment have given a lot of information about the pair correlation function $g(r)$, defined by (7.52). For a homogeneous system, $\rho^{(2)}(\mathbf{r}, \mathbf{r}')$ depends only on $|\mathbf{r} - \mathbf{r}'|$ and (7.52) becomes

$$\rho^{(2)}(|\mathbf{r} - \mathbf{r}'|) = \rho^2 g(|\mathbf{r} - \mathbf{r}'|) \tag{7.76}$$

For small $|\mathbf{r} - \mathbf{r}'|$, the behavior of $g(r)$ is dominated by the interatomic potential $u(r)$. We have shown that $g(r) = \exp[-\beta u(r)]$ when it us unlikely to find more than two particles in close proximity. At the other extreme, the pair correlation function must approach 1 for large $|\mathbf{r} - \mathbf{r}'|$ because there can be no correlation between two particles very far apart.

In many of the integral formulas of the preceding section, the quantity $g(r) - 1$ appears. It is usual to define

$$h(r) \equiv g(r) - 1 \tag{7.77}$$

and call $h(r)$ the total correlation function, since $h(r) = 0$ corresponds to no correlation. The total correlation function approaches 0 at large distances. For low densities, $h(r)$ reflects the correlation due to the interatomic potential. For higher densities, $g(r)$ and $h(r)$ become more complicated because of indirect correlations: the probability of finding particle b at a distance r from a central particle a depends partly on the interaction potential energy between a and b and partly on the interactions between particle a and third particles which also interact with particle b. This suggests the definition of a "direct correlation function" $c(r)$ according to the Ornstein–Zernicke equation:

$$h(\mathbf{r}_1, \mathbf{r}_2) = c(\mathbf{r}_1, \mathbf{r}_2) + \int d\mathbf{r}_2\, h(\mathbf{r}_1, \mathbf{r}_3)\, \rho^{(1)}(\mathbf{r}_3) c(\mathbf{r}_3, \mathbf{r}_2) \tag{7.78}$$

The total correlation function is equal to the direct correlation function for low densities ($\rho^{(1)} \to 0$) but differs from it at higher densities because of indirect correlations through the particle at $\mathbf{r}_2$.

One way to get equation (7.78) is as follows: the probability that a small volume δV_i in the fluid contains a particle is $\rho\, \delta V_i$ on the average ($\rho = N/V$). However, if there is known to be a particle in the nearby small volume δV_j, the probability of a particle in δV_i may be different. The difference is proportional to $\rho h(r_{ij})$. (Of course, the occupation probabilities of other nearby volumes are also affected.) Now suppose it is known that there is a particle in δV_j, *and* none in all nearby volumes $\{\delta V_k\}$. The probability that there is a particle in δV_i again differs from $\rho\, \delta V_i$, but by a different amount from the previous case. The difference is now proportional to $\rho c(r_{ij})$. The difference between $\rho h(r_{ij})$ and $\rho c(r_{ij})$ is that, for the latter, only the particle in δV_j is present; the indirect effect of this particle on the occupation probabilities of other nearby volumes has been removed. This indirect effect is the second term in (7.78).

The Ornstein–Zernicke equation also follows from mathematical manipulation of the exact equation for h. However it is brought forth, though, the introduction of one new function and one new equation adds nothing to our knowledge. The use of the equation stems from the hope that c is easier to understand than h, so that it will be easier to guess at the form of c than at the form of h.

For a homogeneous system, $\rho^{(1)}$ is equal to the number density ρ and independent of position; $c(\mathbf{r}_1, \mathbf{r}_2)$, like $h(\mathbf{r}_1, \mathbf{r}_2)$, depends only on the distance $|\mathbf{r}_1 - \mathbf{r}_2|$. The Ornstein–Zernicke equation for a homogeneous one-component system becomes

$$h(r) = c(r) + \rho \int c(|\mathbf{r} - \mathbf{r}'|) h(r')\, d\mathbf{r}' \qquad (7.79)$$

Note that h and c each depend on a single variable, but the integral is over three coordinates. For a typical dense fluid, $h(r)$ shows shell structure, in the form of the damped oscillations discussed previously (Figure 7.12), but $c(r)$, which can be derived from $h(r)$ using (7.79), shows no oscillations. It has the simple shape of Figure 7.11, although it may take on very large negative values (h can never be less than -1). This suggests that one may be able to get a simple approximation to c, and then solve the Ornstein–Zernicke equation to get h. This has been done in some cases.

Equation (7.78) may be solved iteratively for h, starting from the approximation valid for small ρ: $h \approx c$. Substituting this on the right-hand side of (7.78) gives the next approximation:

$$h(\mathbf{r}_1, \mathbf{r}_2) \approx c(\mathbf{r}_1, \mathbf{r}_2) + \int d\mathbf{r}_3\, c(\mathbf{r}_1, \mathbf{r}_3)\, \rho^{(1)}(\mathbf{r}_3) c(\mathbf{r}_3, \mathbf{r}_2)$$

And substituting this for h on the right-hand side of (7.78) gives the next approximation:

$$h(\mathbf{r}_1, \mathbf{r}_2) \approx c(\mathbf{r}_1, \mathbf{r}_2) + \int d\mathbf{r}_3\, c(\mathbf{r}_1, \mathbf{r}_3)\, \rho^{(1)}(\mathbf{r}_3)\, c(\mathbf{r}_3, \mathbf{r}_2)$$

$$+ \iint d\mathbf{r}_3\, d\mathbf{r}_4\, c(\mathbf{r}_1, \mathbf{r}_3)\, \rho^{(1)}(\mathbf{r}_3)\, c(\mathbf{r}_3, \mathbf{r}_4)\, \rho^{(1)}(\mathbf{r}_4)\, c(\mathbf{r}_4, \mathbf{r}_2)$$

and so on. The total correlation between two particles is given as a sum of chains of correlations, involving $0, 1, 2, \ldots$ intermediate particles.

The relation between h and c looks simpler when one takes Fourier transforms [see equations (7.70)–(7.72), and this is often done in finding approximations to the correlation functions. Fourier transforming the

Ornstein–Zernicke equation (7.79) gives

$$\tilde{h}(\mathbf{k}) = \tilde{c}(\mathbf{k}) + \rho \int d\mathbf{r}\, e^{i\mathbf{k}\cdot\mathbf{r}} \int d\mathbf{r}'\, c(|\mathbf{r} - \mathbf{r}'|)\, h(r')$$

$$= \tilde{c}(\mathbf{k}) + \rho \int d\mathbf{r} \int d\mathbf{r}'\, e^{i\mathbf{k}\cdot(\mathbf{r}-\mathbf{r}')}\, e^{i\mathbf{k}\cdot\mathbf{r}'}\, c(|\mathbf{r} - \mathbf{r}'|)\, h(r')$$

For the variable $\mathbf{r}$, substitute $\mathbf{s} = \mathbf{r} - \mathbf{r}'$; the integration over $\mathbf{r}'$ gives the Fourier transform of h and the integration over $\mathbf{s}$ gives the Fourier transform of c. Therefore

$$\tilde{h}(\mathbf{k}) = \tilde{c}(\mathbf{k}) + \rho\, \tilde{c}(\mathbf{k})\, \tilde{h}(\mathbf{k}) \tag{7.80}$$

That the Fourier transform of the integral in (7.79) is a product of the Fourier transforms of the two functions is an example of the convolution theorem of Fourier analysis. Equation (7.80) is an algebraic relation between the Fourier transforms of h and c, and may be rearranged to

$$\left[1 + \rho\tilde{h}(\mathbf{k})\right]\left[1 - \rho\tilde{c}(\mathbf{k})\right] = 1 \tag{7.81}$$

a sort of reciprocal relation between $\tilde{c}(\mathbf{k})$ and $\tilde{h}(\mathbf{k})$.

In principle, h and c may be calculated in terms of the same N-atom potential function (even though the exact calculation is too difficult), so there must be a relation between them. By rather involved analysis of the integrals involved in the exact calculation (one doesn't get anything without effort), it has been shown that, if there are only two-body potentials,

$$c(r) = h(r) - \ln[1 + h(r)] - \beta u(r) + B(r)$$

where B, an infinite collection of difficult integrals, approaches 0 more rapidly than h for large r. This suggests that, for large r, $c(r)$ approaches $-\beta u(r)$, which suggests in turn that $c(r) \approx -\beta u(r)$ is a reasonable approximation. Applying this to the hard-sphere fluid by putting $c(r) = 0$ for $r > \sigma$ (the hard-sphere diameter), and using the fact that $h(r) = -1$ for $r < \sigma$, one can solve the Ornstein–Zernicke equation analytically. One then has $c(r)$ and $h(r)$ for all r, and can get the equation of state from the compressibility equation. The result is

$$\frac{\beta P}{\rho} = \frac{1 + \eta + \eta^2}{(1 - \eta)^3} \tag{7.82}$$

where $\eta = (\pi/6)\rho\sigma^3$ is called the "packing fraction." Using the same $h(r)$ in the virial equation, one gets the equation of state

$$\frac{\beta P}{\rho} = \frac{1 + 2\eta + 3\eta^2}{(1 - \eta)^2} \tag{7.83}$$

Equations (7.82) and (7.83) are not identical; the difference between them, important for very high values of η, reflects the approximate nature of the approximation, $c(r) \approx -\beta u(r)$.

7.6. DEBYE-HÜCKEL THEORY

Fluids composed of charged particles following the laws of classical mechanics are of much interest to chemists. Examples are ions in an aqueous solution, or a plasma at high enough temperature so that quantum statistics need not be invoked. For such fluids, use of the approximation $c(r) \approx -\beta u(r)$ in the Ornstein-Zernicke equation leads to the well-known Debye-Hückel model, as we now show. For the more usual (and more physically based) derivation, one can skip to (7.98).

A stable system containing electrically charged particles must include more than one species. This means there is a set of correlation functions instead of a single h. Call them $h_{st}(r)$, where the indices s and t label the species, and call the corresponding direct correlation functions, $c_{st}(r)$. The Ornstein-Zernicke equation for a homogeneous system, (7.79), becomes

$$h_{st}(r) = c_{st}(r) + \sum_u \rho_u \int c_{su}(|\mathbf{r} - \mathbf{r}'|)\, h_{ut}(r')\, d\mathbf{r}'$$

or

$$h_{st}(r_{12}) = c_{st}(r_{12}) + \sum_u \int c_{su}(r_{13})\, \rho_u\, h_{ut}(r_{23})\, d\mathbf{r}_3 \tag{7.84}$$

Here, ρ_u is the number density of particles of species u. Assuming that the interaction between particles is electrostatic only,

$$u_{st}(r) = \frac{e_s e_t}{\varepsilon r} \tag{7.85}$$

in electrostatic units. Here, e_s is the charge on a particle of species s (it may be + or −) and ε is the dielectric constant. A more realistic interaction potential includes short-range repulsion between particles of all species in addition to the electrostatic interaction; how this core repulsion may be included will be shown later.

Applying the approximation $c(r) \approx -\beta u(r)$ to this system means assuming

$$c_{st}(r) = -\beta \frac{e_s e_t}{\varepsilon r} \tag{7.86}$$

for all s and t and for all r. Electrostatic repulsion keeps particles apart and electrostatic attraction brings particles together. Equation (7.84) must be solved for $\{h_{st}(r)\}$. This is done using Fourier transforms which, in the one-component system, gave us a simple algebraic equation [(7.80) and (7.81)]. Multiplying (7.84) by $e^{i\mathbf{k}\cdot\mathbf{r}_{12}}$ and integrating over $\mathbf{r}_{12}$ (note that $\mathbf{r}_{12} = \mathbf{r}_{13} + \mathbf{r}_{32}$),

$$\begin{aligned}\tilde{h}_{st}(\mathbf{k}) &= \tilde{c}_{st}(\mathbf{k}) + \sum_u \rho_u \int d\mathbf{r}_{12}\, e^{i\mathbf{k}\cdot\mathbf{r}_{12}} \int c_{su}(r_{13})\, h_{ut}(r_{23})\, d\mathbf{r}_3 \\ &= \tilde{c}_{st}(\mathbf{k}) + \sum_u \rho_u\, \tilde{c}_{su}(\mathbf{k})\, \tilde{h}_{ut}(\mathbf{k})\end{aligned} \tag{7.87}$$

These equations are like (7.80), but for a multi-component system. When solved for $h_{st}(\mathbf{k})$ by successive approximations (like what was done with equation 7.78), they lead to a simple result.

First one requires the Fourier transform of c_{st}:

$$\begin{aligned}\tilde{c}_{st}(\mathbf{k}) &= -\beta \int d\mathbf{r}\, e^{i\mathbf{k}\cdot\mathbf{r}} \frac{e_s e_t}{\varepsilon r} = -\beta \int r^2\, dr \sin\theta\, d\theta\, d\phi\, e^{ikr\cos\theta} \frac{e_s e_t}{\varepsilon r} \\ &= \frac{-\beta e_s e_t}{\varepsilon} \int_0^\infty r\, dr\, 4\pi \frac{\sin kr}{kr}\end{aligned}$$

The value of the integral of $\sin kr$ is determined by considering the integral of $e^{-sr} \sin kr$, $k/(s^2 + k^2)$, and letting s go to zero. Then

$$\tilde{c}_{st}(\mathbf{k}) = \frac{-4\pi\beta e_s e_t}{\varepsilon k^2} \tag{7.88}$$

In (7.87), the first approximation to $\tilde{h}_{st}(\mathbf{k})$ is $\tilde{c}_{st}(\mathbf{k})$; inserting $\tilde{c}_{st}(\mathbf{k})$ for $\tilde{h}_{st}(\mathbf{k})$ in the integral gives the second approximation to $\tilde{h}_{st}(\mathbf{k})$; inserting the second approximation in the integral gives the next; and so on. The result is the sum of chains of direct correlations:

$$\begin{aligned}\tilde{h}_{st}(\mathbf{k}) &= \tilde{c}_{st}(\mathbf{k}) + \sum_u \rho_u\, \tilde{c}_{su}(\mathbf{k})\, \tilde{c}_{ut}(\mathbf{k}) \\ &\quad + \sum_u \rho_u \sum_v \rho_v\, \tilde{c}_{su}(\mathbf{k})\, \tilde{c}_{uv}(\mathbf{k})\, \tilde{c}_{vt}(\mathbf{k}) + \cdots\end{aligned}$$

Using (7.88) makes this

$$\begin{aligned}\tilde{h}_{st}(\mathbf{k}) &= \frac{-4\pi\beta e_s e_t}{\varepsilon k^2} + \left[\frac{-4\pi\beta}{\varepsilon k^2}\right]^2 e_s \sum_u \rho_u e_u^2 e_t \\ &\quad + \left[\frac{-4\pi\beta}{\varepsilon k^2}\right]^3 \sum_u \rho_u e_s e_u^2 \sum_v \rho_v e_v^2 e_t + \cdots \\ &= \frac{-4\pi\beta e_s e_t}{\varepsilon k^2}\left[1 - \frac{\kappa^2}{k^2} + \frac{\kappa^4}{k^4} - \cdots\right]\end{aligned}$$

where

$$\kappa^2 = \frac{4\pi\beta}{\varepsilon} \sum_u \rho_u e_u^2 \tag{7.89}$$

The inverse of κ has physical dimensions of length and is referred to as the Debye length.

Recognizing that $1 - x + x^2 \cdots$ is the power series for $1/(1 + x)$, one sees that

$$\tilde{h}_{st}(\mathbf{k}) = \frac{-4\pi\beta e_s e_t}{\varepsilon k^2}\left(1 + \frac{\kappa^2}{k^2}\right)^{-1} \tag{7.90}$$

Inverting the Fourier transform according to (7.71),

$$\begin{aligned}h_{st}(r) &= \frac{1}{(2\pi)^3}\int e^{-i\mathbf{k}\cdot\mathbf{r}} \frac{-4\pi\beta e_s e_t}{\varepsilon(k^2 + \kappa^2)}\,\mathbf{dk} \\ &= \frac{-2\beta e_s e_t}{\pi\varepsilon r}\int_0^\infty dk \frac{k}{k^2 + \kappa^2} \sin kr\end{aligned}$$

With the help of a table of integrals, the final result is

$$h_{st}(r) = \frac{-\beta e_s e_t}{\varepsilon r} e^{-\kappa r} \tag{7.91}$$

This is the direct correlation function $c_{st}(r)$ multiplied by $e^{-\kappa r}$. Looked at another way, it is $-\beta$ multiplied by a screened Coulomb potential; κ^{-1} is the screening length.

Equation (7.91) is the main result of the Debye–Hückel theory. The total and direct correlation functions between an ion of type s and an ion of type t, h_{st} and c_{st}, differ because of other ions, attracted and repelled by the two ions being looked at. The Coulombic interaction between these two ions is

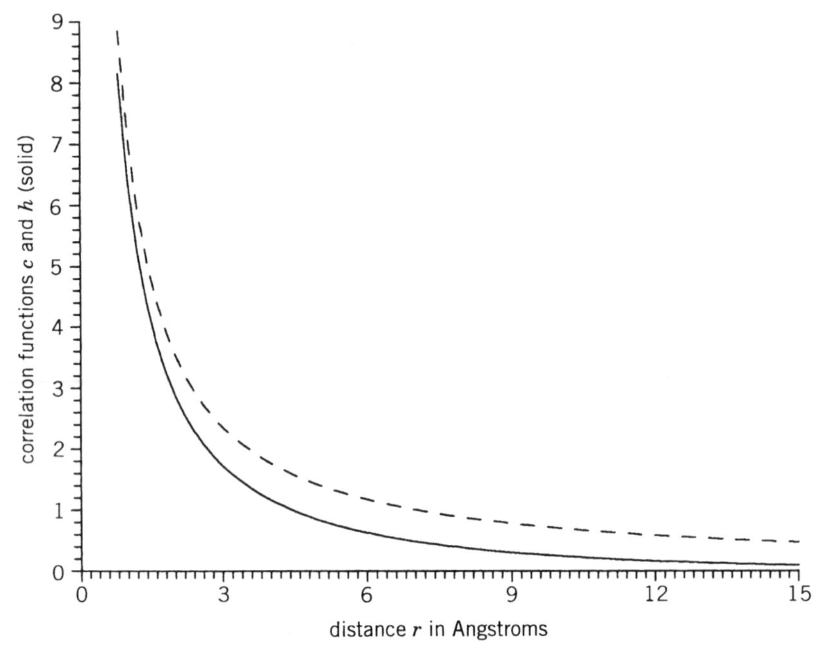

Figure 7.14. Correlation functions between opposite charges in the Debye–Hückel model. The dashed line is the direct correlation function,. $c_{st} = -\beta e_s e_t/\varepsilon r$, for $\beta e_s e_t/\varepsilon = 7.084$ A. (This corresponds to charges of e and $-e$ at temperature 300 K with dielectric constant 78.5.) The solid line is the total correlation function $h_{st} = c_{st}\, e^{-\kappa r}$ for $\kappa = 0.1035$, corresponding to a 1–1 electrolyte at 0.01 M.) Note that c_{st} is $-\beta$ times the Coulombic interaction potential between the ions and h_{st} is $-\beta$ times the screened interaction between the ions.

screened by the other ions. One thinks of the ions around any one ion as constituting its "ionic atmosphere" and κ^{-1} is a measure of the size of the ionic atmosphere, as will be discussed in more detail later. For an 0.1 M solution of a 1–1 electrolyte in water at 20°C, $\kappa^2 = 1.072 \times 10^{14}$ cm^{-2} ($\varepsilon = 80.4$, $\rho_+ = \rho_- = 6.02 \times 10^{19}$ cm^{-3}). The direct and total correlation functions for ions of opposite signs are shown in Figure 7.14 (the Coulomb and screened potentials are just the correlation functions divided by $-\beta$).

The virial equation of state (7.58), with the interparticle potentials (7.85), is

$$P = \sum_s \frac{N_s kT}{V} - \sum_{st} \frac{2\pi \rho_s \rho_t}{3} \int dr r^3 [h_{st}(r) + 1] \frac{e_s e_t}{(-\varepsilon r^2)}$$

Overall electroneutrality, expressed by

$$\sum_s \rho_s e_s = 0 \tag{7.92}$$

means that the second term in brackets makes no contribution, so, with (7.91),

$$\beta P = \sum_s \rho_s - \frac{2\pi}{3} \sum_s \sum_t \frac{\beta^2 \rho_s \rho_t e_s^2 e_t^2}{\varepsilon^2} \int dr e^{-\kappa r}$$

$$= \sum_s \rho_s - \frac{2\pi}{3} \left(\frac{\kappa^2}{4\pi} \right)^2 \kappa^{-1} \tag{7.93}$$

The first term is the ideal-gas contribution to the pressure. The correction to ideality, due to the interparticle interactions, is negative, showing that the net electrostatic force is attractive. It will be noted that the correction is proportional to κ^3, that is, to the 3/2 power of ionic densities. This means that there is no virial expansion of the pressure, since $\rho^{3/2}$ cannot be written as a power series in ρ. The underlying reason is the long range of the Coulomb interaction, which makes it impossible to consider interaction between two, three,... molecules, which lead to terms in $\rho^2, \rho^3, \ldots$ One might say that each molecule feels the presence of all others, even when ρ is small.

The interaction potentials, equation (7.85), are missing the short-range repulsions that must be present between all ions. With a hard-sphere repulsion added to the Coulomb interaction, the model is not much more complicated, and can be treated similarly to (7.86) to (7.93). However, it is more interesting to add the hard-sphere repulsion to a derivation of the Debye–Hückel equations, which emphasizes the physical assumptions involved. This is done later, starting with (7.98).

First, we consider some general properties of the correlation functions, in particular the nth moment of the total correlation functions, defined as

$$M_n \equiv \sum_{st} \rho_s e_s \rho_t e_t \int \mathbf{dr}\, h_{st}(r) r^n \tag{7.94}$$

For the simple Debye–Hückel theory,

$$M_0 = \sum_{st} \rho_s e_s \rho_t e_t 4\pi \int dr r^2 \left(\frac{-\beta e_s e_t}{\varepsilon r} \right) e^{-\kappa r}$$

$$= \frac{-\varepsilon}{4\pi\beta} \kappa^4 \int_0^\infty dr r e^{-\kappa r} = -\sum_s \rho_s e_s^2$$

This result reflects the local electroneutrality condition

$$\sum_t \int \mathbf{dr}\, h_{st}(r) e_t \rho_t = -e_s \tag{7.95}$$

that each ion's atmosphere has a net charge equal and opposite to the ion's charge. Thus it can be expected that it still holds in the presence of short-range repulsions. This is indeed the case (Problem 7.31). Remarkably, the value of the second moment M_2 is also independent of the short-range repulsions, as will be shown in the next two paragraphs.

The moment M_n is proportional to the coefficient of k^n in the power series for $\tilde{H}(\mathbf{k})$, the Fourier transform of

$$H(r) = \sum_{st} e_s \rho_s e_t \rho_t h_{st}(r)$$

To see this, write

$$\begin{aligned}
\tilde{H}(\mathbf{k}) &= 4\pi \int_0^\infty dr\, r^2 \frac{\sin kr}{kr} H(r) \\
&= 4\pi \int_0^\infty dr\, r^2 \sum_j (-1)^j \frac{(kr)^{2j}}{(2j+1)!} \sum_{st} e_s \rho_s e_t \rho_t h_{st}(r) \\
&= \sum_j (-1)^j \frac{k^{2j}}{(2j+1)!} M_{2j}
\end{aligned}$$

The odd moments do not appear. Since it has been assumed that c_{st} is proportional to the interaction potential between particles of species s and t, the short-range repulsions should give rise to an additive term in c_{st}. Then

$$c_{st}(r) = -\beta \frac{e_s e_t}{\varepsilon r} + d_{st}(r)$$

and the Fourier transform is

$$\tilde{c}_{st}(\mathbf{k}) = \frac{-4\pi\beta e_s e_t}{\varepsilon k^2} + \tilde{d}_{st}(\mathbf{k})$$

The expansion of $\tilde{d}_{st}(k)$,

$$\tilde{d}_{st}(k) = \sum_j d_{stj} k^{2j}$$

starts with k^0 because the repulsions are short range (only the long-range Coulomb interaction can give a term in k^{-2}). It includes only even powers of k because $\tilde{d}_{st}(k)$ depends on k and not $\mathbf{k}$. The function $\tilde{h}_{st}(k)$ has a similar expansion to $\tilde{d}_{st}(k)$.

The Ornstein–Zernicke equation (7.87) is now multiplied by $e_s \rho_s e_t \rho_t$ and summed over s and t to give

$$\tilde{H}(k) = -4\pi\beta \sum_{st} e_s^2 e_t^2 \frac{\rho_s \rho_t}{\varepsilon k^2} + \sum_{st} e_s e_t \rho_s \rho_t \sum_{j\geq 0} d_{stj} k^{2j}$$

$$- \frac{4\pi\beta}{\varepsilon k^2} \sum_{stu} e_s^2 e_t e_u \rho_s \rho_t \rho_u \tilde{h}_{ut}$$

$$+ \sum_{stu} e_s e_t \rho_s \rho_t \rho_u \tilde{d}_{su}(k) \tilde{h}_{ut}(k) \qquad (7.96)$$

The terms in k^{-2} must cancel, so that

$$\sum_{st} e_s^2 e_t^2 \rho_s \rho_t = -\sum_s e_s^2 \rho_s M_0$$

which gives immediately the condition on M_0. Now, equating the terms in k^0 on the left- and right-hand sides of (7.96) gives

$$M_0 = \sum_{st} e_s e_t \rho_s \rho_t d_{st0} - \frac{4\pi\beta}{\varepsilon} \sum_{stu} e_s^2 e_t e_u \rho_s \rho_t \rho_u h_{ut1}$$

$$+ \sum_{stu} e_s e_t \rho_s \rho_t \rho_u d_{su0} h_{ut0} \qquad (7.97)$$

In the last sum, h_{ut0} is equal to the value of $\tilde{h}_{ut}(\mathbf{k})$ for $k = 0$ so that

$$\sum_t \rho_t e_t h_{ut0} = \sum_t \rho_t e_t \int \mathbf{dr}\, h_{ut}(r)$$

which is $-e_u$ according to (7.95). Then (7.97) becomes

$$M_0 = \sum_{st} e_s e_t \rho_s \rho_t d_{st0} - \frac{4\pi\beta}{\varepsilon} \sum_s e_s^2 \rho_s \left(\frac{-M_2}{3!} \right)$$

$$+ \sum_{su} e_s \rho_s \rho_u d_{su0}(-e_u)$$

and the short-range terms cancel, leaving the Stillinger–Lovett second-moment condition:

$$M_2 = \frac{-6\varepsilon}{4\pi\beta}$$

This condition, like that on M_0, is a consequence of the long range of the Coulombic interactions. It holds regardless of the form or strength of the short-range interactions.

In deriving the Debye–Hückel model from physical assumptions, we introduce the concept of "potential of mean force," which reflects interparticle correlations. The potential of mean force for species s and t, w_{st}, is defined by

$$g_{st}(r) = e^{-\beta w_{st}(r)} \tag{7.98}$$

To understand its significance and why it is called the potential of mean force, note that $\rho_s \rho_t g_{st}(r)$ is the probability of finding a particle of species s and a particle of species t a distance r apart. According to (7.56),

$$\rho_s \rho_t g_{st}(|\mathbf{r} - \mathbf{r}'|) = \frac{N_s(N_t - \delta_{st})}{Z_N} \int \mathbf{d}^{3N}q\, \delta(\mathbf{q}_{sj} - \mathbf{r})\, \delta(\mathbf{q}_{tk} - \mathbf{r}')\, e^{-\beta U_N}$$

Take the gradient with respect to $\mathbf{r}$ and integrate the integral by parts:

$$\rho_s \rho_t \nabla g_{st}(|\mathbf{r} - \mathbf{r}'|) = \frac{-N_s(N_t - \delta_{st})}{Z_N} \times \int \mathbf{d}^{3N}q\, \delta(\mathbf{q}_{sj} - \mathbf{r})\, \delta(\mathbf{q}_{tk} - \mathbf{r}')(-\beta \nabla U_N)\, e^{-\beta U_N}$$

Using (7.98), this is

$$\rho_s \rho_t g_{st} \nabla w_{st}(|\mathbf{r} - \mathbf{r}'|) = \frac{N_s(N_t - \delta_{st})}{Z_N} \times \int \mathbf{d}^{3N}q\, \delta(\mathbf{q}_{sj} - \mathbf{r})\, \delta(\mathbf{q}_{tk} - \mathbf{r}')(-\nabla U_N)\, e^{-\beta U_N}$$

The right-hand side is the average force exerted by the particles of species t located at $\mathbf{r}'$ on the particles of species s at $\mathbf{r}$. Thus the gradient of w_{st} is the mean force exerted by a particle of species t located at $\mathbf{r}'$ on a particle of species s located at $\mathbf{r}$, and w_{st} is the potential giving the mean force. All correlations between particles are included; the number of particles of any species at point $\mathbf{r}'$ depends on how many particles of each species s are found at $\mathbf{r}$.

In the Debye–Hückel model for particles interacting by Coulomb forces, the potential of mean force is approximated by the charge of a particle multiplied by the electrostatic potential generated by the particle and the particles around it. The electrostatic potential is related to the *average* distribution of these particles, so that some of the correlations between particles are not included in the model. With Φ^s the electrostatic potential around a particle of species s located at $r = 0$, the Poisson equation is

$$\frac{1}{r^2}\frac{d}{dr}\left(r^2\frac{d\Phi^s}{dr}\right) = \frac{-4\pi}{\varepsilon}\,\omega^s(r)$$

The charge density at a distance r is

$$\omega^s(r) = \sum_t e_t\,\rho_t g_{st}(r) \tag{7.99}$$

where ρ_t is the average density of particles of species t.

Using (7.98) for g_{st} and approximating w_{st} by $e_t\Phi^s$, the Poisson equation becomes

$$\frac{1}{r^2}\frac{d}{dr}\left(r^2\frac{d\Phi^s}{dr}\right) = \frac{-4\pi}{\varepsilon}\sum_t e_t\,\rho_t\exp(-\beta\Phi^s e_t)$$

$$= \frac{-4\pi}{\varepsilon}\sum_t e_t\,\rho_t[1 - \beta\Phi^s e_t + \cdots]$$

The first term from the expansion of the exponential gives no contribution because of the electroneutrality condition (7.92). The contribution of terms after the second are neglected (the third term, $\sum \rho_t e_t^3$, is zero for a symmetrical binary electrolyte and probably small for other cases). Then

$$\frac{1}{r^2}\frac{d}{dr}\left(r^2\frac{d\Phi^s}{dr}\right) = \frac{4\pi\beta}{\varepsilon}\sum_t e_t^2\rho_t\Phi^s = \kappa^2\Phi_s \tag{7.100}$$

[see (7.89)]. Equation (7.100) is often called the Poisson–Boltzmann equation, since it comes from the Poisson equation plus the approximation of $g_{st}(r)$ by $\exp[-\beta e_t\Phi^s(r)]$, which implies that ions of type t follow a Boltzmann distribution in the potential Φ_s.

The general solution to (7.100) is that $r\Phi$ must be a linear combination of $e^{\kappa r}$ and $e^{-\kappa r}$. Since Φ must approach 0 for large r, $e^{\kappa r}$ must be rejected. Also, Φ^s must approach $e_s/\varepsilon r$ (the potential of the central ion itself) for small r. Therefore

$$\Phi^s = \frac{e_s}{\varepsilon r}\,e^{-\kappa r} \tag{7.101}$$

which is the screened Coulomb potential. This is the potential which determines the densities of ions around the central ion.

The charge density ω^s around the central ion of species s may be obtained by differentiating Φ^s according to the Poisson equation. The result is

$$\omega^s(r) = \frac{-\varepsilon\kappa^2}{4\pi} \frac{e_s}{\varepsilon r} e^{-\kappa r} \tag{7.102}$$

This charge density is the ionic atmosphere referred to earlier. By integrating (7.102), one finds (see Problem 7.23) that the total charge in the ionic atmosphere is $-e_s$, the negative of the charge of the central ion. The size of the ionic atmosphere, as measured by the average value of r with $\omega^s(r)$ as the weighting factor, is κ^{-1}, confirming our interpretation of the Debye length. Note also that the pair correlation function g_{st}, equal to $e^{-\beta w_{st}}$, is being approximated as $\exp[-\beta e_t \Phi_s]$ which, keeping the leading terms, becomes

$$g_{st}(r) \approx 1 - \beta e_t \Phi_s = 1 - \frac{\beta e_s e_t}{\varepsilon r} e^{-\kappa r}$$

confirming equation (7.91) for $h_{st}(r)$.

The total electrostatic energy per unit volume is calculated using the above g_{st}, the first term (1) making no contribution:

$$\frac{E^{\text{exc}}}{V} = \frac{1}{2} \sum_{st} \rho_s \rho_t \int \mathbf{dr}\, g_{st}(r)\, u_{st}(r)$$

$$= \frac{-2\pi\beta}{\varepsilon^2} \sum_{st} \rho_s e_s^2 \rho_t e_t^2 \int_0^\infty dr\, e^{-\kappa r} = \frac{-\kappa^3}{8\pi\beta} \tag{7.103}$$

or 3 times excess pressure of (7.93). From (7.103) one can obtain the partition function, from which other thermodynamic properties are obtainable. Since

$$\frac{\partial \ln Q}{\partial \beta} = -E^{\text{exc}} - \frac{3N}{2\beta}$$

where $N = \Sigma_s N_s$ = total number of ions, integration from β_0 to β gives

$$\ln Q_N - \ln Q_N^0 = \int_{\beta_0}^{\beta} d\beta \left(\frac{-3N}{2\beta} + \frac{V\kappa^3}{8\pi\beta} \right) = \frac{-3N}{2} \ln\left(\frac{\beta}{\beta_0}\right) + \frac{V}{8\pi} \frac{\kappa^3 - \kappa_0^3}{3/2}$$

Note that κ is proportional to $\beta^{1/2}$, so κ^3/β is proportional to $\beta^{1/2}$. As $\beta_0 \to 0$ (infinite temperature), correlations arising from interionic interactions become unimportant and the collection of ions behaves like an ideal gas, so Q_N^0 is the ideal-gas partition function [see (7.6)] and

$$\ln Q_N = N \ln V - \sum_s \ln\left(\Lambda_s^{3N_s} N_s!\right)_{\beta_0} - \frac{3N}{2} \ln\left(\frac{\beta}{\beta_0}\right) + \frac{V\kappa^3}{12\pi}$$

$$= N \ln V - \sum_s \left(\frac{3N_s}{2} \ln \frac{h^2}{2\pi m_s} + \ln N_s!\right) - \frac{3N}{2} \ln \beta + \frac{V\kappa^3}{12\pi} \quad (7.104)$$

The free energy is

$$A = -\beta^{-1} \ln Q_N$$

$$= \beta^{-1} \sum_s \left(\frac{3N_s}{2} \ln \frac{h^2\beta}{2\pi m_s} + \ln N_s!\right) - \beta^{-1}\left[\frac{V\kappa^3}{12\pi} + N \ln V\right]$$

with $V\kappa^3/12\pi\beta$ obviously being the contribution of interionic interactions.

The chemical potential (per mole of ions) of species s is therefore

$$\mu_s = \mu_s^{\text{ideal}} - \frac{\mathcal{N}_A V}{12\pi\beta} \frac{\partial}{\partial N_s}\left[\frac{4\pi\beta}{\varepsilon} \sum_t \frac{e_t^2 N_t}{V}\right]^{3/2}$$

$$= \mu_s^{\text{ideal}} - \frac{\mathcal{N}_A \kappa e_s^2}{2\varepsilon} \quad (7.105)$$

with $\mathcal{N}_A$ being Avogadro's number. It is usual to write a chemical potential as the chemical potential for the ideal system plus $RT \ln \gamma_s$ with γ_s the activity coefficient. Equation (7.105) thus gives the Debye–Hückel limiting law for activity coefficients:

$$\ln \gamma_s \to -\beta \frac{\kappa e_s^2}{2\varepsilon} \quad (7.106)$$

as $\kappa \to 0$. One does not get ideal behavior for ion s when ρ_s approaches zero, since all ions contribute to the nonideality by affecting the interionic interaction.

The short-range, nonelectrostatic interactions between the ions may easily be included in the model, if represented by hard-sphere repulsions. Only for oppositely charged ions are they of importance, since ions of the same sign of charge are kept apart by electrostatic repulsion. Thus it is a good approximation to use the same hard-sphere radius for all ions, so that $h_{st} = -1$ for $r < \sigma$, the hard-sphere diameter; the value of σ is appropriate for the interaction of oppositely charged ions. The Poisson equation for $r < \sigma$ is simply

$$\frac{1}{r^2}\frac{d}{dr}\left(r^2 \frac{d\Phi^s}{dr}\right) = 0$$

(there is no charge density for r between 0 and σ). The solution is $\Phi^s = C - Dr^{-1}$. For $r \geq \sigma$, the Poisson equation is identical to (7.100), with the identical solution, $\Phi^s = Ar^{-1}e^{-\kappa r}$.

The values of C and D are obtained by demanding that Φ^s and $d\Phi^s/dr$ be continuous at $r = \sigma$. This gives $D = (1 + \kappa\sigma)Ae^{-\kappa\sigma}$ and $C = -\kappa Ae^{-\kappa\sigma}$. To determine A, Φ^s must approach $e_s/\varepsilon r$ for small r. (One might think that since no ion can approach more closely than $r = \sigma$ the same should hold for solvent molecules and, the solvent molecules being responsible for the dielectric constant's differing from unity, Φ^s should approach e_s/r for $r < \sigma$. However, the ions envisaged by this theory are solvated; σ is the outer limit of a solvent sphere, and the dielectric constant should be ε for $r < \sigma$.) Thus $D = e_s/\varepsilon$ and

$$A = \frac{e_s e^{\kappa\sigma}}{\varepsilon(1 + \kappa\sigma)} \tag{7.107}$$

Then

$$C = \frac{-\varepsilon_s \kappa}{\varepsilon(1 + \kappa\sigma)}$$

C is the contribution of the ion atmosphere to the electrostatic potential for $r < \sigma$ (D/r is the contribution of the central ion itself).

The total charge of the ionic atmosphere is easily shown to be $-e_s$. From the expression for C, one can show that the effective radius of the ionic atmosphere is $1/\kappa$ (Problem 7.25). Thus $\kappa\sigma$ is small compared to 1 when the size of the ionic atmosphere is much greater than the size of an ion (for low ionic concentrations), and the expressions for thermodynamic and other properties go over to the expressions for point ions [(7.102)–(7.106)].

Explicitly,

$$g_{st} = e^{-\beta e_t \Phi^s} \approx 1 - \beta e_t A r^{-1} e^{-\kappa r}$$

for $r > \sigma$ and $g_{st} = 0$ for $r \leq \sigma$; A is given by (7.107). The electrostatic interaction energy is therefore [see (7.103)]

$$\frac{E^{\text{exc}}}{V} = \frac{1}{2} \sum_{st} \rho_s \rho_t \int_\sigma^\infty \mathbf{dr} \left[1 - \frac{\beta e_t e_s e^{\kappa(\sigma - r)}}{\varepsilon(1 + \kappa\sigma) r} \right] \frac{e_s e_t}{\varepsilon r}$$

$$= \frac{-2\pi\beta}{\varepsilon^2(1 + \kappa\sigma)} \sum_{st} \rho_s e_s^2 \rho_t e_t^2 \int_\sigma^\infty dr\, e^{\kappa(\sigma - r)}$$

$$= \frac{-\kappa^3}{8\pi\beta(1 + \kappa\sigma)} \tag{7.108}$$

Like (7.103), this is integrated with respect to β to get Q_N. Writing κ^2 as $\alpha^2\beta$, we get

$$\ln Q_N - \ln Q_N^0 + \frac{3N}{2} \ln \frac{\beta}{\beta_0} = \int_0^\beta d\beta \frac{V\kappa^3}{8\pi\beta(1 + \kappa\sigma)}$$

$$= \frac{V}{8\pi} \int_0^{\alpha\sqrt{\beta}} d\kappa\, 2\kappa \frac{\kappa^3}{\kappa^2(1 + \kappa\sigma)}$$

$$= \frac{V}{4\pi} \left(\frac{\kappa^2\sigma^2 - 2\kappa\sigma + 2\ln(1 + \kappa\sigma)}{2\sigma^3} \right)$$

From this

$$A^{\text{exc}} = A - A^{\text{ideal}} = \frac{-V}{4\pi\beta} \left(\frac{\kappa^2\sigma^2 - 2\kappa\sigma + 2\ln(1 + \kappa\sigma)}{2\sigma^3} \right) \tag{7.109}$$

and, differentiating with respect to the number of moles of species s,

$$\mu_s - \mu_s^{\text{ideal}} = \frac{\mathcal{N}_A V}{4\pi\beta\sigma^3} \left[\kappa\sigma^2 - \sigma + \frac{\sigma}{1 + \kappa\sigma} \right] \left(\frac{\partial\kappa}{\partial N_s} \right)_{\beta V \{N_t\}}$$

with $\mathcal{N}_A$ = Avogadro's number. Writing this as $RT \ln \gamma_s$ we have

$$RT \ln \gamma_s = \frac{-\mathcal{N}_A V}{4\pi\beta\sigma^3} \frac{\kappa^2\sigma^3}{1 + \kappa\sigma} \frac{2\pi\beta e_s^2}{\kappa\varepsilon V}$$

so that

$$\ln \gamma_s = \frac{-\kappa\beta e_s^2}{2\varepsilon(1 + \kappa\sigma)} \tag{7.110}$$

As do the other formulas, this goes over to the formula for point ions when $\kappa\sigma$ approaches 0.

Actually, the activity coefficient of a single ion is of little interest, because it is not possible to add appreciable numbers of a single charged species to a system and maintain stability. One always has neutral electrolytes which may dissociate to positively and negatively charged species, preserving electroneutrality. The activity coefficients of interest are averages over the dissociation products. For the binary electrolyte $A_{\nu+}B_{\nu-}$ which dissociates into ν_+ ions of charge q_+e and ν_- ions of charge q_-e, electroneutrality requires $\nu_+q_+ + \nu_-q_- = 0$. The chemical potential of the *electrolyte* at nominal molality m is

$$\begin{aligned}\mu_{\text{el}} &= \nu_+\mu_A + \nu_-\mu_B \\ &= \nu_+\mu_A^0 + \nu_-\mu_B^0 + \nu_+RT\ln(\gamma_+\nu_+m) + \nu_-RT\ln(\gamma_-\nu_-m) \\ &= \mu_{\text{el}}^0 + \nu_+RT\ln\nu_+m + \nu_-RT\ln\nu_-m + \nu RT\ln\gamma_\pm\end{aligned}$$

where $\nu = \nu_+ + \nu_-$ and

$$\gamma_\pm = (\gamma_+\nu_+)^{\nu_+/\nu}(\gamma_-\nu_-)^{\nu_-/\nu}$$

is called the mean ionic activity coefficient. (The standard state is usually taken as the ideal solution at unit molality). Using (7.110) for the individual ionic activity coefficients,

$$\ln \gamma_\pm = \frac{-\kappa\beta e^2}{2\varepsilon\nu(1 + \kappa\sigma)}\left[\nu_+q_+^2 + \nu_-q_-^2\right]$$

With the electroneutrality condition, the square bracket may be written as $(-\nu_-q_-)q_+ + (-\nu_+q_+)q_- = -\nu q_+q_-$. Note that $\ln\gamma_\pm$ is always negative.

The inverse Debye length κ is proportional to the square root of the ionic strength, defined as

$$I = \frac{1}{2}\sum_s e_s^2 c_s \tag{7.111}$$

The ionic strength depends on the concentrations of all the ions in solution, whether or not they come from the electrolyte being considered. For a solution containing only one electrolyte, I is proportional to its molality and

$$\ln \gamma_{\pm} = -|q_{+}q_{-}|\, A\sqrt{m}\,\left(1 + B\sqrt{m}\,\right)^{-1} \tag{7.112}$$

with A and B constants. Thus a plot of $\ln \gamma_{\pm}$ vs. $\sqrt{m}$ becomes linear as m approaches zero.

PROBLEMS

7.1. Repeat the calculation of $(\partial \ln Z/\partial V)$, performed between (7.10) and (7.11), assuming the volume V is not a cube, but a right-angled prism, with edge lengths M, N, and P ($MNP = V$). The scaling must be different in different directions: each x-coordinate, say x_i, should be written as Mt_i, with $0 \le t_i \le 1$; each y_i should be written as Nu_i with $0 \le u_i \le 1$, and each z_i should be written as Pv_i with $0 \le v_i \le 1$.

7.2. Show that, when the series

$$z = \rho + \rho^2\left(V - Z_2V^{-1}\right) + \rho^3\left(2Z_2^2V^{-2} - \tfrac{5}{2}Z_2 + V^2 - \tfrac{1}{2}Z_3V^{-1}\right) + \cdots$$

is substituted into (7.18) (a series for ρ in terms of z), the result is an identity. Also, by substituting the above series in (7.16), find the virial coefficients B_2 and B_3.

7.3. By differentiating (7.5) with respect to T and assuming U_N is independent of T, one gets the excess energy

$$E_{\text{exc}} = \left(Z_N\right)^{-1} \int d^{3N}q U_N\, e^{-\beta U_N}$$

Prove that, even if U_N depends on T, one still has E_{exc} equal to the average value of $[\partial(\beta U_N)/\partial\beta]$.

7.4. A very accurate equation of state for a hard-sphere fluid, derived partly theoretically and partly empirically, is the Carnahan–Starling equation

$$\frac{\beta P}{\rho} = \frac{1 + \eta + \eta^2 - \eta^3}{(1-\eta)^3}; \qquad \eta = \frac{\pi\rho}{6}(r_1)^3$$

with r_1 the hard-sphere diameter. What are the virial coefficients $B_2 \cdots B_6$ for the Carnahan–Starling equation of state?

7.5. B_2 and B_3 for the hard-sphere fluid are $4v$ and $10v^2$ respectively, with v the volume of an atomic hard sphere, $v = \pi r_1^3/6$. Suppose $T = 300$ K and the volume of a mole of molecules is 10 cm^3 (the right size for the heavier rare gases). What are v and r_1? At what pressure will the term $B_2 \rho^2$ be 0.1ρ (10% of the leading term in the virial series)? At what pressure will the term $B_3 \rho^3$ be 0.1ρ? (Since only approximate values are required, one can calculate the densities first and use the ideal-gas equation to get the pressures.)

7.6. The Joule–Thomson coefficient is

$$\mu = \left(\frac{\partial T}{\partial P}\right)_H - V = (C_P)^{-1}\left[T\left(\frac{\partial V}{\partial T}\right)_P - V\right]$$

Derive a virial expansion for the quantity in square brackets in terms of the virial coefficients $\{B_i\}$.

7.7. Many pair potentials are of the form

$$u(r) = Dv(r/\sigma)$$

involving only two parameters, one giving the strength of the interaction and the other the range or scale of distance. This holds for the Lennard–Jones or 6–12 potential [equation (7.23)] and the square-well potential [equation (7.24)]. One might hope that the same pair potential describes a number of different atoms or molecules, with the only difference between different atoms being that they take different values of the two parameters. Show that, for atoms or moecules interacting according to $u(r) = Dv(r/\sigma)$, the partition function may be written as

$$Q_N = (m\sigma^2 D)^{3N/2}\Phi(\phi, \tau)$$

where Φ is the same function, for all the atoms or molecules, of the "reduced volume" $\phi = V/\sigma^3$ and the "reduced temperature" $\tau = kT/D$. Then show that the equation of state may be written as

$$\pi = -\tau(\partial\Phi/\partial\phi)$$

where $\pi = \sigma^3 P/D$ is called the "reduced pressure." This is the principle of corresponding states.

7.8. In the pair potential $u(r) = ar^n$, the constants a and n may be positive or negative. For which signs of a and n will the virial of this potential be positive? For which signs of a and n will the virial be negative?

7.9. The square-well potential [equation (7.24)] may be used to represent a more realistic potential, such as the Lennard–Jones or Morse potential. (For representing the Lennard–Jones potential [equation (7.23)], it is recommended to use the Lennard–Jones σ for r_1, 1.8σ for r_2, and 0.56ε for D.) The square-well parameters for three molecules are

molecule	$r_1(A)$	r_2/r_1	D/k
Ar	3.162	1.85	69.4 K
CCl_3F	4.534	1.545	399 K
CO_2	3.917	1.83	119 K

(a) Calculate B_2 as a function of T for the three molecules. Only for the hard-sphere (infinite) repulsion is B_2 independent of temperature.

(b) The temperature at which $B_2 = 0$ is called the Boyle temperature. At its Boyle temperature, a gas behaves like an ideal gas (P is proportional to ρ) for the largest range of pressure because, in some sense, attractive and repulsive forces balance each other. Find the Boyle temperature for each of the three molecules. (*Answer:* the Boyle temperature is given by $1/kT_B = -D^{-1}\ln(1 - (r_1/r_2)^3)$.)

7.10. Calculate the virial coefficients B_2 and B_3 for the van der Waals equation, (7.34). Represent $2\pi \int_{r_1}^{\infty} dr r^2 u^a(r)$ by the constant a.

7.11. The potential $u = ar^{-m}$ is sometimes called a point center of repulsion (if a is positive) or attraction (if a is negative). Show that, for $m > 3$,

$$B_2 = \frac{2\pi}{3}\Gamma\left(\frac{(m-3)}{3}\right)(\beta a)^{3/m}$$

(*Hint:* Integrate (7.28) for B_2 by parts—the resulting expression will show why m must be greater than 3—and use the integral

$$\int_0^{\infty} dx x^n \exp[-bx^p] = \frac{-\Gamma\left(\dfrac{n+1}{p}\right)}{|p|\, a^{[(n+1)/p]}}$$

where Γ is the gamma-function.)

7.12. The second virial coefficient B_2 for the Lennard–Jones potential (7.23) may be calculated analytically (see Problem 7.13), but it involves an infinite series. However, a reasonably good approximation to B_2 is obtained by writing u as $u_{12} + u_6$ (the repulsive and attractive parts) and, in the integral

$$\int_0^\infty dr\, r^2(1 - e^{-\beta u}),$$

substituting $(1 - e^{-\beta u_{12}} + 1 - e^{-\beta u_6})$ for $1 - e^{-\beta u}$. Then the integral formula of Problem 7.11 can be used. The approximation is best for high temperatures, when $1 - e^{-\beta u} \simeq \beta u$.

(a) Show that, with the approximation above,

$$B_2 = \frac{2\pi}{3}\sigma^3\left[\left(\frac{4D}{kT}\right)^{1/4}\Gamma\left(\frac{3}{4}\right) - \left(\frac{4D}{kT}\right)^{1/2}\Gamma\left(\frac{1}{2}\right)\right]$$

(b) Note that this is temperature dependent. What is the Boyle temperature?

7.13. To get B_2 as a function of temperature for the Lennard–Jones potential without approximation, one introduces the reduced temperature, $t = kT/D$, and the reduced distance, $x = r/\sigma$. Then integrating (7.28) by parts gives

$$B_2 = -2\pi\int_0^\infty dr\,\frac{r^3}{3}\,e^{-\beta u}\,\beta\,\frac{du}{dr}$$

$$= \frac{-8\pi\beta\sigma^3 D}{3}\int_0^\infty dx\,\exp[-4t^{-1}(x^{-12} - x^{-6})]\left(\frac{-12}{x^{10}} + \frac{6}{x^4}\right)$$

Now $\exp[4x^{-6}/t]$ is expanded in a power series, and the result is integrated term by term using the integral from Problem 7.11. Then

$$B_2 = \frac{16\pi\sigma^3}{t}\int_0^\infty dx\sum_{j=0}^\infty\frac{(4t^{-1}x^{-6})^j}{j!}\exp[-4t^{-1}x^{-12}]\left(\frac{2}{x^{10}} - \frac{1}{x^4}\right)$$

$$= \frac{4\pi\sigma^3}{3t}\sum_{j=0}^\infty\frac{(4t^{-1})^j}{j!}\left(\frac{2\Gamma((3+2j)/4)}{(4t^{-1})^{(3+2j)/4}} - \frac{\Gamma((1+2j)/4)}{(4t^{-1})^{(1+2j)/4}}\right)$$

where $\Gamma(x)$ is the gamma function. After some algebra, this can be rearranged to

$$B_2 = \frac{-2\pi\sigma^3}{3}\sum_{j=0}^\infty\frac{t^{-(2j+1)/4}}{4j!}\Gamma\left(\frac{2j-1}{4}\right)2^{(2j+1)/2}$$

Unless t is very small, only a few terms of the series are needed.

(a) Calculate the Boyle temperature for the Lennard–Jones potential.

(b) At what temperature is B_2 a maximum? (The first two terms of the series should be enough to do this calculation.) Assuming that the kinetic energy of a molecule along the direction of approach to another molecule is $kT/2$, what energy corresponds to this temperature, and what distance of closest approach?

7.14. Calculate the potential energy (total energy minus kinetic energy) for a van der Waals gas, using the configurational integral $Z^{hs} = (V - 4vN)^N$ for the hard-sphere gas [see equations (7.33) and following] and the approximations we used in obtaining the pressure [equations (7.33)–(7.36)].

7.15. Graph the Mayer f-function for the square-well potential of (7.24).

7.16. Figure 7.7 shows, for four-particle clusters, twelve of the sixteen clusters with three bonds and eight of the fifteen clusters with four bonds. Draw diagrams for the remaining four clusters with three bonds, for the remaining seven clusters with four bonds, and for the six clusters with five bonds. There is of course only one cluster with six bonds. There are also four ways to draw three bonds which do not produce a four-particle cluster. Draw diagrams for these.

7.17. From (7.41), we derived the density as a power series in the activity for a fluid with pair interactions:

$$\rho = \sum_{n \geq 1} nb_n z^n$$

The $\{b_n\}$ are the cluster integrals. Assume that z may be written as a power series in ρ. Substitute the series into the foregoing equation to find the first four coefficients in this power series. Then, substituting this series in (7.41), derive the virial series (7.42) for the pressure.

7.18. The difference between an irreducible cluster of order n and a reducible cluster of order n is that the contribution of an irreducible cluster to the cluster integral can not be written as a product of integrals over fewer than $n - 1$ coordinates. Thus integrating $f_{12}f_{13}f_{23}$ over $\mathbf{r}_1$, $\mathbf{r}_2$, and $\mathbf{r}_3$ leads to an integral over $\mathbf{r}_{12}$ and $\mathbf{r}_{13}$ but integrating $f_{12}f_{13}$ leads to a product of an integral over $\mathbf{r}_{12}$ and an integral over $\mathbf{r}_{13}$. Show that each type of reducible cluster for $n = 4$ leads to a product of integrals, each of which involves fewer than three coordinates.

7.19. Calculate the compressibility κ_T for the hard-sphere equation of state (7.32):

$$P = kT\rho/(1 - 4v\rho)$$

and show that $\kappa_T \rho/\beta$ is equal to $(1 - 4v\rho)^2$ where v is the volume of an atom, $(4\pi/3)(r_1/2)^3$ with r_1 the hard-sphere diameter. Then, using the compressibility equation of state (7.66), show that $\int_{r_1}^{\infty} (g - 1)r^2\, dr$ is equal to $(2r_1^3/3)\rho v$, so that it becomes less important at lower densities.

7.20. By combining the virial series (7.14) with the compressibility equation of state, derive a formula for $\int h(r)\,\mathbf{dr}$ in terms of the virial coefficients B_2, B_2, and so on. Then, assuming the density is small enough so that $B_3 \rho^2 \ll B_2\rho$ and $B_2\rho \ll 1$, show that the equation is consistent with $g(r) = e^{-\beta u(r)}$.

7.21. The Ornstein–Zernicke equation for a homogeneous fluid, (7.79), may be rewritten by using the law of cosines in the form: $|\mathbf{r} - \mathbf{r}'|^2 = r^2 + (r')^2 - 2rr' \cos\theta$, where θ is the angle between $\mathbf{r}$ and $\mathbf{r}'$, and integrating over the polar coordinates r', θ, and ϕ. The result is

$$h(r) = c(r) + \frac{2\pi\rho}{r} \int_0^{\infty} dr'\, r'\, h(r') \int_{r-r'}^{r+r'} du\, u c(u)$$

where u is $|\mathbf{r} - \mathbf{r}'|$. By multiplying by r^2 and integrating over r (this involves changing the limits of integration—some readers may want to try it), one gets

$$\int_0^{\infty} dr\, r^2\, h(r) = \int_0^{\infty} dr\, r^2\, c(r) + 4\pi\rho \int_0^{\infty} dr'\, r'^2\, h(r') \int_0^{\infty} du\, u^2\, c(u)$$

(a) Show that this equation follows simply from the Fourier transformed version of the Ornstein–Zernicke equation, (7.80).

(b) Use this equation to write the compressibility equation of state (7.66) in terms of an integral over $c(r)$ instead of $h(r)$.

7.22. In (7.82) and (7.83) for the hard-sphere fluid, the packing fraction $\eta = (\pi/6)\rho\sigma^3$ is equal to $v\rho$, so that a power series in η for βP is the virial series. Derive the virial coefficients B_k $(k = 2 \cdots 6)$ from (7.82) and (7.83). For what k do the virial coefficients differ for the two equations of state?

7.23. The charge density in the ionic atmosphere around a point ion of charge e_s is given by the sum of the contributions of all species. The density of ions of species t, with charge e_t, at a distance r is $\rho_t g_{st}(r)$, and, according to (7.91), $g_{st}(r) = 1 - (\beta/\varepsilon r)e_s e_t\, e^{-\kappa r}$. Show that the resulting charge density is as given in (7.102):

$$\omega^s(r) = \frac{-e_s \kappa^2}{4\pi r} e^{-\kappa r}$$

Then calculate the total charge in the ion atmosphere. Calculate the average value of r, and the average value of $1/r$, using $\omega^s(r)$ as weighting function.

7.24. For a spherical distribution of charge with charge density $q(r)$, the electrostatic potential at the origin is given by $\int 4\pi r^2\, dr q(r) r^{-1}$. Calculate the potential at an ion of species s due to its ionic atmosphere; the charge density is given in the preceding problem.

7.25. If the ions in the Debye–Hückel theory have short-range repulsions characterized by a hard-sphere diameter σ, it was shown [see equation (7.107) and following] that the electrostatic potential around an ion of species s is, for $r \geq \sigma$,

$$\Phi^s = \frac{e_s\, e^{\kappa(\sigma - r)}}{\varepsilon(1 + \kappa\sigma) r}$$

and, for $r < \sigma$,

$$\Phi^s = \frac{-\kappa e_s}{\varepsilon(1 + \kappa\sigma)} + \frac{e_s}{\varepsilon r}$$

Use the Poisson equation,

$$\frac{1}{r^2} \frac{d}{dr}\left(r^2 \frac{d\Phi^s}{dr}\right) = \frac{-4\pi}{\varepsilon} q(r)$$

to calculate the charge density q as a function of r. Then integrate q to find the total charge in the ion atmosphere. Calculate the average value of r^{-1}. Multiplying this by the total charge in the ion atmosphere yields the electrostatic potential at the central ion due to the charge in the ion atmosphere.

7.26. Show that (7.109), for the excess free energy of an ionic "gas" in the Debye–Hückel theory, becomes equal to $V\kappa^3/12\pi\beta$ [see equation (7.104) and following] when the hard-sphere diameter σ approaches 0.

7.27. Calculate the entropy associated with the formation of ion atmospheres,

$$S^{\text{exc}} = -\left(\frac{\partial A^{\text{exc}}}{\partial T}\right)_{N,V}$$

7.28. The excess free energy $A^{\text{exc}} = A - A^{\text{ideal}}$ for a system of charged particles is often thought of as the reversible work of charging. One calculates the work required, starting with neutral species which behave ideally, to gradually build up electrical charge on all the species until each ion has its full charge. The work required to add a charge δq_s to an ion s at which the electrostatic potential due to other ions is $\Psi^s(0)$, is $dw_s = \delta q_s \Psi^s(0)$. Suppose that, at some stage of the charging process, the charge on an ion of species s is equal to fe_s, $0 \leq f \leq 1$. Since the charging is supposed to be done reversibly, the ions are at equilibrium at each stage, and

$$\Psi^s(0) = \left[fe_s \frac{e^{-f\kappa r}}{r} - \frac{fe_s}{r} \right]_{r \to 0}$$

(fe_s/r is the part of the potential due to the central ion itself, and contributes to the self energy of the ion rather than to the work of charging). Integrate $\Sigma_s \, \delta q_s \, \Psi^s(0)$ with $\delta q_s = e_s \, \delta f$ to get A^{exc}. Since $A^{\text{exc}} = -\beta \ln(Q/Q^{\text{ideal}})$, this may be compared with what one gets from (7.104).

7.29. The linearization of the exponential used in obtaining the Debye–Hückel results requires $\beta e_s \Phi^s(r)$ to be small compared to 1. The most important values of r are near r_0, for which the radial charge density, the charge between r and $r + dr$, of the ionic atmosphere is largest.

(a) The radial charge density of the ionic atmosphere around an ion of charge e_s is $4\pi r^2$ (area of a sphere of radius r) multiplied by the charge density $\omega^s(r)$ [(7.102)]. What is r_0?

(b) For a 1–1 electrolyte like Na^+Cl^- in water at 298 K ($\varepsilon = 78.5$), what is the maximum concentration for which $\beta e_s \Phi^s(r)$ is less than 0.1 at $r = r_0$? What is the maximum concentration for a 2–2 electrolyte like $Ca^{2+}SO_4^{2-}$?

7.30. Considering a 1–1 electrolyte, at what molar concentration does $\kappa\sigma$ become 0.1 and hence not negligible if the hard-sphere diameter is $10 A$ ($T = 298$ K, solvent water, $\varepsilon = 78.5$)?

7.31. For the Debye–Hückel model when hard-sphere repulsions are included [(7.107)–(7.110)], show that the zeroth moment of the correlation functions M_0 [see (7.94)] is equal to $-\Sigma_s \rho_s e_s^2$. Show that the second moment of the correlation functions M_2 is equal to $-6\varepsilon/4\pi\beta$ for the Debye–Hückel model without the hard-sphere repulsions, and approximately $-6\varepsilon/4\pi\beta$ with them included.

7.32. Calculate an expression for the pressure due to the ions in Debye–Hückel theory, with and without hard-sphere repulsions, from the expression for the Helmholtz free energy A.

7.33. Calculate the concentration needed to give $\kappa^{-1} = 50 A$ for the simple Debye–Hückel theory assuming (i) a 1–1 electrolyte with $\varepsilon = 80$, 10, and 1; (ii) a 2–2 electrolyte with $\varepsilon = 80$, 10, and 1; (iii) a 3–1 electrolyte with $\varepsilon = 80$, 10, and 1. $T = 300$ K. [*Answer:* (i) 0.003801 M, 0.000475 M, 0.0000475 M.]

7.34. When the measured mean ionic activity coefficient for $CaCl_2$ in water at 25°C is plotted vs. the square root of the molality, it apparently passes through a minimum for $\sqrt{m}$ near 0.5. Although the Debye–Hückel theory is not accurate for molalities above 0.1, use the first two terms in a power series for $\ln \gamma_\pm$ in κ to estimate the hard-sphere diameter for $CaCl_2$.

7.35. For dissociation of a weak acid like acetic acid, the apparent equilibrium constant $K_{HA} = m_{H+}m_{A-}/m_{HA}$ is apparently not constant, even at very low concentrations. Show that the Debye–Hückel theory predicts that, for low concentrations in aqueous solution at 25°C,

$$\ln K_{HA} = Q + 2.34\sqrt{c}$$

where c is the molar concentration of the acid and Q is a constant. (Activity coefficients of neutral species are much closer to unity than those for charged species.)

7.36. Calculate the mean ionic activity coefficients for the following aqueous electrolytes at 25°C, at molalities 0.001, 0.005, 0.01, 0.05, and 0.1: HCl, KNO_3, K_2SO_4, $BaCl_2$, and $CuSO_4$. Use the Debye–Hückel theory with $\sigma = 0$. (Experimental values: 0.965, 0.928, 0.904, 0.830, 0.796; 0.965, 0.927, 0.902, 0.823, 0.785; 0.89, 0.78, 0.71, 0.52, 0.43; 0.88, 0.77, 0.72, 0.56, 0.49; 0.74, 0.53, 0.41, 0.21, 0.16). Are the deviations from experiment explainable qualitatively by the fact that σ is not zero?

7.37. The solubility of lead chloride, $PbCl_2$, in water is 0.005 mole/L at 25°C. Calculate the apparent solubility product, $[Pb^{2+}][Cl^-]^2$ (in terms of concentrations) and the thermodynamic solubility product, in terms of activities, taking into account the activity coefficients. From the thermodynamic solubility product, calculate the solubility of $PbCl_2$ at 25°C in an aqueous 0.03 M solution of sodium nitrate, $NaNO_3$. (*Answer:* 0.0063 M.)

CHAPTER 8

TIME DEPENDENCE

With one exception, only systems at equilibrium, for which there are no flows or currents, have been discussed. The exception was the consideration of transport coefficients in gases. Although currents of energy, momentum, and matter were involved, average values did not change with time because a steady state was being analyzed. It is easy to imagine more complicated situations; it can be anticipated that there is a greater variety of time-dependent problems than time-independent. One may also expect time-dependent problems to be more difficult and the methods developed to study them to be more complicated. Because all this is true, only some general ideas and a few simple problems will be discussed here.

The approach is by classical statistical mechanics. An ensemble is described by a cloud of points in the 3N-dimensional phase space. Each point, representing a system in the ensemble, moves around phase space. If the ensemble is an equilibrium ensemble, the cloud of points (meaning the density of points as a function of position in phase space) does not change. Time dependence is represented by a cloud of points which deforms in time. In Sections 8.1 and 8.2 the equations governing movement of phase points are discussed. The application to nonequilibrium situations is found in Section 8.3.

Whether the cloud of phase points deforms or not, the motion of an individual system point is determined by Hamilton's equations of motion. Therefore, the change in a system not at equilibrium is related to how individual points within an equilibrium cloud move. This implies that information about the approach to equilibrium can be derived from study of fluctuations in an equilibrium system, the subject of the very important fluctuation dissipation theorem (Section 8.3). The concepts presented in Sections 8.1 to 8.3 are applied to conduction and diffusion in Sections 8.4 and 8.5, and to chemical reactions in Section 8.6. The general treatment is somewhat formal, so that some readers may want to skip the details of Sections 8.1 and 8.2 and go to Section 8.3.

8.1. LIOUVILLE EQUATION

As discussed in Sections 6.1 and 6.2, a system of N structureless particles in three-dimensional space may be represented by a point in a $6N$-dimensional phase space, that is, by giving values for the $3N$ positions $\{q_i\}$ and the $3N$ momenta $\{p_i\}$. The set of positions and momenta, which are the coordinates in phase space, is abbreviated as Γ_N. Hamilton's equations describe the motion of the point in phase space:

$$\frac{\partial H}{\partial p_i} = \frac{dq_i}{dt} \quad \text{and} \quad \frac{\partial H}{\partial q_i} = -\frac{dp_i}{dt}$$

The Hamiltonian for the system is

$$H = \sum_{i=1}^{3N} \frac{p_i^2}{2m} + U_N(q_i \cdots q_{3N})$$

A dynamical variable is a function of the positions and momenta, and possibly of time as well. The rate of change of the value of the dynamical variable A is

$$\frac{dA}{dt} = \frac{\partial A}{\partial t} + \sum_{i=1}^{3N} \left(\frac{\partial A}{\partial q_i} \frac{dq_i}{dt} + \frac{\partial A}{\partial p_i} \frac{dp_i}{dt} \right) \tag{8.1}$$

with the first term on the right-hand side representing any explicit time dependence in A and the summation representing the change in A due to motion of the phase point. If A is the Hamiltonian which appears in Hamilton's equations, the summation becomes zero.

An ensemble of systems is represented by a large number of points in phase space. Because the number is supposed to be very large, the ensemble is described as a cloud and characterized by giving its density, that is, the number of phase points per unit volume of phase space. For an equilibrium ensemble, the density does not change with time. This is because the density at any Γ_N depends only on the energy (value of the Hamiltonian) at that point. For example, in the canonical ensemble the density is proportional to $\exp[-\beta H]$ where $1/k\beta$ is the temperature. For a nonequilibrium ensemble, the density of phase points may change with time.

If Hamilton's equations are used in (8.1), the rate of change of a dynamical variable $A(\Gamma_N; t)$ is written

$$\frac{dA}{dt} = \frac{\partial A}{\partial t} + \sum_{i=1}^{3N} \left(\frac{\partial A}{\partial q_i} \frac{\partial H}{\partial p_i} - \frac{\partial A}{\partial p_i} \frac{\partial H}{\partial q_i} \right) \equiv \frac{\partial A}{\partial t} + \mathscr{L}A$$

which defines the Liouville operator $\mathscr{L}$. If Cartesian coordinates are used, $(\partial H/\partial p_i) = p_i/m$ and $(\partial H/\partial q_i) = -F_i$ (the force corresponding to position i), so that

$$\mathscr{L} = \sum_{i=1}^{3N} \left(\frac{p_i}{m} \frac{\partial}{\partial q_i} + F_i \frac{\partial}{\partial p_i} \right) \tag{8.2}$$

The Liouville operator, operating on a dynamical variable, produces the rate of change of this variable due to the motion of phase points. If A is a function of the Hamiltonian only, so that A depends on Γ_N only through the Hamiltonian,

$$\mathscr{L}A(\Gamma_N; t) = \sum_{i=1}^{3N} \left(\frac{dA}{dH} \frac{\partial H}{\partial q_i} \frac{\partial H}{\partial p_i} - \frac{dA}{dH} \frac{\partial H}{\partial p_i} \frac{\partial H}{\partial q_i} \right) = 0$$

and $dA/dt = \partial A/\partial t$. The reason for this is that the phase points follow trajectories of constant energy, so that the value of A will not change as the phase point moves through phase space.

The average value of any dynamical variable A is calculated by integrating, over all phase space, the value of A at a point in phase space multiplied by the probability of finding a system in the ensemble at that point. Denoting this probability by $f_N(\Gamma_N)$,

$$\langle A \rangle = \int f_N(\Gamma_N) A(\Gamma_N)\, d\Gamma_N \tag{8.3}$$

[note that $\int f_N(\Gamma_N) = 1$]. In general, f_N is time dependent, and

$$\frac{df_N(\Gamma_N; t)}{dt} = \frac{\partial f_N(\Gamma_N; t)}{\partial t} + \mathscr{L} f_N(\Gamma_N; t)$$

The first term is the change of f_N with time at a particular point in phase space (this term would vanish for an equilibrium distribution) whereas the Liouville term is the change of f_N due to the fact that the motion of system points is changing the location in phase space at which f_N is being evaluated. It should be noted that f_N is not a dynamical variable because it depends on the history of the system.

Since systems in the ensemble are neither created nor destroyed, f_N obeys a continuity equation: $\partial f_N/\partial t$, integrated over a volume of phase space, is equal to the negative of the rate of flow of f_N outward through the boundaries of that volume. With the volume shrunk to zero, the outward rate of flow is equal to the divergence, and the continuity equation reads

$$\frac{\partial f_N}{\partial t} = -\sum_{i=1}^{3N}\left(\frac{\partial}{\partial q_i}\left(f_N\frac{dq_i}{dt}\right) + \frac{\partial}{\partial p_i}\left(f_N\frac{dp_i}{dt}\right)\right) \tag{8.4}$$

The phase space has $6N$ dimensions, $3N$ of which are positions and $3N$ of which are momenta. In (8.4), $f_N(dq_i/dt)$ is the flow rate of f_N along one of the position dimensions (flow rate = density times velocity), and $f_N(dp_i/dt)$ is the flow rate of f_N along one of the momentum dimensions. The right-hand side of (8.4) is the negative of the divergence. It can be simplified using Hamilton's equations: $dq_i/dt = \partial H/\partial p_i$ and $dp_i/dt = -\partial H/\partial q_i$. One has then the Liouville equation

$$\frac{\partial f_N}{\partial t} = -\sum_{i=1}^{3N}\left(\frac{\partial f_N}{\partial q_i}\left(\frac{\partial H}{\partial p_i}\right) + \frac{\partial f_N}{\partial p_i}\left(\frac{-\partial H}{\partial q_i}\right)\right) = -\mathscr{L}f_N(\Gamma_N;t) \tag{8.5}$$

According to the Liouville equation, $df_N/dt = 0$: the total time derivative of f_N vanishes. The meaning of this is that, if one observed the density of phase points from the point of view of a system in the ensemble which moves through phase space, one would see a time-independent density of phase points. This is true for nonequilibrium as well as equilibrium distributions.

The Liouville equation is the most important equation in time-dependent statistical mechanics. It describes the evolution in time of a distribution, and therefore predicts how average values of all properties change with time. Naturally, it is far from easy to solve in any but the simplest systems. The only thing simple about it is that the Liouville operator is a linear operator. Formally, the solution to the equation

$$\frac{dA}{dt} = \mathscr{L}A$$

(for A a dynamical variable with no explicit time dependence) is

$$A(t) = e^{t\mathscr{L}}A(0)$$

To see what this means, expand the exponential in a power series. Sometimes $\exp(t\mathscr{L})$ is called a time-development operator. The inverse, $\exp(-t\mathscr{L})$, is the operator that develops a dynamical variable backwards in time.

To make its meaning clearer, the Liouville equation will be looked at in some detail for a system of particles interacting by momentum-independent pair potentials:

$$U_N = \sum_{j<k} u(r_{jk})$$

as in Chapter 7. If particle i has coordinates $q_{3i-2}q_{3i-1}$ and q_{3i}, and Cartesian coordinates are used,

$$r_{jk}^2 = (q_{3j-2} - q_{3k-2})^2 + (q_{3j-1} - q_{3k-1})^2 + (q_{3j} - q_{3k})^2$$

The Liouville equation becomes

$$\frac{\partial f_N}{\partial t} = -\mathscr{L} f_N = -\sum_{i=1}^{3N} \left(\frac{\partial f_N}{\partial q_i} \frac{p_i}{m} - \frac{\partial f_N}{\partial p_i} \frac{\partial U_N}{\partial q_i} \right)$$

It will be convenient to use, instead of the $3N$ coordinates $\{q_i\}$ and the $3N$ momenta $\{p_j\}$, the N-particle positions $\{\mathbf{r}_i\}$ (each is a set of three coordinates) and the N momenta $\{\mathbf{p}_i\}$. In terms of these, the Liouville equation is written

$$\frac{\partial f_N(\mathbf{r}_1 \cdots \mathbf{r}_N; \mathbf{p}_1 \cdots \mathbf{p}_N)}{\partial t} = \sum_{i=1}^{N} \left(-\frac{\partial f_N}{\partial \mathbf{r}_i} \cdot \frac{\mathbf{p}_i}{m} + \frac{\partial f_N}{\partial \mathbf{p}_i} \cdot \frac{\partial \left[\sum u(r_{jk}) \right]}{\partial \mathbf{r}_i} \right) \tag{8.6}$$

where $\partial f_N / \partial \mathbf{r}_i$ is the gradient, a three-component vector:

$$\frac{\partial f_N}{\partial \mathbf{r}_i} = \left\{ \frac{\partial f_N}{\partial q_{3i-2}}; \frac{\partial f_N}{\partial q_{3i-1}}; \frac{\partial f_N}{\partial q_{3i-1}} \right\}$$

and similarly for $\partial f_N / \partial \mathbf{p}_i$ and $\partial u / \partial \mathbf{r}_i$.

As in Chapter 7, two distribution functions will be of particular interest: the one-particle distribution

$$f^{(1)}(\mathbf{r}, \mathbf{p}) = N \int \mathbf{dr}_2 \cdots \mathbf{dr}_N \, \mathbf{dp}_2 \cdots \mathbf{dp}_N \, f_N(\mathbf{r}, \mathbf{r}_2, \ldots, \mathbf{r}_N, \mathbf{p}, \mathbf{p}_2, \ldots, \mathbf{p}_N)$$

and the two-particle distribution

$$f^{(2)}(\mathbf{r}, \mathbf{r}', \mathbf{p}, \mathbf{p}') = N(N-1) \times \int \mathbf{dr}_3 \cdots \mathbf{dr}_N \, \mathbf{dp}_3 \cdots \mathbf{dp}_N \, f_N(\mathbf{r}, \mathbf{r}', \mathbf{r}_3, \ldots, \mathbf{r}_N, \mathbf{p}, \mathbf{p}', \mathbf{p}_3, \ldots, \mathbf{p}_N)$$

Note that these depend on momenta as well as positions. In our discussion of fluids, we assumed thermal equilibrium for momenta at each point, so that $f^{(1)}$ was a product of the spatial density $\rho^{(1)}$ and the Maxwell–Boltzmann momentum distribution. Similarly, $f^{(2)}$ was assumed to be a product of the two-particle spatial distribution $\rho^{(2)}$ and Maxwell–Boltzmann distributions for the momenta of the two particles.

To see how $f^{(1)}$ changes with time, one integrates (8.6) over the coordinates and momenta of all particles except particle 1 and multiplies by N. Then, separating the terms for $i = 1$, one has

$$\frac{\partial f^{(1)}(\mathbf{r}_1, \mathbf{p}_1)}{\partial t}$$

$$= N \int d\mathbf{r}_2 \cdots d\mathbf{r}_N \, d\mathbf{p}_2 \cdots d\mathbf{p}_N \left(-\frac{\partial f_N}{\partial \mathbf{r}_1} \cdot \frac{\mathbf{p}_1}{m} + \frac{\partial f_N}{\partial \mathbf{p}_1} \cdot \frac{\partial}{\partial \mathbf{r}_1} \sum_k^{k \neq 1} u(r_{1k}) \right)$$

$$+ N \int d\mathbf{r}_2 \cdots d\mathbf{r}_N \, d\mathbf{p}_2 \cdots d\mathbf{p}_N$$

$$\times \sum_{i=2}^{N} \left(-\frac{\partial f_N}{\partial \mathbf{r}_i} \cdot \frac{\mathbf{p}_i}{m} + \frac{\partial f_N}{\partial \mathbf{p}_i} \cdot \frac{\partial}{\partial \mathbf{r}_i} \sum_k^{k \neq i} u(r_{ik}) \right)$$

The integral of $\partial f_N / \partial p_{xi}$ over p_{xi} (the x-component of $\mathbf{p}_i$) gives f_N, the probability of finding a system in the ensemble, for p_{xi} infinite; for any reasonable distribution f_N this will vanish. Similarly, the integral of $\partial f_N / \partial x_i$ over x_i gives the probability of finding a system in the ensemble with x_i infinite; this will also vanish for any reasonable f_N. Thus the terms in the summation over i vanish in the above equation. The sum in the second term (over k with k unequal to 1) actually gives $N - 1$ identical terms because all the particles are equivalent. Thus

$$\frac{\partial f^{(1)}(\mathbf{r}_1, \mathbf{p}_1)}{\partial t} = -\frac{\partial f^{(1)}}{\partial \mathbf{r}_1} \cdot \frac{\mathbf{p}_1}{m}$$

$$+ N(N-1) \int d\mathbf{r}_2 \cdots d\mathbf{r}_N \, d\mathbf{p}_2 \cdots d\mathbf{p}_N \left(\frac{\partial f_N}{\partial \mathbf{p}_1} \cdot \frac{\partial u(r_{12})}{\partial \mathbf{r}_1} \right)$$

$$= -\frac{\partial f^{(1)}}{\partial \mathbf{r}_1} \cdot \frac{\mathbf{p}_1}{m} + \int d\mathbf{r}_2 \, d\mathbf{p}_2 \left(\frac{\partial f^{(2)}}{\partial \mathbf{p}_1} \cdot \frac{\partial u(r_{12})}{\partial \mathbf{r}_1} \right) \tag{8.7}$$

Some physical interpretation can now be given.

Remember that $f_N(\Gamma_N)$ is the density of system points in phase space. If

$$\int f_N(\Gamma_N)\, d\Gamma_N = 1$$

the one-particle density is normalized according to

$$\int f^{(1)}(\mathbf{r}, \mathbf{p})\, \mathbf{dr}\, \mathbf{dp} = N$$

and the two-particle density is normalized according to

$$\int f^{(2)}(\mathbf{r}, \mathbf{p}, \mathbf{r}', \mathbf{p}')\, \mathbf{dr}\, \mathbf{dp}\, \mathbf{dr}'\, \mathbf{dp}' = N(N-1)$$

Thus $f^{(1)}(\mathbf{r}, \mathbf{p})\, \mathbf{dr}\, \mathbf{dp}$ is the number of systems in the ensemble having a particle in the infinitesimal six-dimensional volume element $\mathbf{dr}\,\mathbf{dp}$ at the position $\mathbf{r}$ and with momentum $\mathbf{p}$, or the probability that there is a particle (any particle) with **x**-coordinate between x and $x + dx$, y-coordinate between y and $y + dz$, z-coordinate between z and $z + dz$, x-momentum between p_x and $p_x + dp_x$, y-momentum between p_z and $p_y + dp_z$, and z-momentum between p_z and $p_z + dp_z$. Here, $\mathbf{r} = \{x, y, z\}$ and $\mathbf{p} = \{p_x, p_y, p_z\}$. According to (8.7), the probability that there is a particle at $\mathbf{r}_1$ with momentum $\mathbf{p}_1$ changes with time for two reasons: (1) particles of momentum $\mathbf{p}_1$ arrive at $\mathbf{r}_1$ from other places and leave $\mathbf{r}_1$ for other places, and (2) the momenta of particles at $\mathbf{r}_1$ are changed by the interaction of these particles with other particles.

The flow of particles of momentum $\mathbf{p}_1$ is proportional to their velocity, $\mathbf{v}_1 = \mathbf{p}_1/m$, and to their density, $f^{(1)}(\mathbf{r}_1, \mathbf{p}_1)$. If the density is constant along a line in the $\mathbf{v}_1$ direction, the flow of these particles to $\mathbf{r}_1$ will exactly equal the flow away from $\mathbf{r}_1$. This is the reason for the factor $\partial f^{(1)}/\partial \mathbf{r}_1$, which would vanish if the density were constant. The rate of change of momentum of particles at $\mathbf{r}_1$ is equal to the force exerted on them by the other particles. The total force of $N - 1$ particles on a particle at $\mathbf{r}_1$ with momentum $\mathbf{p}_1$ is

$$F(\mathbf{r}_1, \mathbf{p}_1) = \frac{-\int \mathbf{dr}_2\, \mathbf{dp}_2\, f^{(2)}(\mathbf{r}_1, \mathbf{p}_1, \mathbf{r}_2, \mathbf{p}_2)(\partial u(r_{12})/\partial \mathbf{r}_1)}{f^{(1)}(\mathbf{r}_1, \mathbf{p}_1)}$$

The net increase in the number of particles at $\mathbf{r}_1$ with momentum $\mathbf{p}_1$ will be zero if as many particles have their momenta changed *to* $\mathbf{p}_1$ as have their momenta changed *from* $\mathbf{p}_1$. This is why (8.7) contains the derivative of $f^{(2)}$ with respect to $\mathbf{p}_1$.

To examine the force term in (8.7) more closely, consider a spatially homogeneous and isotropic gas, so that $f^{(1)}$ is independent of $\mathbf{r}_1$ and $f^{(2)}$

depends only on $r_{12} = |\mathbf{r}_1 - \mathbf{r}_2|$ and not on $\mathbf{r}_1$ and $\mathbf{r}_2$ separately. Then (8.7) becomes

$$\frac{\partial f^{(1)}(\mathbf{p}_1)}{\partial t} = \int \mathbf{dr}_2\, \mathbf{dp}_2 \left(\frac{\partial f^{(2)}(r_{12}, \mathbf{p}_1, \mathbf{p}_2)}{\partial \mathbf{p}_1} \cdot \frac{\partial u(r_{12})}{\partial \mathbf{r}_1} \right) \tag{8.8}$$

The interaction potential $u(r)$ normally includes a long-range attractive and a short-range repulsive part. Let R_a and R_r be the ranges over which each differs significantly from zero. For a hard-sphere repulsive potential with hard-sphere diameter σ, R_r is equal to σ. Particles interacting are said to be in collision. The duration of a collision, or collision time, is $\tau_c = R_a/\langle v \rangle$ where $\langle v \rangle$ is the average speed of a particle. For a gas, τ_c is much less than the average time between collisions, $\tau = \lambda/\langle v \rangle$ with λ the mean free path [see equation (6.34)]. In turn, τ is much less than the macroscopic time $T = L/\langle v \rangle$ with L the dimension of the container holding the gas. The behavior of (8.8) will be discussed for times larger than τ_c and smaller than τ, so only a few collisions take place and it is unnecessary to consider collisions with container walls. Since the system is a gas, it suffices to consider pairs of molecules for which $r_{12} < R_a$; it is unlikely that a third molecule will be nearby.

Consider a pair of molecules at time t, separated by a distance $r_{12} < R_a$, one molecule having momentum $\mathbf{p}_1$ and the other momentum $\mathbf{p}_2$. In the past, that is, at a time t_0 before $t - \tau_c$, these particles were too far apart to be interacting, so their positions and momenta were uncorrelated and the two-particle distribution was a product of one-particle distributions. Every pair of particles in collision at time t was an uncorrelated pair at t_0, so

$$f^{(2)}(r_{12}, \mathbf{p}_1, \mathbf{p}_2; t) = f^{(1)}(\mathbf{P}_1; t_0) f^{(1)}(\mathbf{P}_2; t_0) \tag{8.9}$$

The momenta before collision are designated by $\mathbf{P}_1$ and $\mathbf{P}_2$; they can be calculated in terms of $\mathbf{p}_1$ and $\mathbf{p}_2$ by using conservation of energy,

$$\frac{p_1^2 + p_2^2}{2m} + u(r_{12}) = \frac{P_1^2 + P_2^2}{2m} \tag{8.10}$$

and conservation of momentum, $\mathbf{p}_1 + \mathbf{p}_2 = \mathbf{P}_1 + \mathbf{P}_2$. The situation is illustrated in Figure 8.1, which shows calculated trajectories of two particles that are in collision at time t. The potential used for the calculation was $u(r) = 2.8e^{-3r} + 0.9e^{-r}$, so that $R_a \approx 1$ and $R_r \approx 0.3$. For a very low-density gas (no collisions), $f^{(1)}$ does not change with time. When collisions do occur but not too frequently, $f^{(1)}$ changes only slowly. Thus for a dilute gas $f^{(1)}(t_0)$ should

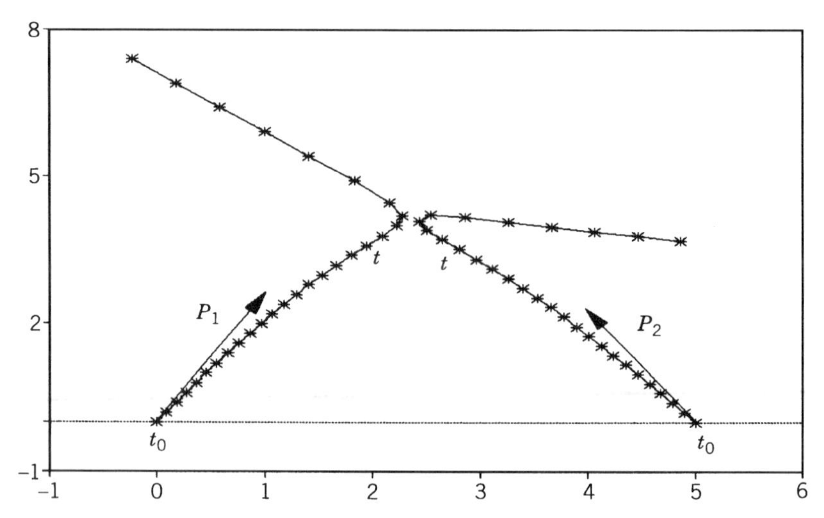

Figure 8.1. Trajectories of colliding particles. Particles of momenta $\mathbf{p}_1$ and $\mathbf{p}_2$ interact with each other at time t. At an earlier time t_0 they had momenta $\mathbf{P}_1$ and $\mathbf{P}_2$ and were too far apart to interact.

be approximately equal to $f^{(1)}(t)$. Using this in (8.9) and substituting for $f^{(2)}$ in (8.8), one has something more tractable:

$$\frac{\partial f^{(1)}(\mathbf{p}_1)}{\partial t} = \int \mathbf{dr}_2\, \mathbf{dp}_2 \left(\frac{\partial \left[f^{(1)}(\mathbf{P}_1) f^{(1)}(\mathbf{P}_2) \right]}{\partial \mathbf{p}_1} \cdot \frac{\partial u(r_{12})}{\partial \mathbf{r}_1} \right)$$

where all the distribution functions are for the same time t.

In the integral, one may introduce the relative coordinate $\mathbf{r}_{12} = \mathbf{r}_2 - \mathbf{r}_1$ and the relative momentum $\mathbf{p}_{12} = \mathbf{p}_2 - \mathbf{p}_1$ so that

$$\frac{\partial f^{(1)}(\mathbf{p}_1)}{\partial t} = \int \mathbf{dr}_{12}\, \mathbf{dp}_{12} \left(\frac{\partial \left[f^{(1)}(\mathbf{P}_1) f^{(1)}(\mathbf{P}_2) \right]}{\partial \mathbf{p}_1} \cdot \frac{\partial u(r_{12})}{\partial r_{12}} \right) \tag{8.11}$$

Since $\mathbf{p}_1 + \mathbf{p}_2 = \mathbf{P}_1 + \mathbf{P}_2$, (8.10) becomes

$$\frac{p_{12}^2}{4m} + u(r_{12}) = \frac{(\mathbf{P}_1 - \mathbf{P}_2) \cdot (\mathbf{P}_1 - \mathbf{P}_2)}{4m}$$

Thus, for given $\mathbf{p}_1$ and $\mathbf{p}_2$, $\mathbf{P}_1$ and $\mathbf{P}_2$ depend on r_{12}. After more algebraic manipulation, (8.11) can be transformed into an equation called the Boltzmann equation, in which the change in the number of particles of momentum $\mathbf{p}_1$ is the difference of two terms. The first describes the creation of particles of momentum $\mathbf{p}_1$ by collisions between particles of momenta (before collision) $\mathbf{P}_3$ and $\mathbf{P}_4$; the second describes the loss of particles of momentum $\mathbf{p}_1$

because they collide with particles having momentum $\mathbf{P}_2$ before collision. The Boltzmann equation can be solved in some important situations. It is valid only when two particles interact at a time (dilute gas). It is also, as seen above, fairly complex. Instead of pursuing this route to time-dependent problems, we return to a consideration of the equilibrium state. Many time-dependent problems of interest involve systems not far from equilibrium.

8.2. TIME-CORRELATION FUNCTIONS

Consider an ensemble appropriate for equilibrium. Let A and B be dynamical variables, functions of the coordinates of phase space, that do not depend explicitly on time. The *values* of A and B for a particular system may depend on time, because the system moves in phase space. If A or B is averaged over an equilibrium ensemble, however, the result should be time independent:

$$\langle A(0)\rangle = \int d\Gamma_N\, f_N^{\text{eq}} A(0) = \langle A(t)\rangle \tag{8.12}$$

Here, f_N^{eq} is the distribution of points in phase space appropriate for the equilibrium ensemble.

Introducing the time-development operator (p. 275)

$$\langle A(t)\rangle = \int d\Gamma_N\, f_N^{\text{eq}} \left[e^{t\mathscr{L}} A(0) \right]$$

One can imagine expanding the time-development operator $e^{t\mathscr{L}}$ as a power series. Considering the kth term in the series,

$$\frac{t^k}{k!} \int d\Gamma_N\, f_N^{\text{eq}}\, \mathscr{L}^k A(0) = -\frac{t^k}{k!} \int d\Gamma_N \left(\mathscr{L} f_N^{\text{eq}} \right) \mathscr{L}^{k-1} A(0)$$

The Liouville operator is a sum of derivatives with respect to coordinates and momenta [see (8.5)], so an integration by parts has been performed. The boundary terms vanish because f_N^{eq} vanishes for infinite values of coordinates or momenta. Repeating the process,

$$\frac{t^k}{k!} \int d\Gamma_N\, f_N^{\text{eq}} \mathscr{L}^k A(0) = \frac{(-t)^k}{k!} \int d\Gamma_N \left(\mathscr{L}^k f_N^{\text{eq}} \right) A(0)$$

and, putting the series back together,

$$\int d\Gamma_N\, f_N^{\text{eq}} \left[e^{t\mathscr{L}} A(0) \right] = \int d\Gamma_N \left[e^{-t\mathscr{L}} f_N^{\text{eq}} \right] A(0)$$

Since f_N^{eq} is an equilibrium distribution, it is unaffected by the Liouville operator, so (8.12) has been proved formally. More generally, the above manipulations show that

$$\int d\Gamma_N\, f_N^{\text{eq}} C(t')\left[e^{u\mathscr{L}} D(t'')\right] = \int d\Gamma_N\, f_N^{\text{eq}}\left[e^{-u\mathscr{L}} C(t')\right] D(t'')$$

or

$$\langle C(t') D(t'' + u)\rangle = \langle C(t' - u) D(t'')\rangle \tag{8.13}$$

where the averages are over an equilibrium ensemble.

In (8.12), $A(t)$ is the value of A at time t for a system for which the value of A was $A(0)$ at time 0. As a function of time, $A(t)$ for any system fluctuates about the ensemble average value $\langle A(t)\rangle$. Remember that the time average is equal to the ensemble average, so

$$\langle A\rangle = \langle A(t)\rangle = T^{-1}\int_0^T A(t)\, dt$$

There is no need to specify the time at which the ensemble average is taken. It is often convenient to define the new dynamical variable δA as $A - \langle A\rangle$. The average value of δA is

$$\langle (A - \langle A\rangle)\rangle = \langle A\rangle - \langle A\rangle = 0$$

for all t. Note that the average value of a constant, like $\langle A\rangle$, is always the constant itself.

The time-correlation function of A and B is the average of $\delta A(0)\, \delta B(t)$ over the equilibrium ensemble:

$$c_{AB}(t) = \langle \delta A(0)\, \delta B(t)\rangle = \langle \delta B(t)\, \delta A(0)\rangle \tag{8.14}$$

One is to find the value of δA for a system at time 0, allow the system to develop until a later time t, find the value of δB at t, and multiply it by δA at 0. Then the product is to be averaged over the systems in the ensemble. Because of the ensemble averaging, $c_{AB}(t)$ also equals $\langle \delta A(T)\, \delta B(T + t)\rangle$ [see (8.13)]. Since the laws of mechanics, which govern the time development of a system, are invariant to time reversal, $c_{AB}(t) = c_{AB}(-t)$. These properties can be derived formally using (8.13). For example, put $\delta A(0)$ for $C(t')$ and $\delta B(t)$ for $D(t'' + u)$, and choose $u = -T$. Then (8.13) reads

$$\langle \delta A(0)\, \delta B(t)\rangle = \langle \delta A(T)\, \delta B(t + T)\rangle$$

This means that the time 0 has no special significance.

If t is very large, the fluctuation of B from its average value cannot be related to the fluctuation of A from its average at time $t = 0$. Then

$$c_{AB}(t) \to \langle \delta A(0) \rangle \langle \delta B(t) \rangle = 0$$

(no correlation). For $t \to 0$, $c_{AB}(t) \to \langle \delta A(0)\, \delta B(0) \rangle$, which may not vanish if the variables A and B are related to each other. Certainly, if A and B are the same dynamical variable, $c_{AB}(0)$ is positive. The "autocorrelation function" $c_{AA}(t)$ still approaches zero for large enough t.

As shown above, $\langle \delta A(0)\, \delta A(t) \rangle$ is equal to $\langle \delta A(T) \delta A(T + t) \rangle$. If T is now taken equal to $-t$,

$$\langle \delta A(0)\, \delta A(t) \rangle = \langle \delta A(-t)\, \delta A(0) \rangle = \langle \delta A(0)\, \delta A(-t) \rangle$$

so $c_{AA}(t) = c_{AA}(-t)$. Also, because the ensemble average value of A is the same at all times,

$$c_{AA}(t) = \langle [A(t) - \langle A \rangle][A(0) - \langle A \rangle] \rangle = \langle A(t) A(0) \rangle - \langle A \rangle^2$$

For large t, $\langle A(t) A(0) \rangle = \langle A(t) \rangle \langle A(0) \rangle = \langle A \rangle^2$ because the value of A at time t will be unrelated to the value of A at time 0, and $c_{AA}(t)$ approaches 0.

As an example of an autocorrelation function, suppose $A = B = v_x$, the x-velocity of a particular particle in an atomic gas. Obviously $\langle A \rangle = 0$, so

$$c_{AA}(t) = \langle v_x(0) v_x(t) \rangle \tag{8.15}$$

All three spatial directions being equivalent,

$$c_{AA}(t) = \tfrac{1}{3} \langle \mathbf{v}(0) \cdot \mathbf{v}(t) \rangle$$

At $t = 0$, $c_{AA}(t)$ is the average squared x-velocity of a particle, which is kT/m. For t less than the average time between collisions, it is likely that the particle will still be moving in the same direction as at $t = 0$ so the ensemble average of $v_x(0)v_x(t)$ will be positive; as the likelihood of a collision increases, the ensemble average will decrease in size. After a long time, the particle will almost certainly have undergone one or more collisions, which changed its direction of motion. Then, for the different systems of the ensemble, $v_x(0)v_x(t)$ will just as likely be negative as positive, and $c_{AA}(t)$ will approach zero.

The approach to zero need not be monotonic. If collisions between atoms were less probable than collisions with the walls of the container, and if collisions with the walls caused an atom to reverse direction, there might be a range of t for which $c_{AA}(t)$ took on negative values. Figure 8.2 shows the form of $c_{VV}(t)$ for a particle in a liquid; the particle is constantly colliding

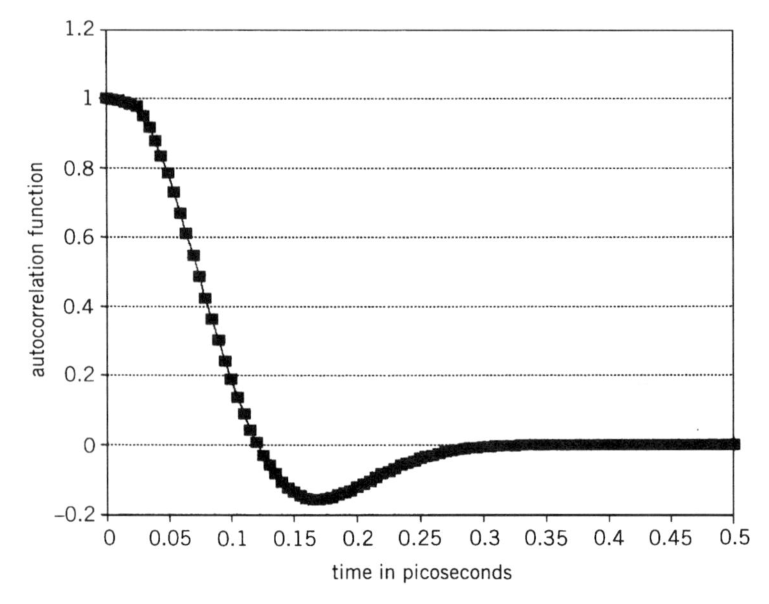

Figure 8.2. Velocity autocorrelation function for a particle in a liquid. It is probable that a particle moving in one direction at time 0 will be moving in the opposite direction 0.15 ps later.

with the walls of the cage formed by neighboring particles. For a perfect oscillator, $x = A\cos(\omega t + b)$,

$$\begin{aligned} v_x(0)v_x(t) &= A^2\omega^2 \sin(\omega t + b)\sin b \\ &= \tfrac{1}{2}A^2\omega^2[\cos \omega t - \cos(\omega t + 2b)] \end{aligned}$$

The microcanonical ensemble includes equal numbers of systems with all values of b between 0 and 2π so that

$$\langle v_x(0)v_x(t)\rangle = \tfrac{1}{2}A^2\omega^2 \cos \omega t$$

and $c_{AA}(t)$ varies sinusoidally between $\langle v_x^2\rangle$ and $-\langle v_x^2\rangle$.

In any case, one can prove that the magnitude of $c_{AA}(t)$ can never exceed $c_{AA}(0)$. The average value of a positive quantity cannot be negative, so

$$0 \le \langle [\delta A(t) \pm \delta A(0)]^2\rangle = 2\langle (\delta A)^2\rangle \pm 2\langle \delta A(t)\delta A(0)\rangle$$

The two signs give two inequalities:

$$-\langle (\delta A)^2\rangle \le c_{AA}(t) \le \langle (\delta A)^2\rangle$$

Since $c_{AA}(0) = \langle(\delta A)^2\rangle$, $|c_{AA}(t)| \leq c_{AA}(0)$. An autocorrelation function is often a decreasing exponential in time:

$$c_{AA}(t) = c_{AA}(0)e^{-t/\tau}$$

Here, τ is called the relaxation time or time constant; $1/\tau$ is the decay rate constant. If a time-correlation function is not a simple exponential, one can define a time constant according to

$$\tau_{AB} = \int_0^\infty dt \, \frac{c_{AB}(t)}{c_{AB}(0)} \tag{8.16}$$

Of course, this definition coincides with the usual definition when the correlation function is an exponential (Problem 8.1).

In general, the variation with time of the dynamical variable A (for a single system) can be represented by a superposition of oscillating functions of different frequencies, that is, as a Fourier transform:

$$A(t) = \int_{-\infty}^{\infty} A_\omega \, e^{i\omega t} \, d\omega \tag{8.17}$$

The Fourier components $\{A_\omega\}$ may be found by inverting the Fourier transform:

$$A_\omega = \frac{1}{2\pi} \int_{-\infty}^{\infty} A(t) \, e^{-i\omega t} \, dt$$

The time-autocorrelation function of A becomes

$$\begin{aligned} c_{AA}(t - t') &= \langle A(t) A(t') \rangle \\ &= \int_{-\infty}^{\infty} \int_{-\infty}^{\infty} \langle A_\omega A_{\omega'} \rangle \, e^{i(\omega t + \omega' t)} \, d\omega \, d\omega' \\ &= \int_{-\infty}^{\infty} \int_{-\infty}^{\infty} \langle A_\omega A_{\omega'} \rangle \, e^{i(\omega + \omega')t} \, e^{i\omega'(t' - t)} \, d\omega \, d\omega' \end{aligned}$$

The right-hand side is to be a function of $t' - t$, which requires that $\langle A_\omega A_{\omega'} \rangle$ be zero unless $\omega + \omega' = 0$, that is,

$$\langle A_\omega A_{\omega'} \rangle = (A^2)_\omega \, \delta(\omega + \omega') \tag{8.18}$$

Here, $(A^2)_\omega = \langle A_\omega A_{-\omega} \rangle$. The meaning of (8.18) is that different Fourier components of $A(t)$ are uncorrelated when averaged over the ensemble. Using (8.18),

$$c_{AA}(t - t') = \int_{-\infty}^{\infty} \int_{-\infty}^{\infty} (A^2)_\omega \, \delta(\omega + \omega') \, e^{i(\omega+\omega')t} \, e^{i\omega'(t'-t)} \, d\omega \, d\omega'$$

$$= \int_{-\infty}^{\infty} (A^2)_{\omega'} \, e^{i\omega'(t'-t)} \, d\omega'$$

In particular, $c_{AA}(0)$ is equal to the integral of $(A^2)_\omega$ over frequency. The inverse of the Fourier transform is

$$(A^2)_\omega = \frac{1}{2\pi} \int_{-\infty}^{\infty} c_{AA}(t'') e^{-i\omega t''} \, dt''$$

An example of the use of these formulas follows.

Suppose a particle in a liquid is displaced from its equilibrium position (or some dynamical variable has a value noticeably different from its equilibrium value), and the rate at which it returns to its equilibrium position is proportional to the displacement:

$$v_x = \frac{dx}{dt} = -kx$$

The equilibrium position (or equilibrium value of the dynamical variable) is taken as 0. Suppose further that, in addition to the restoring force, there are random events that may disturb the return to equilibrium. For a particle in a liquid, these could be collisions with other particles. The correlation function c_{xx} obeys

$$\frac{dc_{xx}(t')}{dt'} = \frac{d}{dt'} \langle x(t) x(t + t') \rangle = -kc_{xx}(t') \tag{8.19}$$

if the random events are assumed to average out. Since $c_{xx}(0) = \langle x^2 \rangle$,

$$c_{xx}(t') = \langle x^2 \rangle \, e^{-kt'}$$

This is true only for $t' > 0$ because it is only for $t' > 0$ that fluctuations from equilibrium diminish as t increases. For $t' < 0$, note that

$$c_{xx}(t') = \langle x(t) x(t + t') \rangle = \langle x(t - t') x(t) \rangle = c_{xx}(-t')$$

so $c_{xx}(-t') = c_{xx}(t')$ and

$$c_{xx}(t) = \langle x^2 \rangle \, e^{-k|t|} \tag{8.20}$$

The discontinuity in $dc_{xx}(t)/dt$ at $t = 0$ is only apparent: (8.19) is not valid for very small times because for small times the effect of random events (collisions) does not average to zero.

The Fourier components of (8.20) are given by

$$\begin{aligned}
(x^2)_\omega &= \frac{1}{2\pi} \int_{-\infty}^{\infty} \langle x^2 \rangle \, e^{-k|t|} \, e^{-i\omega t} \, dt \\
&= \frac{\langle x^2 \rangle}{2\pi} \left(\int_{-\infty}^{0} e^{(k-i\omega)t} \, dt + \int_{0}^{\infty} e^{-(k+i\omega)t} \, dt \right) \\
&= \frac{\langle x^2 \rangle k}{\pi (k^2 + \omega^2)}
\end{aligned} \tag{8.21}$$

Equations (8.19) to (8.21) follow from $dx/dt = -kx$. For any dynamical variable A, $d(\delta A)/dt$ will be a power series in δA, the first term being zero because $d(\delta A)/dt$ vanishes if one is already at equilibrium ($\delta A = 0$). The leading term is linear in δA, so $d(\delta A)/dt \cong -k \, \delta A$ and equations (8.19) to (8.21) should apply approximately.

The time-correlation functions and autocorrelation functions describe the decay of fluctuations in equilibrium ensembles. These fluctuations, deviations in values of dynamical variables from their average values, arise spontaneously. Onsager hypothesized that, in systems not in equilibrium, deviations in values of dynamical variables from their equilibrium values decay in the same way as fluctuations in equilibrium systems. For systems not in equilibrium, the decay to zero of deviations is just the relaxation of the system to equilibrium.

It is not unreasonable that relaxation of a nonequilibrium system to equilibrium and disappearance of spontaneous fluctuations around equilibrium are closely related. The same molecules, acting according to the same laws of mechanics, are responsible for both. For example, the decay of the velocity autocorrelation function in a gas is due to collisions between molecules, which decrease the average velocity in any particular direction to zero. If there is a net flow in a fluid, making the average velocity nonzero in some region of space, the flow will decay to zero because of the same collisions. In Section 6.4 the approximate treatment of transport properties for gases was in terms of the mean free path. "Transport properties" refer to nonequilibrium states, in which nonuniformities exist, and the mean free path is a characteristic of the equilibrium state.

Onsager's hypothesis may be expressed mathematically as follows: let A be a dynamical variable, and let $[A - \langle A \rangle]_t$ be the ensemble average value of

$A - \langle A \rangle$ for a nonequilibrium system at time t. As the system relaxes to equilibrium, $[A - \langle A \rangle]_t \to 0$ as $t \to \infty$, just like $c_{AA}(t)$. Onsager's hypothesis is

$$\frac{[A - \langle A \rangle]_t}{[A - \langle A \rangle]_0} = \frac{c_{AA}(t)}{c_{AA}(0)} \tag{8.22}$$

If the autocorrelation function obeyed (8.20), one would have

$$[A - \langle A \rangle]_t = [A - \langle A \rangle]_0 \, e^{-k|t|}$$

Since Onsager's hypothesis relates fluctuations around equilibrium and relaxation of nonequilibrium systems to equilibrium, which is characterized by dissipation of gradients of density, velocity, and so on, it is sometimes referred to as the fluctuation-dissipation theorem.

8.3. RELAXATION TO EQUILIBRIUM

A system not at thermal equilibrium will undergo changes which will eventually bring it to an equilibrium state. If the system is represented by an ensemble for which the phase-space distribution $f_N(\Gamma_N)$ is not an equilibrium distribution, f_N will change with time, gradually approaching an equilibrium distribution, which is time independent (see Section 8.1). How did the system get to its nonequilibrium state in the first place? As discussed, it could have been due to a spontaneous fluctuation in some dynamical variable or to an imposed fluctuation; the relaxation, or approach to equilibrium, will be the same in either case. The case of an imposed fluctuation is easier to discuss and will be dealt with first.

The Hamiltonian H is written as a sum of two terms, $H_0 + H_1$, with H_0 time independent and H_1 time dependent. A statement that the system is or is not in thermal equilibrium at $t = 0$ refers to thermal equilibrium with respect to H_0. The equilibrium distribution is written f_{N0} to indicate this. For the canonical distribution,

$$f_{N0} = \frac{e^{-\beta H_0}}{\int e^{-\beta H_0} \, d\Gamma_N} \tag{8.23}$$

H_1, which is responsible for taking the system out of thermal equilibrium, is supposed to be small compared to H_0, so it may be considered as a perturbation. Initially, it will be supposed that $H_1(t)$ vanishes except at $t = 0$, so it is an instantaneous impulse:

$$H_1(t) = -A \, \delta t$$

Here, A is a dynamical variable of the system, or a combination of a dynamical variable and an external field. Any more complicated time-dependence of H_1 can be written as a superposition (sum, or integral) of impulses. At some later time, the value of the dynamical variable B is to be found. H_1 affects the value of B because it affects f_N.

The distribution function of the system is f_{N0} for all times up to $t = 0$. For one instant, at $t = 0$, the Hamiltonian changes from H_0 to $H_0 + H_1$, and the distribution function f_N starts to change. Eventually, f_N will return to f_{N0}. The equation of motion is equation (8.5):

$$\frac{\partial f_N}{\partial t} = -\mathscr{L} f_N(\Gamma_N; t) = -\sum_{i=1}^{3N} \left(\frac{\partial f_N}{\partial q_i} \left(\frac{\partial H}{\partial p_i} \right) + \frac{\partial f_N}{\partial p_i} \left(\frac{-\partial H}{\partial q_i} \right) \right)$$

Since H is a sum of two terms, so is the Liouville operator:

$$\mathscr{L} = \mathscr{L}_0 + \mathscr{L}_1$$

where $\mathscr{L}_1$ is small compared to $\mathscr{L}_0$. Similarly,

$$f_N = f_{N0} + f_{N1}$$

where f_{N1} is small compared to f_{N0}. The Liouville equation becomes

$$-(\mathscr{L}_0 f_{N0} + \mathscr{L}_0 f_{N1} + \mathscr{L}_1 f_{N0} + \mathscr{L}_1 f_{N1}) = \frac{\partial f_{N0}}{\partial t} + \frac{\partial f_{N1}}{\partial t}$$

Now $\mathscr{L}_0 f_{N0} = (\partial f_{N0}/\partial t) = 0$. The term $\mathscr{L}_1 f_{N1}$ is small compared to the other terms because it is the product of two small factors. Neglecting it leaves

$$\frac{\partial f_{N1}}{\partial t} + \mathscr{L}_0 f_{N1} = -\mathscr{L}_1 f_{N0} \tag{8.24}$$

which may now be integrated with respect to t, starting before $t = 0$.

The left-hand side of (8.24) is the total time derivative of f_{N1}, corresponding to its change with time as seen from a phase point moving under the influence of H_0. Since $f_{N1} = 0$ for $t < 0$, (8.24) can be integrated from any negative t to some positive t to get

$$f_{N1}[p(t), q(t)] = \int_{-\infty}^{t} dt\, \delta t \sum_{i=1}^{3N} \left(\frac{\partial f_{N0}}{\partial q_i} \left(\frac{\partial A}{\partial p_i} \right) + \frac{\partial f_{N0}}{\partial p_i} \left(\frac{-\partial A}{\partial q_i} \right) \right)$$

$$= -\beta f_{N0} \sum_{i=1}^{3N} \left(\frac{\partial H_0}{\partial q_i} \left(\frac{\partial A}{\partial p_i} \right) + \frac{\partial H_0}{\partial p_i} \left(\frac{-\partial A}{\partial q_i} \right) \right)_{t=0}$$

The last expression arises because of the δ-function in H_1 and from the definition of f_{N0} [equation (8.23)]. The sum in the last expression is just $-dA/dt$ at $t = 0$ [see (8.2) and the equation preceding it; A has no explicit time dependence]. Since it is H_0 which enters the equation, dA/dt refers to the change of A with time under the influence of H_0. Thus one can write

$$f_{N1}[p(t), q(t)] = \beta f_{N0}\left(\frac{dA}{dt}\right)_0, \qquad t > 0 \tag{8.25}$$

where the subscript on (dA/dt) is doing double duty; it means that dA/dt is under the influence of H_0 *and* is evaluated at time 0.

Suppose that the expectation or average value of the dynamical variable B is evaluated at time t.

$$\langle B(t)\rangle = \int [f_{N0}(\Gamma_N) + f_{N1}(\Gamma_N)] B(t)\, d\Gamma_N$$

There are two contributions; the second, involving f_{N1}, is part of the response of the system to the impulse perturbation H_1, while the first would be present even if H_1 never acted. It is the response that is of interest:

$$\begin{aligned}\langle \Delta B(t)\rangle &= \int f_{N1}(\Gamma_N) B(t)\, d\Gamma_N = \beta \int f_{N0}\left(\frac{dA}{dt}\right)_0 B(t)\, d\Gamma_N \\ &= \beta\left\langle \left(\frac{dA}{dt}\right)_0 B(t)\right\rangle, \qquad t > 0 \end{aligned} \tag{8.26}$$

using (8.25). The last expression, sometimes referred to as a response function, is an average over the equilibrium distribution. It will be recognized as β multiplied by a time-correlation function.

The result (8.26) is the response of a system to an impulse at $t = 0$ ($H_1 = -A\,\delta t$). If the impulse had occurred at $t = t'$, $(dA/dt)_{t'}$ would replace $(dA/dt)_0$. Now consider a perturbation H_1 extending over time, say

$$H_1 = -A(p, q)F(t) = -A(p, q)\int_{-\infty}^{\infty} dt'\, F(t')\,\delta(t - t')$$

The reason for writing $F(t)$ as an integral over t' is to show that $F(t)$ may be considered a sum of an infinite number of impulses. Each impulse being treated as a small perturbation [see (8.23)–(8.25)], the effect of a sum of impulses is the sum of the effects of the individual impulses. Then

$$\langle \Delta B(t)\rangle = \beta \int_{-\infty}^{t} dt'\, F(t')\left\langle \left(\frac{dA}{dt}\right)_{t'} B(t)\right\rangle \tag{8.27}$$

Only the impulses for times t' before the observation time t are considered; cause always precedes effect. For the integral to converge, $F(t')$ must approach zero as t' approaches $-\infty$. The physical meaning of this requirement is that the system must have been in an equilibrium state before the perturbation began. The time-correlation function in (8.27) is a function of $t - t'$.

After what may seem like involved manipulations, the result (8.27) is quite simple. The effect of a perturbation on the property B is proportional to the time-correlation function of B and the time derivative of the perturbation. Equation (8.27) is sometimes called Kubo's theorem (after one of its developers) or the linear-response theorem (in the linear regime, the response is proportional to the strength of the perturbation). Its significance should become clearer when it is applied to particular problems.

The first application is to electrical conductivity by charged particles in a fluid. In this case, the perturbation H_1 is that of an electric field and the measured property B is the electrical current. The law of conductivity, equivalent to the familiar Ohm's law, is $\mathbf{j} = \sigma \mathbf{E}$, with $\mathbf{j}$ being the current density, σ the conductivity, and $\mathbf{E}$ the electric field. For an isotropic medium, $\mathbf{j}$ is in the direction of $\mathbf{E}$. Let particles of the fluid carry a charge q and hence be accelerated by the electric field (to have overall electrical neutrality, there should also be particles of opposite sign of charge which cannot be moved by the field). For a field E along the x-direction, the perturbing Hamiltonian is

$$H_1 = -AF(t) = -qE \sum_{i=1}^{N} x_i \, e^{\alpha t}, \qquad t \leq 0$$

The choice of $e^{\alpha t}$ for $F(t)$ ensures that $F(t) \to 0$ as $t \to -\infty$ (later, α will be allowed to approach 0). The field is built up slowly to its full value at $t = 0$, at which time the current in the x-direction is measured. Thus the operator A for (8.27) is qE multiplied by the sum of the x-coordinates of the particles, and the operator B is

$$J_x = \sum_{i=1}^{N} q \frac{dx_i}{dt} = \frac{q}{m} \sum_{i=1}^{N} p_{xi}$$

Of course, the average current in any direction vanishes in the absence of the field.

Now the Kubo equation (8.27) becomes

$$\langle J_x(0) \rangle = \beta \int_{-\infty}^{0} dt' \, e^{\alpha t'} \left\langle qE \frac{d}{dt} \left(\sum_{i=1}^{N} x_i \right)_{t=t'} J_x(t=0) \right\rangle$$

$$= \beta E \int_{-\infty}^{0} dt' \, e^{\alpha t'} \langle J_x(t') J_x(0) \rangle \qquad (8.28)$$

The current is proportional to the electric field E, in accordance with Ohm's law. The conductivity involves the current autocorrelation function, an average of $J_x(t')J_x(0)$ over the equilibrium ensemble. This autocorrelation function depends on $|t'|$, like the velocity autocorrelation function, approaching 0 when $|t'|$ becomes large. Thus one can put α equal to zero in the integral, which will still converge because of the behavior of the autocorrelation function. Also,

$$\langle J_x(t')J_x(0)\rangle = q^2 \sum_{i,j} \langle v_{xi}(t')v_{xj}(0)\rangle \approx q^2 \sum_{i} \langle v_{xi}(t')v_{xi}(0)\rangle$$

Correlations between velocities of different particles ($i \neq j$) have been neglected in reducing the double sum to a single sum over particles. The neglected terms are proportional to the square of the number density, whereas those kept are first power in the density, so that the derived results are correct in the limit of small density. The last sum consists of N identical terms, so the current density is

$$j_x = V^{-1}J_x(0) = V^{-1}\beta E q^2 \int_{-\infty}^{0} dt'\, N\langle v_{x1}(t')v_{x1}(0)\rangle$$

$$= \left[\frac{\rho q^2}{kT} \int_{-\infty}^{0} dt' \langle v_{x1}(t')v_{x1}(0)\rangle\right] E \tag{8.29}$$

with ρ the density of charged particles. The conductivity σ, which is the ratio of the current density to the electric field, is the quantity in square brackets in (8.29). The velocity autocorrelation function in (8.29) has been discussed above [see (8.15)]. Further discussion and interpretation of the conductivity formula will be given in Section 8.4.

What has been calculated, starting with (8.24), is the response of a system in thermal equilibrium to an external force (due to the electric field in the case of conductivity). The external force was treated as a small change in the Hamiltonian, which was written as $H_0 + H_1$. The system was assumed to be originally in thermal equilibrium with respect to H_0, and the change in the phase-space distribution function f_N due to H_1 was calculated. More common, and perhaps of greater interest, than the effect of an external force, is a system which is not in thermal equilibrium, and which returns to equilibrium. It is this situation, sometimes referred to as a "thermal perturbation," that is referred to in the Onsager hypothesis.

The treatment of thermal perturbations involves a reformulation of the problem so a thermal perturbation can be thought of as if it were created by a change in the Hamiltonian, similar to an external field. Thus, consider a system with Hamiltonian H_0 which is not a state of thermal equilibrium, so that $f_N(\Gamma_N)$ differs from the equilibrium distribution

$$f_{N0}(\Gamma_N) = \frac{e^{-\beta H_0}}{\int d\Gamma_N \, e^{-\beta H_0}}$$

Under the influence of H_0, f_N will change and eventually approach f_{N0}. One can imagine that f_N is the equilibrium distribution for some Hamiltonian H_T, which differs from H_0 by a term H_1:

$$f_{NT}(\Gamma_N) = \frac{e^{-\beta(H_0+H_1)}}{\int d\Gamma_N \, e^{-\beta(H_0+H_1)}}$$

The Hamiltonian H_1, which may be thought of as an external field, is then turned off. The system finds itself under the rule of the Hamiltonian H_0 and, not being at equilibrium with respect to H_0, it starts to change its distribution. Since the Hamiltonian which governs the change is H_0, the equation of motion is the Liouville equation with H_0 as Hamiltonian:

$$\frac{\partial f_N}{\partial t} = -\mathscr{L}_0 f_N(\Gamma_N;t) = -\sum_{i=1}^{3N}\left(\frac{\partial f_N}{\partial q_i}\left(\frac{\partial H_0}{\partial p_i}\right) + \frac{\partial f_N}{\partial p_i}\left(\frac{-\partial H_0}{\partial q_i}\right)\right)$$

The average value of any time-dependent dynamical variable, which would be time-independent if the Hamiltonian remained H_T, changes as well.

If the system is not very far from equilibrium (the only case to be considered here), f_{N1} is not very different from f_{N0}, meaning βH_1 is small. Then $f_N = f_{N0} + f_{N1}(t)$ with f_{N1} small and $\mathscr{L}_0 f_{N0} = 0$. The expectation value of a dynamical variable A at time 0 is

$$\int A(\Gamma_N) f_{NT}(\Gamma_N)\, d\Gamma_N = \frac{\int e^{-\beta(H_0+H_1)} A \, d\Gamma_N}{\int e^{-\beta(H_0+H_1)}\, d\Gamma_N}$$

$$\approx \frac{\int e^{-\beta H_0}(1-\beta H_1) A \, d\Gamma_N}{\int e^{-\beta H_0}\, d\Gamma_N - \beta \int e^{-\beta H_0} H_1 \, d\Gamma_N}$$

Since βH_1 is small, $\exp(-\beta H_1)$ has been approximated as $1 - \beta H_1$. Furthermore, the expression $(a-b)^{-1}$ with $b \ll a$ may be approximated as $[a(1-b/a)]^{-1} \approx a^{-1}(1+b/a)$, so

$$\int A(\Gamma_N) f_{NT}(\Gamma_N)\, d\Gamma_N \approx \frac{\int e^{-\beta H_0}(1-\beta H_1) A \, d\Gamma_N}{\int e^{-\beta H_0}\, d\Gamma_N}\left(1 + \beta \frac{\int e^{-\beta H_0} H_1 \, d\Gamma_N}{\int e^{-\beta H_0}\, d\Gamma_N}\right)$$

$$\approx \langle A \rangle - \beta \langle H_1 A \rangle + \beta \langle A \rangle \langle H_1 \rangle \tag{8.30}$$

A term in $(\beta H_1)^2$ has been dropped because it is very small. The fences indicate averages over the equilibrium ensemble, with phase-space distribution f_{N0}. According to (8.30), the difference between the expectation value of A at $t \to \infty$, when the system is at equilibrium and the distribution has become f_{N0}, and the expectation value of A at $t = 0$, when H_1 has just been turned off, is $\beta\langle H_1 A\rangle - \beta\langle A\rangle\langle H_1\rangle$.

Now the expectation value of A at $t > 0$ is to be evaluated. To distinguish expectation values for the nonequilibrium system from expectation values for the equilibrium system, which were indicated above by fences, we use square brackets for the former. The solution to $\partial f_N/\partial t = -\mathscr{L}_0 f_N$ may be written formally (see Section 8.1) as

$$f_N(t) = e^{-\mathscr{L}_0 t} f_N(0) \approx e^{-\mathscr{L}_0 t}\{f_{N0}(0)[1 - \beta H_1(0) + \langle \beta H_1\rangle]\}$$

Since $\mathscr{L}_0$ has no effect on f_{N0}, the expectation value of A at time t is

$$[A]_t = \int A(\Gamma_N) f_{N0}\left[1 - \beta\, e^{-\mathscr{L}_0 t} H_1(0) + \langle \beta H_1\rangle\right] d\Gamma_N$$

$$= \langle A\rangle - \beta\langle A(0) H_1(t)\rangle + \beta\langle A\rangle\langle H_1\rangle$$

One has therefore

$$[A]_{t'} - [A]_t = \beta\left[\langle A(0) H_1(t)\rangle - \langle A(0) H_1(t')\rangle\right] \tag{8.31}$$

In terms of the time-correlation functions defined by (8.14),

$$[A]_{t'} - [A]_t = \beta\left[c_{AH_1}(t) - c_{AH_1}(t')\right] \tag{8.32}$$

(see Problem 8.5). Note that $c_{AH_1}(t) \to 0$ as $t \to \infty$. With $A = H_1$, (8.32) becomes

$$[H_1]_{t'} - [H_1]_t = \beta\left[c_{H_1 H_1}(t) - c_{H_1 H_1}(t')\right]$$

The left-hand side gives the change in the value of the imposed fluctuation H_1 with time, and the right-hand side is proportional to the change of the corresponding spontaneous fluctuation in the equilibrium system. If $t' \to \infty$, $[H_1]_{t'} = \langle H_1\rangle$ and $c_{H_1 H_1}(t') \to 0$, so

$$[H_1]_t - \langle H_1\rangle = -\beta c_{H_1 H_1}(t) \tag{8.33}$$

for any H_1.

Since H_1 is arbitrary except that it is a possible Hamiltonian term, meaning that it has the units or dimensions of energy, kH_1 can be anything, provided that the constant k takes care of the units. Then $A = -kH_1$ is an arbitrary operator and (8.33) may be rewritten

$$[A]_t - \langle A \rangle = \beta k^{-1} c_{AA}(t)$$

If this equation is divided by the corresponding equation for $t = 0$, one gets

$$\frac{[A]_t - \langle A \rangle}{[A]_0 - \langle A \rangle} = \frac{c_{AA}(t)}{c_{AA}(0)} \tag{8.34}$$

This is the same as (8.22), Onsager's hypothesis. Onsager was right.

As an example of the use of (8.34), let A be the operator for number density at point $\mathbf{r}$:

$$A = \sum_{i=1}^{N} \delta(\mathbf{r}_i - \mathbf{r}) = N\delta(\mathbf{r}_1 - \mathbf{r})$$

Then (8.34) describes how an imposed deviation of the number density from its average value decreases with time. If extra particles are put at a point $\mathbf{r}$ in a fluid, it is expected that the excess will decrease with time; because of the random motion of particles, more particles leave the point $\mathbf{r}$ than come to it. One is dealing here with diffusion.

The autocorrelation function involved is the density–density autocorrelation function, but it turns out to be useful to examine the more general density–density time-correlation function

$$\langle A(\mathbf{r} + \mathbf{R}, t) A(\mathbf{r}, 0) \rangle$$

This function is proportional to the probability of having a particle at $\mathbf{r}$ at time 0 and a particle at $\mathbf{r} + \mathbf{R}$ at time t. The connection to the autocorrelation function $c_{AA}(t)$ is through the function

$$\begin{aligned} G(\mathbf{R}, t) &\equiv \langle [A(\mathbf{r} + \mathbf{R}, t) - \rho][A(\mathbf{r}, 0) - \rho] \rangle \\ &= \langle A(\mathbf{r} + \mathbf{R}, t) A(\mathbf{r}, 0) \rangle - \rho^2 \end{aligned} \tag{8.35}$$

where ρ is the average density; $G(\mathbf{R}, t)$ becomes equal to $c_{AA}(t)$ when $\mathbf{R} = 0$. Properties of $\mathbf{G}(\mathbf{R}, t)$ include: it approaches 0 for $R \to \infty$ (t finite) and 0 for $t \to \infty$ (R finite), while its value for $t = 0$ is proportional to $\delta(\mathbf{R})$ (a particle cannot be in two places at the same time). In an isotropic fluid, G depends only on $\mathbf{R}$, the magnitude of $\mathbf{R}$.

As discussed in the next section, a particle in a fluid is constantly undergoing collisions which change its velocity, so that its trajectory is a series of steps of varying length and apparently random direction (see Fig. 8.3). The result of this "random walk" is that, for a particle observed for a time t, the

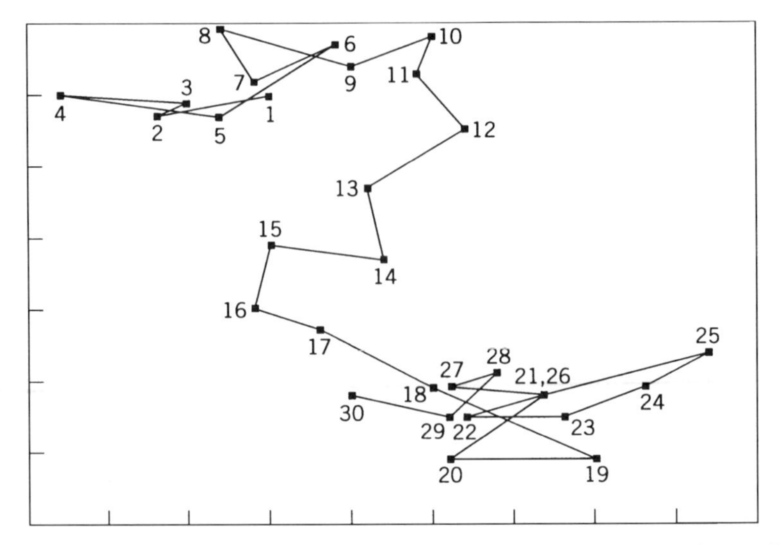

Figure 8.3. Random walk in two dimensions. Numbers indicate sucessive positions after steps of random length in x and y directions.

average distance it gets from its position at $t = 0$ is proportional to $\sqrt{t}$. Usually, one writes $\langle R^2 \rangle_{\text{av}} = 6Dt$ where D is the diffusion coefficient. The probability that at time t the particle is at a point a distance R from where it was at time 0 is

$$P(R,t) = (4\pi Dt)^{-3/2} e^{-R^2/4Dt} \tag{8.36}$$

(P is normalized so that the integral over all space is 1). The probability of having a particle at $\mathbf{r}$ at time 0 and a particle at $\mathbf{r} + \mathbf{R}$ at time t is the sum of (a) the probability that the two particles are not the same, which is ρ^2 (no correlation) and (b) the probability that it is the same particle that was at $\mathbf{r}$ and has migrated to $\mathbf{r} + \mathbf{R}$ during the time t, which is $\rho P(r,t)$. Therefore,

$$G(\mathbf{R},t) = \rho^2 + \rho P(R,t) - \rho^2 \tag{8.37}$$

At $t \to 0$, $G(\mathbf{R},t) \to 0$ for all $\mathbf{R} \neq 0$ and $G(\mathbf{R},0)$ is infinite at $\mathbf{R} = 0$; because $\int G(\mathbf{R},t)\,\mathbf{dR}$ is equal to ρ for all t, $G(\mathbf{R},0)$ is just $\rho\delta(\mathbf{R})$. This unphysical behavior arises because (8.36) is not valid for very short times, but only when the migrating particle has undergone enough collisions to make statistical averaging meaningful.

Now we can return to our problem, the return to equilibrium of a fluid which is not at equilibrium because the particle density is not uniform.

According to equations (8.35) to (8.37), the density–density autocorrelation function is

$$c_{AA}(t) = G(0,t) = (4\pi Dt)^{-3/2}$$

except for very small values of t. Therefore an imposed density fluctuation at a single point, denoted by F, decreases according to

$$\frac{F(t) - F(t')}{F(t)} = \frac{c_{AA}(t) - c_{AA}(t')}{c_{AA}(t)} = 1 - \left(\frac{t}{t'}\right)^{3/2}$$

The formalism may be extended to consider more complicated situations, in which the deviation from equilibrium is not localized (a delta-function), but a sum or integral of delta-functions, which, as has been seen earlier, is quite general.

The next application of this formalism also relates to diffusion. Suppose the Hamiltonian which takes the system out of equilibrium is

$$H_1 = g \sum_i x_i$$

with g small. This looks like a gravitational field in the x-direction, and will make the density of particles in a fluid at constant temperature follow the barometric formula

$$\rho(x) = \rho_0\, e^{-\beta g x} \approx \rho_0(1 - \beta g x)$$

For small g, this is essentially a constant density gradient, $d\rho/dx = -\beta\rho_0 g$. If H_1 is turned off at $t = 0$, allowing the system to relax, there will be a flow or flux of particles in the positive x-direction. The flux density at $x = 0$ is the expectation value of the operator

$$j(0) = \sum_{i=1}^{N} \delta(x_i) v_{xi}(0)$$

Then (8.32) is

$$[j(0)]_{t'} - [j(0)]_t = \beta\left[c_{jH_1}(t) - c_{jH_1}(t')\right]$$

The correlation function is

$$c_{jH_1}(t) = g\left\langle \left[\sum_{i=1}^{N} \delta(x_i) v_{xi}(0)\right]\left[\sum_{j=1}^{N} x_j(t)\right]\right\rangle \tag{8.38}$$

For $t' \to \infty$, the system approaches equilibrium and the flux vanishes; also, $c_{jH_1}(t') \to 0$ for $t' \to \infty$. Thus $[j(0)]_t = -\beta c_{jH_1}(t)$.

The time-correlation function is a property of the equilibrium system, in which the velocity of one particle is uncorrelated with the position of another. Thus i and j must be the same in (8.38). Since the density of particles at $x = 0$ is always ρ_0,

$$c_{jH_1}(t) = g\rho_0 \langle v_x(0)x(t)\rangle$$

where v_x and x are the x-velocity and x-position of a single particle. Since $\langle v_x(0)x(t)\rangle$ is identical to $\langle v_x(-t)x(0)\rangle$, one may write

$$[j(0)]_t = -\beta g\rho_0 \langle v_x(-t)x(0)\rangle = -\beta g\rho_0 \frac{d}{dt}\langle x(-t)x(0)\rangle \qquad (8.39)$$

The flux is proportional to the density gradient and to the time derivative of the position–position autocorrelation function. To evaluate this factor, we use the fact that the mean-square distance traveled by a particle in an isotropic fluid in time t is $6Dt$ with D the diffusion coefficient. The x-, y-, and z-directions being equivalent, the mean-square distance traveled in the x-direction is

$$2Dt = \langle [x(t) - x(0)]^2\rangle = \langle x^2(t)\rangle + \langle x^2(0)\rangle - 2\langle x(0)x(t)\rangle$$

Now $\langle x^2(t)\rangle$ is really independent of t so differentiating gives

$$2D = -2\frac{d}{dt}\langle x(-t)x(0)\rangle \qquad (8.40)$$

and (8.39) becomes

$$[j(0)]_t = \beta g\rho_0 D = -D(d\rho/dx)$$

The flux of particles is proportional to the negative of the density gradient, which is Fick's law of diffusion. The constant of proportionality is the diffusion coefficient.

In Sections 8.1 to 8.3, we have tried to show how results of time-dependent statistical mechanics are proved. Some properties of distribution functions in phase space and time-correlation functions were studied. Two questions were answered: (1) If an external field or other perturbation acts on a system at equilibrium, what is the response of the system to first order (response proportional to the perturbation)? (2) As a system not at equilibrium returns to equilibrium, how do its properties change? As an example of (1), we considered electrical conduction; as an example of (2), we considered diffusion. In Section 8.4, we return to conduction and diffusion for more discussion.

8.4. CONDUCTION, DIFFUSION, AND THE RANDOM WALK

According to (8.29), the current density produced by an electric field E in a fluid is given by

$$j_x = \left[\frac{\rho q^2}{kT} \int_{-\infty}^{0} dt' \langle v_{x1}(t') v_{x1}(0) \rangle \right] E$$

where ρ is the density of particles of charge q, so the quantity in square brackets is the conductivity σ. The velocity autocorrelation function has the value kT/m at $t' = 0$ (average value of v_x^2 at thermal equilibrium), and approaches 0 for large t'. Introducing the time constant τ according to (8.16), we have for the conductivity

$$\sigma = \frac{\rho q^2}{m} \tau \tag{8.41}$$

Equation (8.41) can be given a simple physical interpretation. The current produced in a fluid by an electric field is the result of competition between the acceleration of particles by the field and their deceleration by collisions with other particles. Just after collision, charged particles are supposed to have an average x-velocity v_x of zero. They are accelerated by the electric field until undergoing another collision, at which time the average x-velocity is again reduced to zero. The force qE of the electric field on a charged particle is equal to $m(dv_x/dt)$. If T is the average time between collisions for a particle, an observed particle would undergo acceleration, on the average, for a time T, and have an average x-velocity of

$$m^{-1} \int_{t}^{t+T} Eq\, dt = \frac{EqT}{m}$$

The average current density would then be $\rho q(EqT/m)$ and the conductivity $\rho q^2 T/m$. Thus τ in (8.41) is interpreted as the average time between collisions, which is the mean free path divided by the average relative speed (see Section 6.3).

That a particle loses all memory of which way it was going whenever it collides with another particle is an exaggeration. Probably it takes more than one collision per particle to reduce its average x-velocity to zero. The time constant τ which appears in (8.41) should be somewhat longer than, but of the same magnitude as, the average time between collisions. Note also that the charged particles whose motion is responsible for electric current collide with particles of other kinds, charged or uncharged. The average time between collisions must be calculated taking all the collision probabilities into account.

The obvious application of (8.41) is to metals, in which the current-carrying particles are the conduction electrons ($m = 9.1 \times 10^{-31}$ kg and $q = 1.6 \times 10^{-19}$ C). The particles of opposite charge, the ion cores, are essentially immobile and do not carry current. With one conduction electron per metal atom and one atom for 3 cubic Angstroms, ρ would be 3.3×10^{29} m^{-3}. The conductivity of silver is 6.7×10^{7} (ohm m)$^{-1}$. According to (8.41), the time constant τ is

$$\frac{6.1 \times 10^{-23}\ \mathrm{C\ kg\ V^{-1} s^{-1}}}{(3.3 \times 10^{29}\ \mathrm{m^{-3}})(1.6 \times 10^{-19}\ \mathrm{C})^2} = 7.2 \times 10^{-15}\ \mathrm{kg\ m^2\ J^{-1}}$$

or 7.2×10^{-15} s. This implies, if the average speed of electrons at room temperature is estimated as $\sqrt{3kT/m} = 1.2 \times 10^{5}$ m/s, that the average distance an electron travels between collisions is of the order of 10^{-9} m or 10 Angstroms. (The estimates used for the effective mass and the average speed are quite crude, but suffice to show how small is the mean free path.) It turns out that the "particles" with which the electrons collide, limiting the mean free path, are the phonons of vibrational energy. This is why increasing the temperature, by raising the amount of vibrational energy, lowers τ and hence the conductivity.

Other metals may have conductivities one or two orders of magnitude smaller than that of silver. Conductivities of intrinsic semiconductors are much lower, ranging from 0.001 to 2 (ohm m)$^{-1}$. The reason for the low conductivities of semiconductors is, of course, the low density of current-carrying species. Since this density increases exponentially as the temperature increases, the conductivity of intrinsic semiconductors increases rapidly with temperature.

It is not difficult to generalize the conductivity formula to a fluid containing several different charged species. The perturbing Hamiltonian would be

$$H_1 = -E \sum_{i=1}^{N} q_i x_i e^{\alpha t}$$

and the current

$$J_x = \sum_{i=1}^{N} q_i \frac{dx_i}{dt}$$

allowing for the possibility that different particles have different charges. Equation (8.28) is unchanged, except for the new definition of the current:

$$\langle J_x(0) \rangle = \beta E \int_{-\infty}^{0} dt'\, e^{\alpha t'} \langle J_x(t') J_x(0) \rangle$$

where

$$\langle J_x(t')J_x(0)\rangle \approx \sum_i q_i^2 \langle v_{xi}(t')v_{xi}(0)\rangle \tag{8.42}$$

Correlations between velocities of different particles are neglected, as was done in going from (8.28) to (8.29); the results are strictly valid only for low densities of charged particles. The sum over particles may be written as a sum over species. If there are N_k particles of species k, one has instead of (8.29)

$$j_x = V^{-1}J_x(0) = \left[V^{-1}\beta \sum_k N_k q_k^2 \int_{-\infty}^0 dt' \langle v_{xk}(t')v_{xk}(0)\rangle\right] E$$

where v_{xi} is the x-velocity of a particle of species i. The conductivity is

$$\sigma = \beta \sum_k \rho_k q_k^2 \int_{-\infty}^0 dt' \langle v_x(t')v_x(0)\rangle_k \tag{8.43}$$

where $\rho_k = N_k/V$. It should be noted that the electric field E in the above expressions is the local electric field which acts on the charged particles; it may differ from the field outside the medium because of such effects as dielectric screening.

For each species i, one may define a time constant τ_i according to [see (8.16)]

$$\frac{kT}{m_i}\tau_i = \int_{-\infty}^0 dt' \langle v_{xi}(t')v_{xi}(0)\rangle$$

The velocity autocorrelation function differs for different species of particles. Then the total current density is

$$j_x = \left[\sum_k \rho_k q_k^2 \tau_k m_k^{-1}\right] E \tag{8.44}$$

This is a sum of contributions of the different charged species, the contribution of species i being the product of the charge q_i, the number density ρ_i, and the average velocity of an ion of species i. The average velocity is proportional to the strength of the field E. It is common in electrochemistry to define the mobility of an ionic species as the average velocity (sometimes referred to as the "drift velocity") of that ion in unit electric field. According to (8.44), the average velocity is $q_i\tau_i E/m_i$, so the mobility is

$$U_i = |q_i|\frac{\tau_i}{m_i}$$

(mobilities are always taken to be positive).

Outside of electrochemistry, it is common to define the friction coefficient ζ_i as the ratio of the average velocity of a particle to the average force acting on it. Since the force exerted on an ion of species i by an electric field E is $q_i E$, the friction coefficient is related to the electrochemical mobility by

$$\zeta_i = \frac{U_i}{|q_i|} = \frac{\tau_i}{m_i} \tag{8.45}$$

The friction coefficient can be defined for any force, not only electrostatic. In the case of diffusion caused by a concentration gradient, we showed that the flux of a species is equal to D, the diffusion coefficient, multiplied by the negative of the concentration gradient. Since the flux is the particle density ρ multiplied by the velocity, the average velocity of a diffusing particle is

$$-D\rho^{-1}\nabla\rho = \left[\frac{d}{dt}\langle x(t)x(0)\rangle\right]\nabla \ln \rho$$

Equation (8.40) has been used for D. The quantity $kT\nabla \ln \rho$ has the physical dimensions of energy divided by distance, or force. One may calculate the friction coefficient for diffusion as the velocity divided by the "generalized force":

$$\zeta = -(kT)^{-1}\frac{d}{dt}\langle x(t)x(0)\rangle \tag{8.46}$$

This friction coefficient is obviously related to the friction coefficient of (8.45), which is

$$\frac{\tau_i}{m_i} = (kT)^{-1}\int_{-\infty}^{0} dt'\langle v_{xi}(t')v_{xi}(0)\rangle \tag{8.47}$$

It will be shown that they are identical. The origin of the friction coefficient is the same in both cases: interparticle collisions prevent a particle from moving in a straight line.

The calculation showing (8.46) is the same as (8.47) is a bit involved. Remember that (8.46) was derived [see (8.40)] by assuming that the mean square displacement of a particle in one direction is $2Dt$ for large t, that is,

$$\zeta = (kT)^{-1}\left\{\lim_{t\to\infty}\frac{\langle[x(0)-x(t)]^2\rangle}{2t}\right\} \tag{8.48}$$

For any particle,

$$x(t) = x(0) + \int_0^t dt'\, v_x(t')$$

so the bracket in (8.48) is

$$\lim_{t\to\infty} \frac{\left\langle \int_0^t dt'\, v_x(t') \int_0^t dt''\, v_x(t'')\right\rangle}{2t}$$

$$= \lim_{t\to\infty} \frac{\left(\int_{-t}^{0} dw \int_{-(1/2)w}^{(1/2)w+t} du + \int_0^t dw \int_{(1/2)w}^{-(1/2)w+t} du\right)\left\langle v_x\left(u-\frac{1}{2}w\right) v_x\left(u+\frac{1}{2}w\right)\right\rangle}{2t}$$

where $u = \frac{1}{2}(t' + t'')$ and $w = t'' - t'$. Figure 8.4, showing the area of integration and the equations of some of the boundary lines, may clarify this change of variables. From the properties of autocorrelation functions $\langle v_x(u - \frac{1}{2}w) v_x(u + \frac{1}{2}w)\rangle$ is equal to $\langle v_x(0) v_x(w)\rangle$, which permits the integrations over u to be carried out. Then

$$\zeta = (kT)^{-1}\left\{\lim_{t\to\infty} \frac{\int_{-t}^{t} dw\, t\langle v_x(0) v_x(w)\rangle - \int_{-t}^{t} dw\, |w| \langle v_x(0) v_x(w)\rangle}{2t}\right\}$$

Because $\langle v_x(0) v_x(w)\rangle$ is an even function of w, neither integral is zero, but because the autocorrelation function approaches zero rapidly as $|w|$ gets

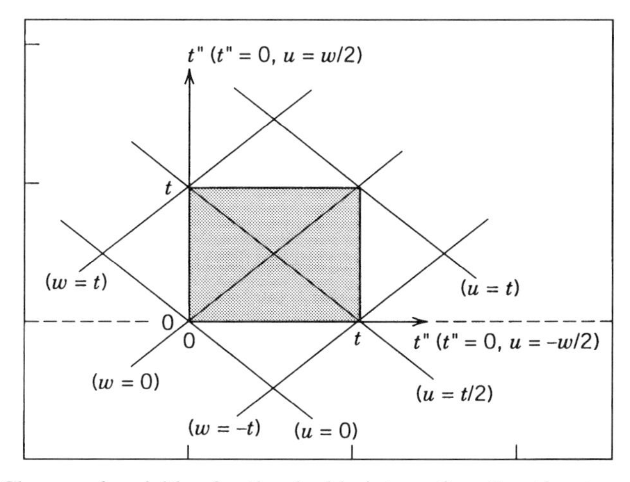

Figure 8.4. Change of variables for the double integration, $0 \le t' \le t$ and $0 \le t'' \le t$ (hatched region). With $u = \frac{1}{2}(t' + t'')$ and $w = t'' - t'$, the integrals become

$$\int_{-t}^{0} dw \int_{-(1/2)w}^{(1/2)w+t} du + \int_0^t dw \int_{(1/2)w}^{-(1/2)w+t} du$$

The lines $w = -t$, $w = 0$, and $w = t$ are shown, as well as the lines $u = 0$, $u = \frac{1}{2}t$, and $u = t$.

large, the value of the second integral remains finite as $t \to \infty$. After division by t, it makes no contribution to ζ, and

$$\zeta = (kT)^{-1}\left\{\frac{1}{2}\int_{-\infty}^{\infty} dw\langle v_x(0)v_x(w)\rangle\right\}$$

which is identical to (8.47).

Thus the particles of species i react to the generalized force $kT\nabla \ln \rho$ in exactly the same way as they react to the true force $q_i E$. This is expressed by the Nernst–Einstein relation between mobility and diffusion coefficient:

$$D_i = \frac{U_i kT}{|q_i|} \tag{8.49}$$

which is valid for a dilute solution. The flux of particles of species i in an electrolyte solution may be considered to consist of two contributions, the diffusive flux due to the concentration gradient and the migration due to the electric field. Using the Nernst–Einstein relation, one may write

$$\mathbf{j}_i = -D_i \nabla \rho_i + \rho_i q_i U_i \mathbf{E} = -\frac{\rho_i D_i}{kT}(kT \nabla \ln \rho_i + q_i \nabla\phi) \tag{8.50}$$

where ϕ is the electrical potential ($\mathbf{E} = -\nabla\phi$). In a dilute solution, approaching ideality, the chemical potential μ_i is equal to $\mu_i^0 + kT \ln \rho_i$ (μ_i^0 refers to the standard state of infinitely dilute solution with unit number density of solute species i) so that (8.50) may be written

$$\mathbf{j}_i = -\frac{\rho_i D_i}{kT}\nabla(\mu_i + q_i\phi) = -\frac{\rho_i D_i}{dT}\nabla\tilde{\mu}_i$$

The electrochemical potential $\tilde{\mu}_i$ is the sum of chemical and electrical parts. This equation is valid even when solute concentrations are too high for the ideal dilute solution law to hold for the chemical potential.

The existence of a friction constant implies that on average a particle reaches a terminal velocity and that this velocity is proportional to the imposed force. The fact that the velocity is constant (no acceleration) means that the net force on the particle is zero. Thus, in addition to the imposed force $\mathbf{F}$, there is an opposing force proportional to particle velocity so that

$$\mathbf{F} - \zeta\mathbf{v} = 0$$

and $\zeta = \mathbf{F}/\mathbf{v}$. In a gas or liquid, the opposing force is the average effect of collisions, as will be discussed in more detail in the next section. In a liquid, the opposing force is usually referred to as a viscous drag and explained by assuming that the liquid at the surface of the moving particle moves with the particle, whereas liquid far away from the particle remains stationary. Thus a particle moving at velocity $\mathbf{v}$ sets up a velocity gradient in the liquid, and the force $\mathbf{F}$ is required to maintain a velocity gradient because of viscosity (see Section 6.4). The friction constant is proportional to the viscosity of the liquid. This explanation of the friction constant is most appropriate when the particle is much larger than the molecules of the liquid, and when the density is so high that it is inconvenient to consider individual collisions between the moving particle and the liquid molecules.

In addition to viscosity, two effects involving interionic interactions are important to the friction coefficient of an ion. Both make the friction coefficient dependent on total ion concentration, and both involve ions of different signs of charge, which are always present, moving in opposite directions. The "electrophoretic effect" refers to the tendency of an ion to carry solvent molecules with it so that a nearby ion of opposite charge must move through solvent molecules which are moving in the opposite direction. The "relaxation effect" relates to the ion atmosphere (Section 7.6): the electric field which moves an ion in one direction moves its ion atmosphere, which is charged oppositely to the ion, in the other direction. When the ion is no longer at the center of the ion atmosphere, the atmosphere exerts a restoring force on the ion and vice versa. It takes some time for the ion atmosphere to relax to its new equilibrium position, which slows down the moving particle and increases the friction constant.

As shown, the friction constant relates to diffusion as well as to migration in an electric field. An important general result about diffusion has been used: on the average, the distance a diffusing particle gets from its starting point in time t is proportional to $\sqrt{t}$. This result will now be derived, using a model in which the diffusing particle takes a series of steps in random directions (Figure 8.3). In a gas, the length of a step depends on the frequency of collisions; in a liquid or solid, the step length is related to the structure (distance between possible equilibrium positions for a particle).

Motion in one dimension will be considered first. Most physical situations are concerned with two- or three-dimensional space, but it will be assumed that randomness includes the independence of different spatial directions. A further simplification is to assume each step is of the same length. This is not a serious limitation as long as the result of many steps is the interest; the step length λ represents the average step length of a more realistic system (perhaps the mean free path in a gas). In one dimension, the step can be to the left or to the right. Let the probability of stepping right be p, the probability of stepping left be $1 - p$. The same probabilities obtain for each step; the moving particle has no memory of its history. For an unbiased walk, $p = 1 - p = \frac{1}{2}$.

Suppose n steps of length λ are taken. If r of these are to the right, the distance moved by the particle (positive to the right) is

$$x = r\lambda - (n - r)\lambda = (2r - n)\lambda$$

The probability that r steps are to the right and $n - r$ are to the left is

$$W(r, n) = p^r(1 - p)^{n-r} \frac{n!}{r!(n - r)!} \tag{8.51}$$

The quotient of factorials is a binomial coefficient, equal to the number of ways of choosing r of n distinguishable objects (i.e., which of the n steps are to the right). The average value of r is

$$\begin{aligned}
\sum_{r=0}^{n} rW(r, n) &= \sum_{r=0}^{n} p^r(1 - p)^{n-r} \frac{rn(n - 1)!}{r!(n - r)!} \\
&= np \sum_{r=1}^{n} p^{r-1}(1 - p)^{n-r} \frac{(n - 1)!}{(r - 1)!(n - r)!} \\
&= np \sum_{s=0}^{n-1} p^s(1 - p)^{n-1-s} \frac{(n - 1)!}{s!(n - 1 - s)!} = np
\end{aligned}$$

because the last sum is $(p + 1 - p)^{n-1}$. The most probable value of r is also np (Problem 8.7). Similarly, the average value of r^2 is

$$\begin{aligned}
&\sum_{r=0}^{n} [r(r - 1) + r]W(r, n) \\
&\quad = n(n - 1)p^2 \sum_{r=2}^{n} p^{r-2}(1 - p)^{n-r} \frac{(n - 2)!}{(r - 2)!(n - r)!} + \langle r \rangle_{\text{av}} \\
&\quad = n(n - 1)p^2 \sum_{s=0}^{n-2} p^s(1 - p)^{n-2-s} \frac{(n - 2)!}{s!(n - 2 - s)!} + np \\
&\quad = n(n - 1)p^2 + np
\end{aligned}$$

The mean-square fluctuation in r is $\langle r^2 \rangle - \langle r \rangle^2 = np(1 - p)$. As n increases, the root-mean-square fluctuation in r decreases relative to the average value of r:

$$\frac{\sqrt{\langle r^2 \rangle - \langle r \rangle^2}}{\langle r \rangle} = \sqrt{\frac{(1 - p)}{np}} \tag{8.52}$$

This indicates that the probability function $W(r, n)$ becomes narrower, lowering the probability that r differs significantly from its average value.

For n large, therefore, only large values of r and $n - r$ are of interest (unless p is very close to 0 or to 1). Thus the factorials in $W(r, n)$ may be approximated by the Stirling approximation (Section 1.1 and Problem 1.2)

$$\ln N! \approx \ln\sqrt{2\pi} + \left(N + \tfrac{1}{2}\right)\ln N - N$$

Using this in (8.51) gives

$$W(r,n) = p^r(1-p)^{n-r}\frac{n^{n+1}}{\sqrt{2\pi n}\; r^{r+(1/2)}(n-r)^{n-r+(1/2)}} \equiv \frac{p^r(1-p)^{n-r}}{\sqrt{2\pi n}}T$$

We want to treat r as a continuous variable. Since the difference between r and np will be small compared to n, define ε by

$$\varepsilon n = r - np$$

so that ε will be small compared to 1. Now

$$\begin{aligned} T &= \left(\frac{n}{r}\right)^{r+(1/2)}\left(\frac{n}{n-r}\right)^{n-r+(1/2)} \\ &= (p+\varepsilon)^{-n(p+\varepsilon)-(1/2)}(1-p-\varepsilon)^{-n(1-p-\varepsilon)-(1/2)} \end{aligned}$$

The case of $p = 1 - p = \frac{1}{2}$ will be considered. The algebra is easier to deal with if one looks at $\ln T$, expanding in terms through ε^2.

$$\begin{aligned} \ln T &= -\tfrac{1}{2}(n+1+2n\varepsilon)\ln\left(\tfrac{1}{2}+\varepsilon\right) - \tfrac{1}{2}(n+1-2n\varepsilon)\ln\left(\tfrac{1}{2}-\varepsilon\right) \\ &\approx -(n+1)\ln\left(\tfrac{1}{2}\right) - 2(n-1)\varepsilon^2 \end{aligned}$$

Inserting the definition of ε,

$$T \cong 2^{n+1}\exp\left[-2(n-1)\left(\frac{2r-n}{2n}\right)^2\right]$$

and

$$W(r,n) = \frac{2}{\sqrt{2\pi n}}\exp\left[-2n\left(\frac{2r-n}{2n}\right)^2\right] \tag{8.53}$$

Since n is large, $n - 1$ in the exponential has been replaced by n.

This is normalized in the sense that, when the sum over r is approximated by an integral,

$$\int_{-\infty}^{\infty} W(r,n)\,dr = 1$$

The approximation of the binomial distribution (8.51) by the Gaussian,

$$W(r, n) = \frac{2}{\sqrt{2\pi n}} e^{-2(r - \langle r \rangle)^2/n} \tag{8.54}$$

is a good one even for small n, if one considers values of r near the average or most probable value. Figure 8.5 shows the comparison for $n = 11$. For $n = 15$ and $r = 5$, 6, and 7, the Gaussian (8.54) gives $W(r, n)$ equal to 0.08953, 0.15262, and 0.19926, respectively. The binomial distribution gives 0.09164, 0.15274, and 0.19638, so the errors are 2.4%, 0.1%, and 1.4%, respectively.

The probability of taking between r and $r + dr$ steps to the right is $W(r, n)\, dr$. The probability of moving a distance between x and $x + dx$, $P(x, n)\, dx$, obeys

$$P(x, n)\, dx = W(r, n)\, dr$$

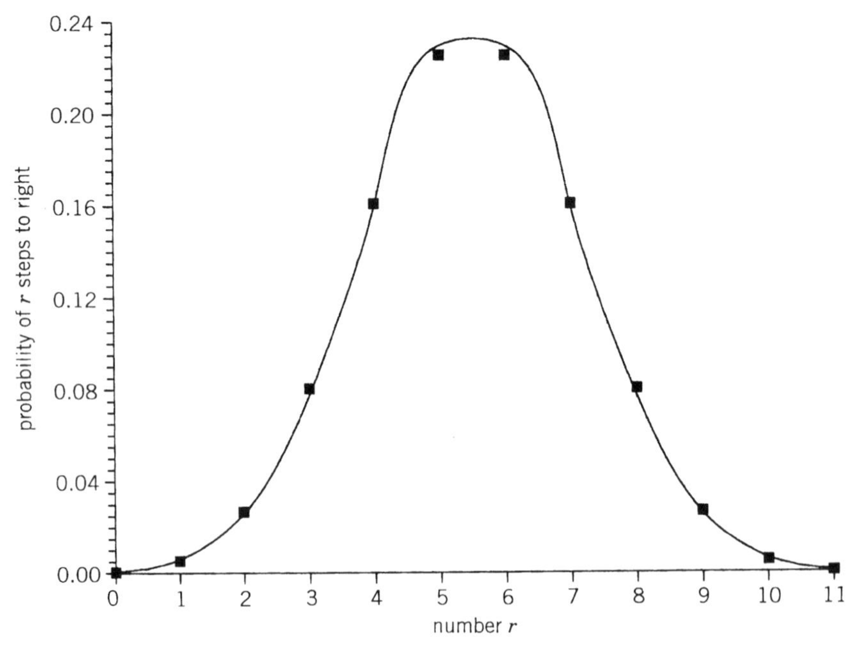

Figure 8.5. Binomial distribution for $n = 11$ and large-n approximation. Points are the binomial distribution $W(r, n) = (\frac{1}{2})^{11}(11!)/[r!(11 - r)!]$; the curve is the approximation to it,

$$W(r, n) = \sqrt{\frac{2}{11\pi}} \exp\left[-22\left(\frac{2r - 11}{22}\right)^2\right]$$

Since the net distance x moved after n steps of length λ, r of them to the right, is given by $(2r - n)\lambda$,

$$P(x, n) = W(r, n)\frac{dr}{dx} = \frac{1}{2\lambda}\frac{2}{\sqrt{2\pi n}}e^{-x^2/2n\lambda^2} \tag{8.55}$$

The average square distance traversed by a particle is

$$\int_{-\infty}^{\infty} P(x, n)x^2\,dx = \frac{1}{\sqrt{2\pi n\lambda^2}}\int_{-\infty}^{\infty} e^{-x^2/2n\lambda^2}x^2\,dx = n\lambda^2$$

and the root-mean-square distance is proportional to $\sqrt{n}$. If one is interested in distance traversed in time t, one replaces n by st, where s is the number of steps taken per unit time. The root-mean-square distance traversed is $\lambda\sqrt{st}$. If the diffusion coefficient D is defined as $\frac{1}{2}s\lambda^2$, the root-mean-square distance is $\sqrt{2Dt}$.

The reason for this definition of D is Fick's law of diffusion, that the flux of particles is given by $-D$ multiplied by the concentration gradient. Instead of considering $P(x, n)$ or $P(x, t)$ to give the probability that one particle moves a distance x, consider a collection of particles, each independently executing a random walk in one dimension. Then $P(x, t)\,dx$, multiplied by the number of particles, gives the number of particles found between x and $x + dx$ after time t, if all particles were at $x = 0$ at time 0. The density or concentration of particles is

$$P(x, t) = \frac{1}{\lambda\sqrt{2\pi st}}e^{-x^2/2st\lambda^2}$$

Fick's law requires

$$j_x = -D\frac{\partial P(x, t)}{\partial x} = \frac{Dx}{\sqrt{2\pi st}\;\lambda^3 st}e^{-x^2/2st\lambda^2} \tag{8.56}$$

and the flux is related to $\partial P/\partial t$ by the equation of continuity

$$\frac{\partial j_x}{\partial x} = -\frac{\partial P}{\partial t}$$

Differentiating j_x of (8.56) with respect to x, one sees that the continuity equation is obeyed (Problem 8.9) if $D = \frac{1}{2}s\lambda^2$. Thus the one-dimensional random walk model verifies the diffusion law, and gives the diffusion constant equal to $\frac{1}{2}s\lambda^2$.

For a random walk in three dimensions, it is assumed that each step moves the particle a distance λ in the positive or negative x-direction, a distance λ

in the positive or negative y-direction, and a distance λ in the positive or negative z-direction. The distance moved in each step is thus $\lambda\sqrt{3}$. If the x-, y-, and z-directions are independent, the probability of being between x and $x + dx$, between y and $y + dy$, and between z and $z + dz$ after n steps is

$$P(x, y, z)\,dx\,dy\,dz = P(x, n)P(y, n)P(z, n)\,dx\,dy\,dz$$

$$= \frac{1}{\lambda^3(2\pi n)^{3/2}}\,e^{-(x^2+y^2+z^2)/2n\lambda^2}\,dx\,dy\,dz$$

$$= \frac{1}{\lambda^3(2\pi n)^{3/2}}\,e^{-r^2/2n\lambda^2}r^2\,dr\,\sin\theta\,d\theta\,d\theta$$

Spherical polar coordinates have been introduced so that the probability of moving a distance between r and $r + dr$, regardless of direction, can be obtained by integrating over angles:

$$P(r, n) = \frac{4\pi}{\lambda^3(2\pi n)^{3/2}}\,e^{-r^2/2n\lambda^2}r^2\,dr$$

If the diffusion constant is defined, as above, by $D = \frac{1}{2}s\lambda^2$ so that $Dt = \frac{1}{2}n\lambda^2$, the three-dimensional radial probability distribution is

$$P(r, t)\,dr = \frac{1}{\sqrt{4\pi}\,(Dt)^{3/2}}\,e^{-r^2/4Dt}r^2\,dr \tag{8.57}$$

The mean square distance traveled from the origin is

$$\int_0^\infty P(r, t)r^2\,dr = \frac{1}{\sqrt{4\pi}\,(Dt)^{3/2}}\int_0^\infty e^{-r^2/4Dt}r^4\,dr = 6Dt$$

This is three times the result for one dimension. It must be remembered that the step size in three dimensions is actually $\sqrt{3}$ times the step size in one dimension.

Diffusion in a liquid or solid may also be calculated from a random walk, but the physical model is somewhat different. A diffusing particle is at an equilibrium position with respect to the forces exerted by neighboring particles. Nearby, there are m other similar positions or sites, separated from the original position by a potential barrier, which determines the probability per unit time that the particle jumps to one of these sites. Let λ_i be the distance of the ith new position from the original equilibrium position, let p_i be the probability that a particular jump is to the ith new position, and let θ_i be the angle between the x-axis and the vector from the original position to the ith position. The average distance a particle moves in the x direction as a result of the kth jump is

$$d_k = \left(\sum_{i=1}^{m} p_i \lambda_i \cos \theta_i \right)_k \equiv \langle \lambda \cos \theta \rangle \tag{8.58}$$

This is zero if there is no external field to make the probabilities differ for jumps in the positive or negative x-directions.

The average distance moved in the x-direction after n jumps is n times $\langle \lambda \cos \theta \rangle$ and vanishes. The average square of the distance moved in the x-direction is

$$\langle x^2 \rangle = \sum_{h;\,k} d_h \, d_k = \sum_{h} d_h^2 = n \langle \lambda^2 \cos^2 \theta \rangle$$

since there is no correlation between distance moved on different jumps ($\langle d_h \, d_k \rangle = 0$ for $h \neq k$). If there is no correlation between λ_i and $\cos \theta_i$,

$$\langle x^2 \rangle = n \langle \lambda^2 \rangle \langle \cos^2 \theta \rangle$$

The average value of $\cos^2 \theta$ is $1/3$, and the average value of λ^2 is the square of the average step length. Of course, $\langle y^2 \rangle$ and $\langle z^2 \rangle$ are equal to $\langle x^2 \rangle$. Thus the model again leads to the result that the average square distance moved by a particle is proportional to the number of steps taken and to the square of the average step length.

So far the laws of diffusion have been derived from the time dependence of the phase space distribution function for a nonequilibrium system (Sections 8.1 and 8.2), and from the random walk. Using these laws, we show how particle density varies with time. Combining Fick's first law, that the flux of particles is equal to $-D$ multiplied by the concentration gradient, with the equation of continuity gives Fick's second law,

$$\frac{\partial \rho}{\partial t} = \frac{\partial}{\partial x} \left(D \frac{\partial \rho}{\partial x} \right) \tag{8.59}$$

in one dimension. Here, $\rho = \rho(x, t)$ is the number density or concentration of particles, $\rho(x, t)\, dx$ being the number of particles between x and $x + dx$ at time t. It is normalized according to

$$\int_{-\infty}^{\infty} dx \, \rho(x, t) = N$$

where N is the total number of particles. If D can be assumed independent of x, (8.59) simplifies to

$$D \frac{\partial^2 \rho}{\partial x^2} = \frac{\partial \rho}{\partial t} \tag{8.60}$$

Normally, D depends on concentration, and diffusion occurs in the presence of a concentration gradient, so D will depend on x. If the concentration is low, this dependence may be unimportant. Another situation in which one may take D to be a constant is tracer diffusion, in which one observes labeled particles which, except for their labeling, are indistinguishable from other particles. The total concentration of particles, labeled plus unlabeled, is independent of x, and so is D.

The solution to (8.60) depends on initial conditions. If $\rho = N\delta(x)$ at $t = 0$ (all particles are at $x = 0$ to start), the solution (see Problem 8.10) is

$$\rho(x,t) = N(4\pi Dt)^{-1/2} \exp\left[\frac{-x^2}{4Dt}\right] \tag{8.61}$$

It is easy to show that this satisfies (8.60), and that $\langle x^2 \rangle = 2Dt$. Another interesting initial condition is the step-function: $\rho = \rho_0$ for $x \le 0$ and $\rho = 0$ for $x > 0$ (all diffusing particles are on one side of the plane at $x = 0$). The solution can be found by considering this initial density profile as a superposition of δ-function profiles, so that for $t > 0$ one can just superpose density profiles of the form of (8.61). [This is the way the result of a general time-dependent perturbation was written as a sum of δ-function impulses in going from (8.26) to (8.27).] If the step-function profile is written

$$\rho(x,0) = \rho_0 \int_{-\infty}^{0} \delta(x-y)\, dy$$

the density at $t > 0$ is

$$\rho(x,t) = \rho_0 \int_{-\infty}^{0} (4\pi Dt)^{-1/2} \exp\left[\frac{-(x-y)^2}{4Dt}\right] dy$$

On substituting $z = x - y$ as the variable of integration, this becomes

$$\rho(x,t)/\rho_0 = (4\pi Dt)^{-1/2} \int_{x}^{\infty} \exp\left[\frac{-z^2}{4Dt}\right] dz$$

$$= \frac{1}{\sqrt{\pi}} \int_{x/\sqrt{Dt}}^{\infty} e^{-y^2}\, dy = \frac{1}{2}\left[1 - \operatorname{erf}\left(\frac{x}{\sqrt{Dt}}\right)\right] \tag{8.62}$$

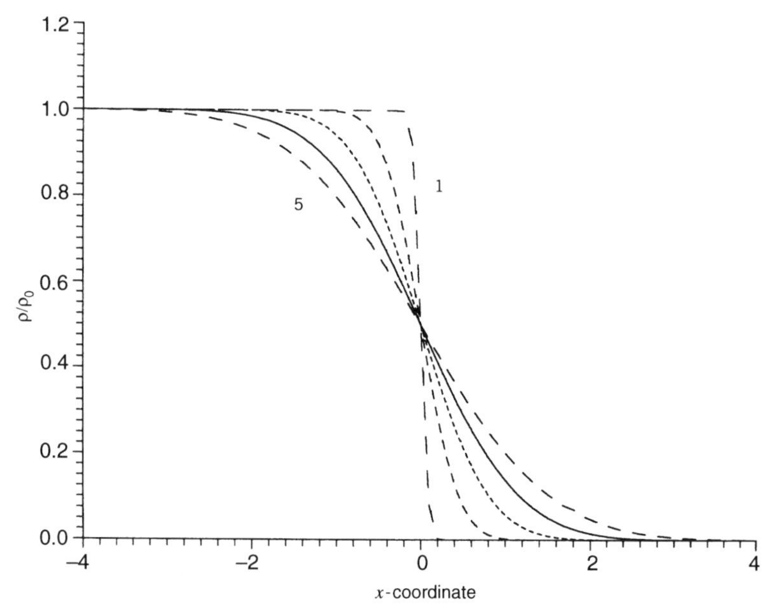

Figure 8.6. Solution to diffusion equation for various times, starting with step profile. At $t = 0$, the profile is $\rho_0[1 - S(x)]$ where $S(x) = 0$ for $x < 0$ and 1 otherwise. Plots 1 through 5 correspond to $\sqrt{Dt} = 0.1, 0.5, 0.9, 1.3$, and 1.7, respectively.

The error function erf(a) is defined as $2/\sqrt{\pi}$ times the integral from 0 to a of $\exp(-y^2)$ so that $\text{erf}(\infty) = 1$. Equation (8.62) is plotted in Figure 8.6 for $\sqrt{Dt} = 0.1, 0.5, 0.9, 1.3$, and 1.7. Note how the originally sharp boundary at $x = 0$ blurs out as t increases, the density becoming more and more uniform.

The diffusion equation in three dimensions is

$$\frac{\partial \rho}{\partial t} = \nabla \cdot (D\, \nabla \rho)$$

where ∇ is the gradient, so that (8.60) is replaced by

$$D\, \nabla^2 \rho = D\left(\frac{\partial^2 \rho}{\partial x^2} + \frac{\partial^2 \rho}{\partial y^2} + \frac{\partial^2 \rho}{\partial z^2}\right) = \frac{\partial \rho}{\partial t}$$

The solution, if $\rho(\mathbf{r}, 0) = N\delta(\mathbf{r})$, is

$$\rho(\mathbf{r}, t) = \frac{1}{(4\pi Dt)^{3/2}}\, e^{-r^2/4Dt} \tag{8.63}$$

that is, identical to (8.57) divided by $4\pi r^2$. Here, $\delta(\mathbf{r})$ is the three-dimensional delta-function, which is 0 when $\mathbf{r}$ is not the origin, and infinite at the origin in such a way as to make $\int \delta(\mathbf{r})\,d\mathbf{r} = 1$.

8.5. LANGEVIN EQUATION

Diffusion and migration in a field involve a random process; in a gas or liquid, this involves collisions of the moving particle with other particles. If one followed a dynamical variable such as velocity of a particle in a fluid, one would see rapid and apparently random fluctuations in its value, due to the random process. The velocity and position correlation functions are smooth functions of time because they are averages over an ensemble of particles, but they also reflect the random process. As noted above, these correlation functions are not correct for small times, because the averaging is not valid. In this section, we look at the random process in more detail.

Suppose that the property monitored is the velocity of a particle in a fluid, v. Thus random variable could be characterized by giving (a) its average value, $\langle v \rangle$, (b) its spread, $\langle v^2 \rangle - \langle v \rangle^2$, and (c) its persistence in time. The third property, sometimes referred to as the coherence time or time constant, would be

$$\int_0^\infty dt \, \frac{\langle \delta v(t)\, \delta v(0) \rangle}{\langle (\delta v)^2 \rangle} = \int_0^\infty dt \, \frac{c_{vv}(t)}{c_{vv}(0)}$$

where $\delta v = v - \langle v \rangle$ [see (8.15) and (8.16)]. Other properties may be necessary to completely characterize the variable. However, there are certain random processes, called Gaussian–Markov processes, which are completely characterized by giving the three properties (a) to (c).

The motion of a particle in a fluid, under the influence of collisions, is such a process. If the particle is big enough to observe directly, its motion is called "Brownian motion" (see Problem 8.12). The collisions that cause Brownian motion are also responsible for the apparently random motion of the molecules of the fluid themselves. The collisions change the velocity of the observed particle, which is sometimes referred to as a "brownon" even if it is not big enough to see, in a random way. Because collisions on the front of a moving particle are harder than collisions on the rear, they always slow the brownon down on the average. The force of collisions can thus be thought of as having two components: a force proportional to the brownon velocity and in the opposite direction, and a random force. The former is the frictional or viscous force, or the drag.

If the two components are put into Newton's equation of motion (force = mass times acceleration) in one dimension, one obtains a form of the Langevin equation:

$$m \frac{d^2x}{dt^2} = \frac{-m}{\tau} \frac{dx}{dt} + R(t) \tag{8.64}$$

The viscous force has been written as (m/τ) times the velocity to make the physical dimensions correct when τ is a time. The random force $R(t)$ is characterized by its average value,

$$\langle R(t) \rangle = 0$$

by its autocorrelation function,

$$\langle R(t)R(0) \rangle = K\delta(t)$$

(the coherence time approaches 0), and by the fact that it is not correlated with the velocity of the particle,

$$\langle R(t)(dx/dt)(0) \rangle = 0, \qquad t \geq 0 \tag{8.65}$$

where $(dx/dt)(0)$ is the velocity at $t = 0$.

If there is an external force F, such as the electric field causing migration, it may be added to the Langevin equation, which becomes

$$m \frac{d^2x}{dt^2} = \frac{-m}{\tau} \frac{dx}{dt} + R(t) + F$$

(F is time independent). Averaging this over an ensemble of systems and noting that the average acceleration of a particle must vanish, one finds that

$$\frac{1}{F} \left\langle \frac{dx}{dt} \right\rangle = \frac{\tau}{m} \tag{8.66}$$

The average velocity divided by the force is just the friction coefficient [see (8.45) and following] so τ can be identified with the time constant of the velocity autocorrelation function [(8.44) through (8.47)].

Returning to the Langevin equation in the absence of an external field [equation (8.64)], we solve it by multiplying by $(dx/dt)(0)$ and averaging over the ensemble. Using (8.65), the result is

$$\left\langle \left(\frac{dx}{dt}\right)_0 \left(\frac{d^2x}{dt^2}\right)_t \right\rangle = \frac{-1}{\tau}\left\langle \left(\frac{dx}{dt}\right)_0 \left(\frac{dx}{dt}\right)_t \right\rangle = -\tau^{-1}\langle v_x(0)v_x(t)\rangle$$

The ensemble average on the left-hand side is the derivative with respect to t of the ensemble average on the right-hand side (which would be the velocity autocorrelation function if the average velocity were zero). Therefore

$$\langle v_x(0)v_x(t)\rangle = Ce^{-t/\tau}, \qquad t \geq 0 \tag{8.67}$$

C, the constant of integration, is the value of $\langle v_x(0)v_x(t)\rangle$ at $t = 0$. For thermal equilibrium, this is just kT/m. For $t < 0$, $|t|$ should be used in the exponential of (8.67) because the correlation function should depend only on the magnitude of t.

It was shown earlier that the integral over t of $\langle v_x(0)v_x(t)\rangle$ from 0 to ∞ is the diffusion coefficient D [see (8.40), (8.46), (8.47) and the discussion following]. Equation (8.67) integrates to $C\tau$; with $C = kT/m$, $D = kT\tau/m$. This result also comes from the definition of the diffusion coefficient,

$$D = \lim_{t\to\infty} \frac{\langle [x(t) - x(0)]^2\rangle}{2t} \tag{8.68}$$

Remembering that the time-correlation function of operators A and B depends only on the difference in times at which A and B are evaluated, one may write

$$\begin{aligned}\frac{d^2}{dt^2}\langle [x(t) - x(0)]^2\rangle &= 2\frac{d}{dt}\left\langle [x(0) - x(-t)]\frac{dx}{dt}(0)\right\rangle \\ &= 2\left\langle \frac{dx}{dt}(-t)\frac{dx}{dt}(0)\right\rangle = 2\frac{kT}{m}e^{-t/\tau}\end{aligned}$$

Therefore

$$\begin{aligned}\langle [x(t) - x(0)]^2\rangle &= 2\frac{kT}{m}\int_0^t dt' \int_0^{t'} dt''\, e^{-t''/\tau} \\ &= 2\frac{kT}{m}\int_0^t dt'\, \tau[1 - e^{-t'/\tau}] = \frac{2kT\tau}{m}[t + \tau(e^{-t/\tau} - 1)]\end{aligned} \tag{8.69}$$

Dividing by $2t$ and taking the limit of $t \to \infty$ [equation (8.68)] gives $D = kT\tau/m$ again.

For small t, the mean square displacement is not proportional to t. According to (8.69),

$$\frac{\langle [x(t) - x(0)]^2 \rangle}{2D} = t + \tau e^{-t/\tau} - \tau \tag{8.70}$$

For times t small compared to τ, the exponential may be written as a power series in t/τ; the leading term on the right-hand side of (8.70) is $t^2/2\tau$. Thus for small time, the average square displacement is proportional to t^2, and the average displacement is proportional to t; this is free-particle or ballistic behavior. For times large compared to τ (essentially the collision time) the average displacement is proportional to $\sqrt{t}$; this is the diffusive regime. The mean-square displacement is plotted as a function of time in Figure 8.7.

To summarize what has been shown from the Langevin equation for a particle moving in one dimension: a time-independent force F (such as would arise from an external field) leads to an average or drift velocity of a particle of $F\tau/m$, where the time constant τ is a measure of the strength of the frictional or viscous force opposing F. In the absence of the force F, the velocity–velocity time-correlation function (8.67) is equal to $C\exp(-t/\tau)$, so

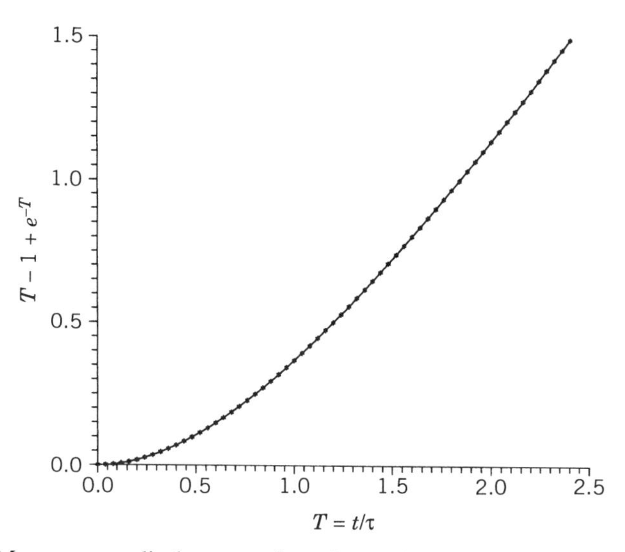

Figure 8.7. Mean-square displacement from Langevin equation as a function of time. For small t, mean-square displacement is proportional to t^2 (ballistic regime); for large t, it is proportional to t (diffusive regime).

that τ may also be found from the fluctuations in velocity in the equilibrium ensemble:

$$\langle v_x(0)v_x(0)\rangle^{-1}\int_0^\infty dt\langle v_x(0)v_x(t)\rangle = \tau \tag{8.71}$$

This is an illustration of Kubo's fluctuation-dissipation theorem: the linear response of a system to a force can be calculated from the fluctuatións occuring in the system at equilibrium, in the absence of the force.

It was also shown [see (8.70)] that the mean-square displacement of a particle in time t is proportional to t when t is large compared to τ and proportional to t^2 when t is small compared to τ. This is an improvement, for small times, on the law of diffusion,

$$\langle[x(t) - x(0)]^2\rangle = 2Dt$$

However, (8.70) is not correct for *very* small values of t, because it was derived assuming that R, the random force due to collisions, was instantaneous, that is,

$$\langle R(t)R(0)\rangle = K\delta(t)$$

In fact, a collision has a nonzero duration τ_c, roughly equal to the effective range of the interparticle force divided by the velocity of a particle. The above equation should be replaced by

$$\langle R(t)R(0)\rangle = \Psi(t) \tag{8.72}$$

where $\Psi(t)$ is a peaked function of time (see figure 8.8) which approaches 0 for $|t| > \tau_c$. Equation (8.64) is now solved using (8.72) for $R(t)$.

Consider a particle which has velocity v_0 at time 0 and velocity $v = dx/dt$ at time t. The solution to the homogeneous version of (8.64) [without R] is $v = v_0 \exp(-t/\tau)$. A standard method for solution of inhomogeneous differential equations gives the solution of (8.64) in terms of the solution to the homogeneous equation:

$$v = v_0\, e^{-t/\tau} + m^{-1}\, e^{-t/\tau}\int_0^t dt'\, e^{t'/\tau}R(t')$$

[This can be shown to be a solution of (8.64) by substitution into that equation.] If v is averaged over an ensemble of brownons having velocity v_0 at time 0, the last term vanishes because $\langle R\rangle = 0$, showing that the average velocity of the brownons decays exponentially with time constant τ. The

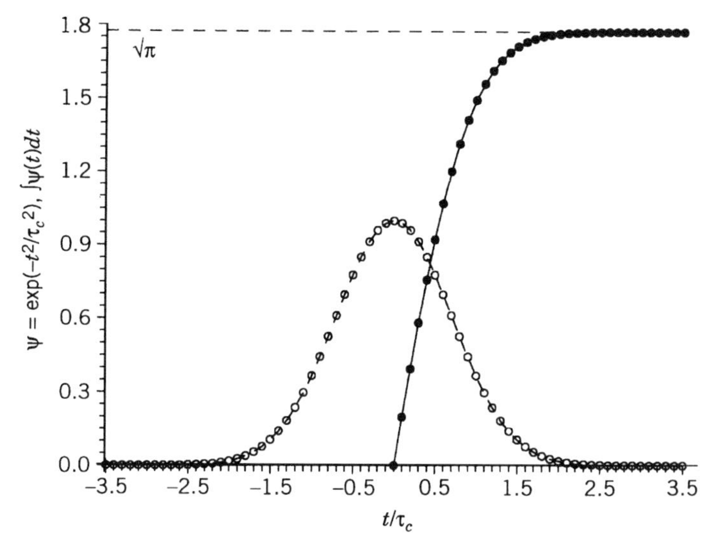

Figure 8.8. Correlation function for random force (solid circles) and its integral (open circles). $\langle R(t)R(0)\rangle$ is assumed to be $\exp(-t^2/\tau_c^2)$ with τ_c the duration of a collision. The integral $A(t) = \int_{-T}^{T}\Psi(t)\,dt$ approaches $A(\infty)$ when T exceeds τ_c.

average value of v^2 is

$$\langle v^2\rangle = v_0^2\, e^{-2t/\tau} + m^{-2}\, e^{-2t/\tau}\int_0^t dt' \int_0^t dt''\; e^{t'/\tau} e^{t''/\tau}\langle R(t')R(t'')\rangle \quad (8.73)$$

(There is no correlation between R and v_0 so there are no cross terms.) In the integrals, make the change of variables, $t_+ = t' + t''$ and $t' - t''$, so

$$\int_0^t dt' \int_0^t dt''\; e^{t'/\tau} e^{t''/\tau}\langle R(t')R(t'')\rangle$$

$$= -\frac{1}{2}\left\{\int_0^t dt_+ \int_{-t_+}^{t_+} dt_- + \int_t^{2t} dt_+ \int_{t_+-2t}^{2t-t_+}\right\} e^{t_+/\tau}\langle R(t_-)R(0)\rangle$$

$$= \frac{1}{2}\int_0^t dt_+\, e^{t_+/\tau}A(t_+) + \frac{1}{2}\int_t^{2t} dt_+\, e^{t_+/\lambda}A(2t - t_+)$$

where

$$A(T) = \int_{-T}^{T} dt'\,\Psi(t')$$

and $\Psi(t)$ is given by (8.72).

If $\Psi(t)$ approaches 0 for $|t| > \tau_c$, as shown in Figure 8.8, $A(T)$ becomes independent of T and essentially equal to $A(\infty)$ when T becomes larger than τ_c. Figure 8.8 shows $A(T)$ as a function of T, assuming $\Psi = \exp(-t^2/\tau_c^2)$. As long as t is large compared to τ_c, (8.73) for the average square velocity at time t is

$$\begin{aligned}\langle v^2 \rangle &= v_0^2 e^{-2t/\tau} \\ &\quad + m^{-2} e^{-2t/\tau} \left[\frac{1}{2} \int_0^t dt_+ \, e^{t_+/\tau} A(\infty) + \frac{1}{2} \int_t^{2t} dt_+ \, e^{t_+/\tau} A(\infty) \right] \\ &= v_0^2 e^{-2t/\tau} + \frac{\tau A(\infty)}{2m^2} [1 - e^{-2t/\tau}]\end{aligned}$$

Note that when $2t/\tau$ is small compared to unity, the first term is large and the second term is small, so that $\langle v^2 \rangle$ is the square of the initial velocity multiplied by a decay factor. When $2t/\tau$ is large compared to unity, the first term is unimportant compared to the second and it is the collisions that determine $\langle v^2 \rangle$, the initial velocity being forgotten. In the limit $t \to \infty$, the ensemble should approach thermal equilibrium, with $\langle v^2 \rangle = kT/m$, so $\tau A(\infty)/2m^2$ must equal kT/m and

$$\langle v^2 \rangle = v_0^2 e^{-2t/\tau} + \frac{kT}{m} [1 - e^{-2t/\tau}] \tag{8.74}$$

as long as t is larger than τ_c. To calculate $\langle v^2 \rangle$ for t as small as τ_c would require knowledge of the details of a collision.

However, is should be emphasized that the collision process has not been ignored in deriving (8.74). The duration of a collision is in $A(\infty)$, which is the integral over time of the correlation function of the random force, $\Psi(t) = \langle R(t)R(0) \rangle$. Remember that the "drag" force, proportional to the velocity dx/dt and in the opposite direction, is also due to collisions. If $\Psi(t)$ is nonzero for $t > 0$, meaning that collisions have a finite duration, this should also be reflected in the drag force. The drag force at time t should depend on the velocity of the particle at earlier times and not just on the velocity at time t. This may be referred to as a "memory" effect.

The Langevin equation (8.64) can be generalized to take into account memory effects by replacing the drag force $(-m/\tau)(dx/dt)$ by an integral of the velocity over times preceding t. The result is the generalized Langevin equation

$$m \frac{d^2x}{dt^2} = -m \int_0^t dt' \, M(t - t') \frac{dx}{dt'} + R(t)$$

where dx/dt' is the velocity at time t' earlier than t and the memory function $M(t - t')$ approaches 0 as $t - t'$ becomes large. In terms of momentum $p = m(dx/dt)$,

$$\frac{dp}{dt} = -\int_0^t dt'\, M(t - t')p(t') + R(t) \tag{8.75}$$

If $M(t - t')$ is $\tau^{-1}\,\delta(t - t')$, (8.75) is the Langevin equation (8.64). The memory function should be related to the correlation function of the random force, because the drag force and the random force both originate in the collision process. The relation will now be derived from the generalized Langevin equation.

Equations expressing the properties of the momentum–momentum correlation function are first derived from (8.75). Note first that the random force is not correlated with the velocity or momentum of the particle:

$$\langle R(t)p(0)\rangle = 0 \quad \text{for } t \geqq 0 \tag{8.76}$$

Then, multiplying (8.75) by $p(0)$ and averaging over an ensemble of systems,

$$\left\langle p(0)\frac{dp}{dt}\right\rangle = \frac{d}{dt}\langle p(0)p(t)\rangle = -\int_0^t dt'\, M(t - t')\langle p(0)p(t')\rangle$$

$$= -\int_0^t dt'\, M(t')\langle p(0)p(t - t')\rangle \tag{8.77}$$

An expression for the second derivative of the momentum correlation function may be obtained by differentiating (8.77).

$$\frac{d^2}{dt^2}\langle p(0)p(t)\rangle$$

$$= -\int_0^t dt'\, M(t')\frac{d}{dt}\langle p(0)p(t - t')\rangle - M(t)\langle p(0)p(0)\rangle \tag{8.78}$$

The last term is $M(t)\langle p^2\rangle$ for an equilibrium ensemble (the average value is independent of time).

Another expression for the second derivative is now obtained.

$$\frac{d^2}{dt^2}\langle p(0)p(t)\rangle = \frac{d}{dt}\left\langle p(0)\left(\frac{dp}{dt}\right)_t\right\rangle = \frac{d}{dt}\left\langle p(-t)\left(\frac{dp}{dt}\right)_0\right\rangle$$

$$= -\left\langle\left(\frac{dp}{dt}\right)_{-t}\left(\frac{dp}{dt}\right)_0\right\rangle = -\left\langle\left(\frac{dp}{dt}\right)_0\left(\frac{dp}{dt}\right)_t\right\rangle \tag{8.79}$$

It follows from (8.75) that $(dp/dt)_0 = R(0)$ so, multiplying (8.75) by $(dp/dt)_0$ and averaging over the ensemble gives

$$\left\langle \left(\frac{dp}{dt}\right)_0 \left(\frac{dp}{dt}\right) \right\rangle = -\int_0^t dt'\, M(t-t') \left\langle \left(\frac{dp}{dt}\right)_0 p(t') \right\rangle + \langle R(0) R(t) \rangle$$

Combining this with (8.79),

$$\begin{aligned}
\frac{d^2}{dt^2} \langle p(0) p(t) \rangle &= \int_0^t dt'\, M(t-t') \left\langle \left(\frac{dp}{dt}\right)_0 p(t') \right\rangle - \langle R(0) R(t) \rangle \\
&= \int_0^t dt'\, M(t-t') \left\langle \left(\frac{dp}{dt}\right)_{-t'} p(0) \right\rangle - \langle R(0) R(t) \rangle \\
&= -\int_0^t dt'\, M(t') \frac{d}{dt} \langle p(t-t') p(0) \rangle - \langle R(0) R(t) \rangle
\end{aligned}$$

Setting this expression for the second derivative equal to (8.78) yields

$$M(t) \langle p(0) p(0) \rangle = \langle R(0) R(t) \rangle \tag{8.80}$$

This is the relation sought, between the memory function and the time correlation of a collision.

Earlier [see (8.72)–(8.74)] the following equation was derived for an equilibrium ensemble:

$$\frac{\tau A(\infty)}{2m^2} = \frac{kT}{m}$$

where

$$A(\infty) = \int_{-\infty}^{\infty} dt \langle R(0) R(t) \rangle$$

Using (8.80), this becomes

$$\frac{\tau \langle p(0) p(0) \rangle}{m} \int_0^{\infty} dt\, M(t) = kT$$

The infinite limit on the integral really means a value of t larger than the duration of a collision, or large enough for the memory function $M(t)$ to fall to zero. Then the integral is equal to τ^{-1} and one recovers the equipartition result $\langle p^2 \rangle = mkT$.

8.6. CHEMICAL REACTIONS

To chemists, chemical reactions are among the most important time-dependent problems. The reaction of chemical species A and B to form chemical species C and D includes the approach of A and B so they are close enough to react (the collision process) and the reaction itself, in which the electrons rearrange so that A and B are converted to C and D. The collision process has been touched on in our discussion of molecular trajectories in gases and in our discussion of diffusion.

Suppose that A contains n_A nuclei and B contains n_B nuclei. The relative positions of the $n_A + n_B$ nuclei are specified by $3(n_A + n_B) - 6$ coordinates $\{\zeta_j\}$, combinations of position coordinates for individual nuclei. For each set of values for these coordinates, consider the ground-state electronic energy and electronic wave function. This electronic energy becomes the potential energy surface on which the nuclei move (see section 4.2). For some values of $\{\zeta_j\}$, the electronic wave function corresponds to the species A and B; for other values, the electronic wave function corresponds to the species C and D. In the region of the potential energy surface corresponding to $A + B$, there is a minimum energy corresponding to the equilibrium configuration of A and B; in the region corresponding to $C + D$, there is a minimum energy corresponding to the equilibrium configuration of $C + D$. The two regions are separated by a region in which the energy is higher than either of these minimum energies, constituting a barrier to reaction.

On the potential energy surface, there are many paths leading from one minimum to the other. Since these paths cross the barrier, a plot of energy vs. distance along any path will include a maximum. The path having the lowest maximum is sometimes considered the reaction coordinate. Along the reaction coordinate, the energy must have the shape shown in Figure 8.9. Diagrams (1) and (2) correspond to exothermic and endothermic reactions. More correctly, the surface should involve free energy, rather than energy, and the plots should be of free energy vs. distance along the reaction coordinate. The free energy includes, in addition to the electronic energy, such contributions as the vibrational energy associated with coordinates other than the reaction coordinate, the entropy associated with rotational and vibrational motions, and, in condensed phases, the free energy of coordinated solvent molecules.

The free-energy plots will have the shape of the energy plots shown in Figure 8.9, including two minima (the stable species) separated by a maximum (the barrier to reaction). The position of the maximum (c in Figure 8.9) is usually called the "transition state;" it is a nuclear configuration which can be assigned neither to reactants nor to products. For simplicity of calculation, a continuous potential curve like (1) or (2) in the figure may be represented by (3), a superposition of parabolas. The problem is to describe the motion of

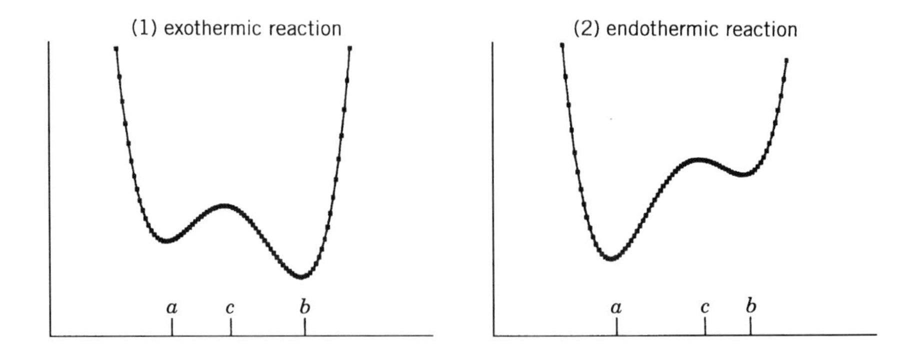

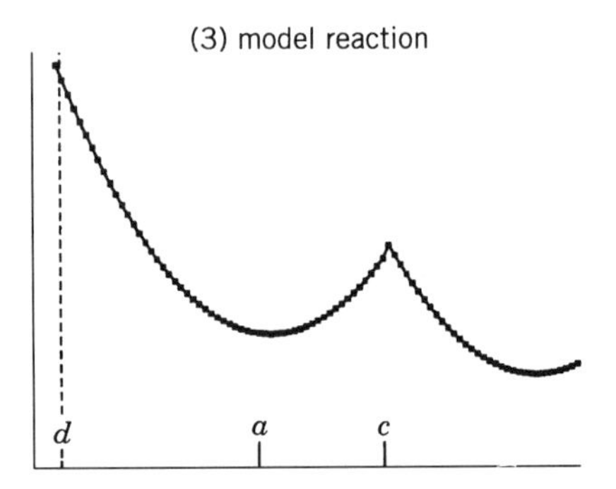

Figure 8.9. Potential energy (or free energy) along the reaction coordinate. Plots (1) and (2) are continuous curves; plot (3) is a superposition of two parabolas and is easier to calculate with. On each plot, point a is the minimum for reactants, point b the minimum for products, and point c the maximum for the transition state.

the system along the reaction coordinate, starting from the minimum corresponding to reactants. If a system reaches the transition state at $x = c$, there is some probability that it will change its electronic state to that of products. If it changes its state, it will proceed rapidly to the minimum at $x = b$.

The simplest reaction is an isomerization, $A \rightleftarrows B$, for which $x < c$ corresponds to species A and $x > c$ corresponds to species B. One can define the dynamical variable $S(x)$ such that $S(x) = 1$ for $x < c$ and $S(x) = 0$ for $x > c$. Then $S(x)$, averaged over an ensemble, is the probability that a molecular system is in a state corresponding to A, which corresponds to the mole fraction of A:

$$\langle S(x) \rangle = x_A = \frac{c_A}{c_A + c_B}$$

(c_A and c_B are concentrations). At thermal equilibrium, c_B/c_A is equal to the equilibrium constant K, assuming ideality, and

$$\langle S(x) \rangle_{\text{eq}} = x_A^{\text{eq}} = (1 + K)^{-1}$$

Assuming the forward and reverse reactions to be first order with rate constants k_f and k_r ($K = k_f/k_r$),

$$\frac{dc_A}{dt} = -k_f c_A + k_r c_B = \frac{-dc_B}{dt}$$

This is rewritten

$$\frac{dc_A}{dt} = k_r(c_A + c_B) - (k_f + k_r)c_A \tag{8.81}$$

which, since $c_A + c_B$ is constant in time, has the solution (Problem 8.20)

$$c_A(t) = \frac{k_r(c_A + c_B)}{k_f + k_r}[1 - e^{-(k_f + k_r)t}] + c_A(0)\, e^{-(k_f + k_r)t}$$

For $t \to \infty$, c_A approaches $(c_A + c_B)/(1 + K)$, which is the equilibrium value.

The deviation of c_A from its equilibrium value decreases exponentially with time:

$$\Delta c_A(t) = c_A(t) - c_A^{\text{eq}} = \left[c_A(0) - \frac{k_r(c_A + c_B)}{k_f + k_r} \right] e^{-(k_f + k_r)t}$$

According to the Onsager hypothesis or fluctuation-dissipation theorem [(8.22) or (8.34)], spontaneous fluctuations in the dynamical variable S in an equilibrium system decay in the same way as imposed fluctuations. Since S corresponds to x_A, it follows that

$$\frac{c_{SS}(t)}{c_{SS}(0)} = \frac{\Delta x_A(t)}{\Delta x_A(0)} = \frac{\Delta c_A(t)}{\Delta c_A(0)} = e^{-(k_f + k_r)t} \tag{8.82}$$

where c_{SS} is for the equilibrium ensemble. It has thus been shown that in an equilibrium system fluctations in concentration of an isomer decay exponentially with time constant $\tau_c = 1/(k_f + k_r)$; the autocorrelation function of S is an exponential. Also, information about chemical rate constants may be derived from the way that fluctuations in concentration decay in an equilibrium system.

Writing the autocorrelation functions in terms of the dynamical variable S, the decay equation (8.82) is

$$\frac{\langle (S(t) - \langle S \rangle)(S(0) - \langle S \rangle) \rangle}{\langle (S(0) - \langle S \rangle)^2 \rangle} = e^{-(k_f + k_r)t} = e^{-t/\tau_c}$$

Note that the averages are over an equilibrium ensemble; $\langle S \rangle = x_A^{\text{eq}}$. Also, $S(x) = 1$ for $x < c$ and $S(x) = 0$ for $x > c$, so that $S(0)^2 = S(0)$ and the above equation is

$$\frac{\langle S(t) S(0) \rangle - \left(x_A^{\text{eq}}\right)^2}{x_A^{\text{eq}} - \left(x_A^{\text{eq}}\right)^2} = e^{-t/\tau_c}$$

To interpret this equation, take the time derivative to get

$$\left\langle \frac{dS}{dt}(t) S(0) \right\rangle = x_A^{\text{eq}} x_B^{\text{eq}} \frac{e^{-t/\tau_c}}{-\tau_c}$$

(note that $x_B = 1 - x_A$). Since the dynamical variable S is a step function at $x = c$,

$$\frac{dS}{dt} = \left(\frac{dS}{dx}\right)\left(\frac{dx}{dt}\right) = -\left(\frac{dx}{dt}\right)\delta(x - c)$$

with dx/dt being $v(t)$, the velocity of the systems. Combining this with the previous equation yields

$$\langle v(t) \delta(x - c) S(0) \rangle = x_A^{\text{eq}} x_B^{\text{eq}} \frac{e^{-t/\tau_c}}{\tau_c}$$

for the rate of reaction. The left-hand side represents the average velocity of systems crossing the barrier at $x = c$ at time t, considering only systems that have $x < c$ (i.e., were A molecules) at time 0. For times larger than τ_c, some of the systems making the crossing have crossed previously.

For the purpose of calculating the reaction rate, the system may be considered as a mass point or particle moving on a potential surface U. Values of x, the reaction coordinate, less than c correspond to reactants; values of x greater than c correspond to products. The (forward) reaction rate is the rate at which particles with $x < c$ cross to $x > c$.

Consider an ensemble of systems such that at $t = 0$ all systems are on the reactant side. Let $P(x, t)\, dx$ be the fraction of systems in the ensemble with positions between x and $x + dx$ at time t, and $Q(x, t)\, dx$ be the fraction of systems which have positions between x and $x + dx$ *and* have not crossed the barrier ($x = c$) since $t = 0$. For $x < c$, $Q(x,t)$ is less than $P(x,t)$ since $P(x,t)$ includes systems which have crossed the barrier from left to right and, later, recrossed from right to left. The surface of Figure 8.9 (3) is used; at $x = d$ the energy is so high that $P(x,t) \approx 0$. The fraction of systems that have not reacted is given by

$$A(t) = \int_{-\infty}^{c} dx\, Q(x,t) < 1$$

where $A(t)$ is a decreasing function of time. A characteristic time may be defined by

$$\tau_f = \int_0^{\infty} dt\, A(t) \tag{8.83}$$

(the subscript f refers to first crossing of the transition barrier). If $A(t)$ is an exponential, $A(t) = e^{-kt}$, the characteristic time τ_f is equal to $1/k$.

It is assumed that there is an equilibrium distribution of systems for $t \leq 0$,

$$Q(x,t) = Q_0(x) \equiv \frac{e^{-\beta U(x)}}{\int_d^c e^{\beta U(x)}\, dx}, \qquad t \leq 0 \tag{8.84}$$

The change of $Q(x,t)$ with time is associated with the current $j(x,t)$:

$$\partial Q(x,t)/\partial t = -\left[\frac{\partial j(x,t)}{\partial x}\right] \tag{8.85}$$

The current is proportional to the sum of the diffusive and external forces. The equivalent between the diffusive force $-kT(\partial \ln Q/\partial x)$ and the true force $-dU/dx$ was discussed earlier in connection with the Nernst–Einstein relation (8.49). It also follows from the Langevin equation (8.64) when an external force is added: (8.66) shows the ratio of drift velocity to force is τ/m which [see (8.68) and (8.69)] is equal to D/kT. Thus in the pressure of the force $-dU/dx$

$$j(x,t) = -D(x)\left[\frac{\partial Q(x,t)}{\partial x} + \frac{Q(x,t)}{kT}\frac{\partial U}{\partial x}\right] \tag{8.86}$$

A position-dependent diffusion coefficient $D(x)$ is used. Boundary conditions on j are: $j(d,t) = 0$ (no particles enter from the left) and $j(c,t) = \kappa Q(c,t)$ for $t > 0$ (κ is the mean velocity of systems arriving at the transition state). Much calculation is now required to get τ_f in terms of Q_o (equation 8.90).

Equation (8.85) may be integrated from d to x to get

$$j(x,t) = -\int_d^x \frac{\partial Q(y,t)}{\partial t}\, dy \tag{8.87}$$

since $j(d,t) = 0$. Now multiply (8.86) by $e^{\beta U}$, obtaining

$$\begin{aligned} e^{\beta U} j(x,t) &= -D(x)\left[e^{\beta U}\frac{\partial Q(x,t)}{\partial x} + Q(x,t)\frac{\partial(e^{\beta U})}{\partial x}\right] \\ &= -D(x)\frac{\partial(Q e^{\beta U})}{\partial x} \end{aligned}$$

which may be rewritten

$$\frac{j(x,t)}{Q_0(x)} = -D(x)\frac{\partial[Q(x,t)/Q_0(x)]}{\partial x} \tag{8.88}$$

Obviously,

$$\frac{Q(x,t)}{Q_0(x)} - \frac{Q(c,t)}{Q_0(c)} = \int_c^x dy\, \frac{\partial}{\partial y}\left[\frac{Q(y,t)}{Q_0(y)}\right]$$

which, using $j(c,t) = \kappa Q(c,t)$ and (8.88), is

$$\frac{Q(x,t)}{Q_0(x)} = \frac{j(c,t)}{\kappa Q_0(c)} - \int_c^x dy\left[\frac{j(y,t)}{D(y)Q_0(y)}\right]$$

Next, (8.87) is used and d is allowed to approach $-\infty$. The result is

$$\frac{Q(x,t)}{Q_0(x)} = \frac{\int_{-\infty}^{c} dy\, \frac{\partial Q(y,t)}{\partial t}}{\kappa Q_0(c)} + \int_c^x dy\, \frac{1}{D(y)Q_0(y)} \int_{-\infty}^{y} dz\, \frac{\partial Q(z,t)}{\partial t} \tag{8.89}$$

This is an equation for the decrease of Q with time, which is the rate that systems cross the barrier from reactants to products.

Now the time constant τ_f [see (8.83)] is obtained by integrating $Q(x,t)/Q_0(x)$ over x from $-\infty$ to c, and over t from 0 to ∞. Since $Q(x,t) \to 0$ for $t \to \infty$,

$$\int_0^\infty dt\, \frac{\partial Q(y,t)}{\partial t} = -Q_0(y)$$

so that it is easier to carry out the t integration of (8.89) first. The result is

$$\frac{\int_0^\infty dt\, Q(x,t)}{Q_0(x)} = \frac{\int_{-\infty}^c dy\, Q_0(y)}{\kappa Q_0(c)} - \int_c^x \frac{dy}{D(y)Q_0(y)} \int_{-\infty}^y dz\, Q_0(z)$$

and the first integral on the right is unity. Now, multiplying through by $Q_0(x)$ and integrating over x gives

$$\tau_f = \int_{-\infty}^c dx \frac{Q_0(x)}{\kappa Q_0(c)} + \int_{-\infty}^c dx\, Q_0(x) \int_x^c \frac{dy}{D(y)Q_0(y)} \int_{-\infty}^y dz\, Q_0(z)$$

In the second term, integrating first over y from x to c and then over x from $-\infty$ to c may be replaced by integrating first over x from $-\infty$ to y and then over y from $-\infty$ to c. Then the time constant is

$$\tau_f = \frac{1}{\kappa Q_0(c)} + \int_{-\infty}^c \frac{dy}{D(y)Q_0(y)} \left[\int_{-\infty}^y dx\, Q_0(x)\right]^2 \tag{8.90}$$

The time constant, which is the reciprocal of a rate constant, is a sum of two terms. The first is related to the rate at which systems cross the barrier at $x = c$. The second relates to the rate at which systems reach $x = c$ by diffusion. The systems are diffusing particles or brownons.

To get explicit results, assume that the potential or free-energy function is parabolic: $U(x) = \frac{1}{2}Kx^2$. Assume also that the diffusion constant is independent of position, and that the barrier height is very large: $\frac{1}{2}Kc^2$ is large compared to β^{-1} or kT. This means that

$$\int_{-\infty}^c dx\, e^{-\beta K x^2/2} \approx \sqrt{\frac{2\pi}{K\beta}}$$

since the upper limit of the integral is essentially infinite. Then (8.90) becomes [see (8.84) for the definition of Q_o]

$$\tau_f = \sqrt{\frac{2\pi}{\beta K}} \frac{e^{\beta K c^2/2}}{\kappa} + \sqrt{\frac{\beta K}{2\pi}} \int_{-\infty}^c dy \frac{e^{\beta K y^2/2}}{D} \left[\int_{-\infty}^y dx\, e^{-\beta K x^2/2}\right]^2$$

The integral over y is dominated by the largest values of y because the exponential increases so rapidly with y; for the largest values of y, the integral over x is essentially $\sqrt{(2\pi/K\beta)}$. The second term in τ_f is then

$$\sqrt{\frac{2\pi}{\beta K}} \frac{1}{D} \int_{-\infty}^c dy \left[e^{\beta K y^2/2} y\right]\left[y^{-1}\right] = \sqrt{\frac{2\pi}{\beta K}} \frac{1}{D} \frac{e^{\beta K c^2/2}}{\beta K c} + \cdots$$

on integration by parts. Successive terms have βKc^2 (a large quantity) to successively higher negative powers and are neglected.

All together,

$$\tau_f = \sqrt{\frac{2\pi}{\beta K}}\, e^{\beta Kc^2/2}\left(\frac{1}{\kappa} + \frac{1}{D\beta Kc}\right)$$

Here, $Kc^2/2$ is the difference in energy (or free energy) between the top of the barrier and the bottom of the harmonic potential well; denote it by E^*. The solution to Newton's equation of motion for a particle of mass m in the potential $\frac{1}{2}Kx^2$ is $x = x_0 \cos(\omega t)$ where $\omega^2 = K/m$. For an ensemble of particles at temperature $1/k\beta$, the spread in position (average value of x^2) is $1/(K\beta) = (\beta\omega^2 m)^{-1}$ (see Problem 8.21). If the particles are subject to a frictional force $-m(dx/dt)/\tau$, the correlation time (see Problem 8.21) is

$$\tau_0 = \int_0^\infty dt\, \frac{\langle x(0)x(t)\rangle}{\langle x(0)x(0)\rangle} = \frac{1}{\omega^2\tau}$$

In this case the diffusion coefficient is $D = \tau/\beta m$. Thus

$$\tau_f = e^{\beta E^*}\left(\frac{\sqrt{2\pi\langle x^2\rangle}}{\kappa} + \sqrt{\frac{\pi}{\beta E^*}}\,\tau_0\right)$$

Introducing τ_t, the time it takes a particle moving with velocity κ to travel through the distance $\sqrt{\langle x^2\rangle}$, the rate constant k is

$$k = \tau_f^{-1} = \frac{e^{-\beta E^*}\pi^{-1/2}}{2^{1/2}\tau_t + \tau_0(\beta E^*)^{-1/2}} \tag{8.91}$$

The two terms in the denominator correspond to the times for particles (brownons) to reach $x = c$ with energy E^* by transport and by diffusion.

Which term is larger depends on the strength of the interaction between the brownons and the particles of the medium. For very strong interaction, diffusion on the free-energy surface is slow and limits the rate of reaction. Then $\tau_t(\beta E^*)^{1/2} \ll \tau_0$, and

$$k = \frac{e^{-\beta E^*}\pi^{-1/2}}{\tau_0(\beta E^*)^{-1/2}}$$

In the other limit, $\tau_t \gg \tau_0$, diffusion is rapid because there is weak interaction between brownons and the medium: D and τ are large so τ_0 is small.

The rate constant is

$$k = e^{-\beta E^*}(2\pi)^{-1/2}\tau_t^{-1}$$

as in transition-state theory. The reaction rate is controlled by the height of the energy barrier, which limits the number of brownons at $x = c$, and the velocity κ at which the barrier is crossed. Because diffusion is rapid, the Boltzmann distribution of brownons over position [equation (8.84)] is maintained, the loss of brownons at $x = c$ by reaction being supplied by other brownons arriving by diffusion.

According to (8.91), k decreases monotonically with the strength of the interaction between the brownons and the medium, since stronger interaction increases τ_0. However, if the interaction is very weak (and τ_0 is very small), there is no mechanism for the brownons to gain, from the medium, the energy necessary to get to $x = c$. Then the rate constant must fall to zero. The reason that the theory given above fails to predict this is the use of the diffusion equation (8.86), which implies an average displacement proportional to $\sqrt{t}$. As we know from the Langevin equation, this holds only for times larger than the collision time [see (8.70)]. The diffusion equation is not correct for smaller times because it takes into account the viscous force only, neglecting the random force, due to individual collisions [see (8.64)]. The strength of the random force must decrease when the strength of the viscous force decreases, since they have the same origin, and it is the random force which produces energy transfer from medium to brownon.

PROBLEMS

8.1. Show that, if a correlation function is a simple exponential ($c_{AB}(t) = A e^{-t/T}$), the general definition of the time constant, equation (8.16), agrees with the usual definition, that is, $\tau_{AB} = T$.

8.2. Show that for any dynamical variables A and B, $\langle B(t+t')A(t')\rangle = \langle B(0)A(-t)\rangle$ for any t'. From this result, show that

$$\left\langle \left(\frac{dB}{dt}\right)_{t=t} A(0) \right\rangle = -\left\langle B(t)\left(\frac{dA}{dt}\right)_{t=0} \right\rangle$$

8.3. Calculate $(A^2)\omega$ for the correlation function shown in Figure 8.2, which is represented approximately by $c_{AA}(t) = b\cos(at)\, e^{-ct^2}$ where a, b, and c are constants.

8.4. Equation (8.28) gives the electric current at time $t = 0$ for a system which was in equilibrium at $t = -\infty$ and subjected to an electric field in the x-direction $Ee^{\alpha t}$;

$$\langle J_x(0) \rangle = \beta E \int_{-\infty}^{0} dt'\, e^{\alpha t'} \langle J_x(t') J_x(0) \rangle$$

Assuming the current autocorrelation function is equal to $A \exp[-|t'|/\tau]$, calculate the current at $t = 0$, letting $\alpha \to 0$. Then suppose the electric field is shut off at a time $-T$, so that $F(t)$ [see (8.27)] is $Ee^{\alpha t}S(t + T)$, where the unit step function $S(x) - 1$ for $x < 0$ and $= 0$ for $x \geq 0$. What is the current at $t = 0$?

8.5. In (8.31), we showed that the difference of expectation values of A at times 0 and t (for a nonequilibrium system) was equal to $\langle A(0)H_1(0) \rangle - \langle A(t)H_1(0) \rangle$ with the fences denoting average values for an equilibrium ensemble. Show that this expression is equal to

$$\langle (A(0) - \alpha)(H_1(0) - \beta) \rangle - \langle (A(t) - \alpha)(H_1(0) - \beta) \rangle$$

where α and β represent the equilibrium-ensemble average values of A and H_1.

8.6. For a particle executing a random walk in three dimensions, the probability that after a time t it reaches a point a distance R from where it was at time 0 is

$$P(R, t) = (4\pi Dt)^{-3/2} e^{-R^2/4Dt}$$

This is normalized such that

$$\int P(R, t)\, \mathbf{dR} = \int_0^{\infty} 4\pi R^2\, dR\, P(R, t) = 1$$

and leads to an average value of R^2 equal to $6Dt$.

(a) Verify the normalization and the average value of R^2.

(b) If the particle executes its walk in two dimensions instead of three, $P(R, t)$ remains proportional to $\exp(-R^2/4Dt)$, but the normalizing factor is no longer $(4\pi Dt)^{-3/2}$. Calculate the proper normalizing factor and the average value of R^2 in this case.

8.7. It was shown that, for a particle executing a random walk in one dimension with p the probability that any step is to the right, the average number of steps taken to the right after n steps is np. Prove that this is also the most probable number of steps to the right.

8.8. Suppose that a particle in a gas moves with constant velocity until it collides with another particle, and suppose that there is a probability $p\,d\ell$ that a collision takes place when the particle moves through the infinitesimal distance $d\ell$. Then the probability $P(\ell)$ that the particle goes through a distance ℓ without suffering a collision obeys

$$P(\ell + d\ell) = P(\ell)(1 - p\,d\ell)$$

so that $dP/d\ell = -pP$ and $P = \exp(-p\ell)$. Suppose further that the velocity of a particle after a collision is unrelated to the velocity before: on the average, $\mathbf{v}_{\text{after}} \cdot \mathbf{v}_{\text{before}} = 0$.

(a) Calculate the average distance a particle travels without colliding with another particle and, assuming the velocity is $\mathbf{v}_a$, calculate the average time for which it travels at constant velocity.

(b) Calculate an expression for the velocity autocorrelation function, $\langle \mathbf{v}(0) \cdot \mathbf{v}(t) \rangle$.

(c) Suppose that, after a collision, the particle retains some memory of its velocity before the collision, so that on the average $\mathbf{v}_{\text{after}} \cdot \mathbf{v}_{\text{before}} = f(v_{\text{before}})^2$. The velocity after a second collision is unrelated to the velocity before the first. Calculate an expression for the velocity autocorrelation function.

8.9. For the one-dimensional random walk, the flux or current density of particles is given by (8.56). Show it obeys the equation of continuity, $\partial j_x / \partial x = -\partial P / \partial t$, where $P = P(x, t)$ [equation above (8.56)], provided that $D = \frac{1}{2} s \lambda^2$.

8.10. Show that the function (8.61) satisfies the diffusion equation (8.60). Show that (8.61) reduces to $N\delta(x)$ for $t = 0$. (This requires showing that the function vanishes except for $x = 0$, and that the integral over x of the function is N even though the function becomes infinite at $x = 0$.)

8.11. Show that the function (8.63) satisfies the three-dimensional diffusion equation. Calculate the concentration gradient and the flux of particles (direction and magnitude).

8.12. In Section 6.3, the number of collisions per second undergone by a particle of type 1 in a gas of particles of type 2 was shown to be

$$z_{12} = \frac{N_2}{V} \sqrt{\frac{8kT}{\pi\mu}}\, \pi (r_1 + r_2)^2$$

If the first particle is much bigger than the others, $r_1 + r_2 \approx r_1$ and the reduced mass μ is approximately m_2, so that the average relative speed is approximately $\sqrt{(8kT/\pi m_2)}$. This situation corresponds to Brownian motion: a particle large enough to be visible, such as a grain of dust, apparently moves randomly in short steps, due to collisions with the unseen particles of the gas which surround it. Derive an expression for the mean free path or step size λ, and, from this, an expression for the root-mean-square distance moved in time t by a particle undergoing Brownian motion.

8.13. A particle executes a one-dimensional random walk with step size λ. What are the probabilities that, after taking two, four, and eight steps, the particle is found four steps to the left of its original position? See (8.51).

8.14. It was shown that a concentration gradient is equivalent to a force, in the sense that the average velocity of a particle in the presence of a concentration gradient ∇c is

$$\langle \mathbf{v} \rangle = -\zeta \beta^{-1} c^{-1} \nabla c$$

and the average velocity in the presence of a force $\mathbf{F}$ is $\zeta \mathbf{F}$ (ζ is the friction coefficient). Consider a gas at constant temperature in the presence of a gravitational field, so that the force in the z direction on a particle of mass m is mg, $g = -0.98$ N/kg $= -0.98$ m/sec^2. The average velocity in the z direction being 0, the gravitational force must be balanced by an upward force due to the concentration gradient.

(a) Show that this leads to the barometric formula, $c/c_0 = \exp(-\beta mgz)$.

(b) Calculate the concentration of molecules of air at $T = 298$ K, assuming there is only one kind of molecule, with molar mass $M = 0.029$ kg/mol. Calculate the concentration gradient due to the gravitational force.

8.15. The mobility U_{Zn} of Zn^{2+} ion in water at 298 K is 5.47×10^{-8} m^2 sec^{-1} V^{-1}.

(a) What is the diffusion coefficient D_{Zn}? What is the friction constant ζ_{Zn}?

(b) The friction constant for a spherical neutral particle in liquid solution is given by $6\pi\eta a$, where η is the viscosity of the solvent (0.0010 kg m^{-1} sec^{-1} for water at 298 K) and a is the radius of the particle. Assuming this formula works approximately for Zn^{2+} even though it is charged, estimate the radius of the (solvated) ion.

(c) Diffusion of a molecule in a liquid may be considered to occur by jumps from one equilibrium position to another [see (8.58) and following discussion]. If the distance jumped is equal to the molecular diameter, how many jumps does a Zn^{2+} ion make per second?

8.16. Suppose a gas contains, per unit volume, ρ_i ions of mass m_i and charge q, and ρ_n neutral molecules of mass m_n, with m_i small compared to m_n. If ρ_n is much larger than ρ_i, most collisions of an ion are with neutral molecules; let the effective cross-section for these collisions be σ_{in}. Show that the electrical conductivity is

$$\sigma = \sqrt{\frac{\pi}{8m_i kT}}\,\frac{n_i q^2}{n_n \sigma_{in}}$$

8.17. Show that the generalized Langevin equation (8.75) becomes the Langevin equation (8.64) when $M(t-t')$ is $\tau^{-1}\,\delta(t-t')$. Then show that, with this choice of M, (8.80) gives $\langle R(0)R(t)\rangle = (mkT/\tau)\,\delta(t)$ [in the Langevin equation, this correlation function was assumed to be $K\delta(t)$ with K an unknown constant].

8.18. From hydrodynamics it can be shown that the force necessary to drag a sphere of radius r with velocity v through a continuous medium of viscosity η is $n\pi r\eta v$. The numerical factor n is 4 when there is no friction between the sphere and the viscous medium (complete slippage) and 6 when the friction is maximal (so the medium sticks to the surface of the sphere). The Langevin equation, assuming constant velocity, is then

$$0 = -\frac{m}{\tau}v + R(t) + n\pi r\eta v$$

We showed that the diffusion constant D was $\tau/m\beta$. Derive an expression for the diffusion constant in terms of the viscosity and the particle radius. This is valid when the diffusing particle (brownon) is large compared to the particles of the medium (so the medium may be considered continuous) and when the concentration of brownons is low (interactions between them are ignored).

8.19. It was shown from the Langevin equation that a particle having velocity v_0 at time 0 has a velocity of

$$v = v_0\,e^{-t/\tau} + m^{-1}\,e^{-t/\tau}\int_0^t dt'\,e^{t'/\tau}R(t')$$

at time t. Derive an expression for the velocity–velocity correlation function $\langle v(t)v(0)\rangle$ for such particles. This function should equal v_0^2 for $t = 0$ and approach 0 for t much larger than τ. By integration of the velocity–velocity correlation function, derive an expression for $\langle (x(t) - x(0))^2\rangle$ for such particles.

8.20. Equation (8.81) is a first-order differential equation with constant coefficients (because $c_A + c_B$ is independent of t); such an equation has a solution of the form

$$c_A(t) = a + b\,e^{ct}$$

with a, b, and c constants. By substituting this form into (8.81), find a and c in terms of k_r and k_f. Then find b in terms of the other constants and $c_A(0)$.

8.21. Newton's equation of motion for a particle of mass m moving in the potential $\frac{1}{2}Kx^2$ is

$$m\left(\frac{d^2x}{dt^2}\right) = -Kx$$

A solution is $x = x_0 \cos \omega t$ with $\omega^2 = K/m$.

(a) Calculate the average value of x^2, $\langle x^2 \rangle$, for a single such particle.

(b) Consider an ensemble of such particles at temperature $T = 1/k\beta$; the probability of finding a particle at position x is

$$P(x) = e^{-\beta Kx^2/2}\left[\int_{-\infty}^{\infty} dx\, e^{-\beta Kx^2/2}\right]^{-1}$$

Calculate $\langle x^2 \rangle$ for the ensemble.

(c) If a frictional term is added to the harmonic potential, so that the equation of motion is

$$m\,\frac{d^2x}{dt^2} = -Kx - \frac{m}{\tau}\,\frac{dx}{dt}$$

the correlation time for the ensemble may be calculated in terms of K, m, and τ as follows [see equations (8.64) and following). Multiply by $(dx/dt)(0)$ and average over the ensemble, so

$$\left\langle \left(\frac{dx}{dt}\right)_0 \left(\frac{d^2x}{dt^2}\right)_t \right\rangle + \frac{K}{m}\left\langle \left(\frac{dx}{dt}\right)_0 (x)_t \right\rangle + \frac{1}{\tau}\left\langle \left(\frac{dx}{dt}\right)_0 \left(\frac{dx}{dt}\right)_t \right\rangle = 0$$

If the second average value is called $u(t)$, the first average value is d^2u/dt^2 and the third is du/dt. The differential equation has the general solution

$$u(t) = A_+\,e^{b_+t} + A_-\,e^{b_-t}$$

where A_+ and A_- are arbitrary constants. Find the two values of the exponential parameters, b_+ and b_-. Since $u(0)$ should be 0, A_+ must equal $-A_-$. Integrate u to get $\langle x(0)x(t)\rangle$ as a sum of exponentials plus a constant of integration, and choose the constant so that $\langle x(0)x(t)\rangle$ approaches 0 for infinite time.

(d) Show that

$$\frac{\langle x(0)x(t)\rangle}{\langle x(0)x(0)\rangle} = \frac{b_- e^{b_+ t} - b_+ e^{b_- t}}{b_- - b_+}$$

The time constant associated with the correlation function is obtained by integrating the above quantity over time from 0 to ∞. Show that the result is $m\tau/K$.

BIBLIOGRAPHY*

Balescu, R. *Equilibrium and Non-Equilibrium Statistical Mechanics*. Wiley, New York, 1975. (Chapter 1, 1.1; Chapter 2, 2.1–2.4; Chapter 5, 5.1–5.3; Chapter 6, 6.1 and 6.2; Chapter 7, 7.1–7.5; Chapter 8, 8.1–8.4)

Bard, A. J. and Faulkner, L. R. *Electrochemical Methods*. Wiley, New York, 1980. (Chapter 7, 7.6; Chapter 8, 8.5)

Bockris, J. O'M. and Khan, S. U. M. *Surface Electrochemistry*. Plenum, New York, 1993. (Chapter 5, 5.3; Chapter 6, 6.5, 6.6; Chapter 7, 7.4)

Böttcher, C. J. F. *Theory of Electric Polarisation.* Elsevier, Amsterdam, 1952. (Chapter 6, 6.5, 6.6)

Boublik, T., Nezbeda, I. and Hlavaty, K. *Statistical Thermodynamics of Simple Liquids and Their Mixtures*. Elsevier, Amsterdam, 1980. (Chapter 7, 7.1–7.6)

Chandler, D. *Introduction to Modern Statistical Mechanics*. Oxford University Press, New York, 1987. (Chapter 1, 1.1, 1.2; Chapter 2, 2.1–2.4; Chapter 3, 3.1–3.3; Chapter 6, 6.1–6.4; Chapter 7, 7.4, 7.5; Chapter 8, 8.1–8.6)

Crawford, F. H. *Heat Thermodynamics and Statistical Physics*. Harcourt Brace and World, New York, 1963. (Chapter 3, 3.3; Chapter 6, 6.1–6.5; Chapter 5, 5.1–5.3, 5.5)

Croxton, C. A. *Statistical Mechanics of the Liquid Surface*. Wiley, Chichester, 1980. (Chapter 7, 7.3–7.5)

Davis, H. T. *Statistical Mechanics of Phases, Interfaces, and Thin Films*. VCH, New York, 1996. (Chapter 2, 2.1, 2.2; Chapter 3, 3.1, 3.2; Chapter 4, 4.5; Chapter 7, 7.1–7.5)

* Material in parentheses refers to sections of the present book for which the source is especially relevant.

Davidson, N. *Statistical Mechanics*. McGraw-Hill, New York, 1962. (Chapter 2, 2.2–2.4; Chapter 3, 3.1–3.6; Chapter 4, 4.1–4.4; Chapter 5, 5.1–5.3; Chapter 6, 6.1–6.6; Chapter 7, 7.1–7.3, 7.6)

Everdell, M. H. *Statistical Mechanics and Its Chemical Applications*. Academic Press, London, 1975. (Chapter 3, 3.1, 3.2; Chapter 4, 4.2–4.5; Chapter 7, 7.2, 7.4)

Eyring, H., Henderson, D., Stover, B. J. and Eyring, E. M. *Statistical Mechanics and Dynamics*, 2nd ed. Wiley, New York, 1982. (Chapter 2, 2.1–2.4; Chapter 3, 3.1–3.3; Chapter 5, 5.1–5.3; Chapter 6, 6.1–6.6; Chapter 7, 7.1–7.3, 7.6; Chapter 8, 8.1–8.4)

Fisher, I. Z. *Statistical Theory of Liquids*. University of Chicago, Chicago, 1964. (Chapter 6, 6.1–6.3; Chapter 7, 7.1–7.6)

Flygare, W. H. *Molecular Structure and Dynamics*. Prentice-Hall, Englewood Cliffs, NJ, 1978. (Chapter 3, 3.4; Chapter 4, 4.1)

Friedman, H. L. *A Course in Statistical Mechanics*. Prentice-Hall, Englewood Cliffs, NJ, 1985. (Chapter 2, 2.1–2.4; Chapter 3, 3.1, 3.2; Chapter 4, 4.1–4.3; Chapter 5, 5.1, 5.2; Chapter 6, 6.1–6.3; Chapter 7, 7.1–7.6; Chapter 8, 8.1, 8.2, 8.5, 8.6)

Goodisman, J. *Electrochemistry: Theoretical Foundations*. Wiley, New York, 1987. (Chapter 5, 5.2–5.4; Chapter 7, 7.6)

Grossman, L. M. *Theormodynamics and Statistical Mechanics*. McGraw-Hill, New York, 1969. (Chapter 3, 3.1–3.3; Chapter 4, 4.1–4.3; Chapter 5, 5.1–5.3; Chapter 7, 7.1–7.3)

Harrison, W. A. *Solid State Theory*. Dover, New York, 1979. (Chapter 3, 3.3; Chapter 5, 5.1–5.3; Chapter 6, 6.4; Chapter 8, 8.4)

Hayman, H. J. G. *Statistical Thermodynamics*. Elsevier, Amsterdam, 1967. (Chapter 1, 1.1, 1.2; Chapter 2, 2.1–2.3; Chapter 5, 5.1–5.3, 5.5)

Herzberg, G. *Atomic Spectra and Atomic Structure*. Dover, New York, 1944. (Chapter 3, 3.4; Chapter 4, 4.1)

Herzberg, G. *Spectra of Diatomic Molecules*. Van Nostrand, Toronto, 1950. (Chapter 3, 3.4; Chapter 4, 4.2; Chapter 6, 6.5)

Hill, T. L. *Introduction to Statistical Thermodynamics*. Addison-Wesley, Reading, MA, 1960. (Chapter 1, 1.1, 1.2; Chapter 2, 2.1–2.4; Chapter 3, 3.1–3.3; Chapter 4, 4.1–4.3; Chapter 6, 6.3–6.6; Chapter 7, 7.6)

Hill, T. L. *Statistical Mechanics*. McGraw-Hill, New York, 1956. (Chapter 2, 2.1–2.4; Chapter 6, 6.1–6.3; Chapter 7, 7.1–7.5)

Huang, K. *Statistical Mechanics*. Wiley, New York, 1963. (Chapter 5, 5.1–5.3, 5.5; Chapter 6, 6.1–6.4; Chapter 7, 7.1–7.5)

Hukins, D. W. L. *X-Ray Diffraction by Disordered and Ordered Systems*. Pergamon, Oxford, 1981. (Chapter 7, 7.4, 7.5)

Kauzmann, W. *Kinetic Theory of Gases*. Benjamin, New York, 1966. (Chapter 6, 6.1–6.4; Chapter 7, 7.2)

Kauzmann, W. *Theormodynamics and Statistics: With Applications to Gases*. Benjamin, New York, 1967. (Chapter 1, 1.1, 1.2; Chapter 2, 2.1–2.3; Chapter 4, 4.1, 4.2; Chapter 6, 6.2–6.4)

Koryta, J., Dvořák, J. and Boháčková, V. *Electrochemistry*. Methuen, London, 1970. (Chapter 7, 7.6; Chapter 8, 8.4)

Kubo, R. *Statistical Mechanics*. North-Holland, Amsterdam, 1964 (Chapter 2, 2.2–2.4; Chapter 3, 3.1, 3.2; Chapter 5, 5.1–5.3, 5.5; Chapter 6, 6.1–6.3; Chapter 7, 7.1, 7.2; Chapter 8, 8.1–8.4)

Landau, L. and Lifchitz, E. *Statistical Physics*. Pergamon Press, Oxford, 1969. (Chapter 1, 1.1, 1.2; Chapter 2, 2.1–2.4; Chapter 3, 3.1, 3.2; Chapter 5, 5.1–5.4; Chapter 7, 7.1–7.5)

Levich, B. G. *Quantum Statistics and Physical Kinetics*. North-Holland, Amsterdam, 1973. (Chapter 6, 6.1, 6.2; Chapter 7, 7.4; Chapter 8, 8.1–8.3)

March, N. H. and Tosi, M. P. *Coulomb Liquids*. Academic Press, London, 1984. (Chapter 7, 7.3–7.6; Chapter 8, 8.1–8.4)

Mayer, J. E. and Mayer, M. G. *Statistical Mechanics*, 2nd ed. Wiley, 1977. (Chapter 1, 1.1; Chapter 2, 2.1, 2.2; Chapter 3, 3.1–3.3; Chapter 5, 5.1–5.5; Chapter 6, 6.1–6.6; Chapter 7, 7.1–7.3; Chapter 8, 8.1, 8.2)

McQuarrie, D. A. *Statistical Thermodynamics*. Harper and Row, New York, 1973. (Chapter 2, 2.1–2.4; Chapter 3, 3.1–3.3; Chapter 4, 4.1–4.4; Chapter 5, 5.1–5.3, 5.5; Chapter 6, 6.1–6.3, 6.5; Chapter 7, 7.5)

Reif, F. *Fundamentals of Statistical and Thermal Physics*. McGraw-Hill, New York, 1965. (Chapter 1, 1.1; Chapter 2, 2.1–2.3; Chapter 3, 3.1, 3.2; Chapter 4, 4.1–4.4; Chapter 5, 5.1, 5.2; Chapter 6, 6.5; Chapter 7, 7.1, 7.2; Chapter 8, 8.1–8.4)

Rice, O. K. *Statistical Mechanics Thermodynamics and Kinetics*. Freeman, San Francisco, 1967. (Chapter 2, 2.1–2.4; Chapter 3, 3.1, 3.2, 3.5; Chapter 4, 4.1–4.3; Chapter 5, 5.1–5.3; Chapter 7, 7.1–7.3; Chapter 8, 8.1)

Sonntag, R. E. and VanWylen, G. J. *Introduction to Thermodynamics, Classical and Statistical*. Wiley, New York, 1991. (Chapter 1, 1.1; Chapter 2, 2.1–2.4; Chapter 3, 3.1)

Teichmann, H. *Semiconductors*. Butterworth's, Washington, 1964. (Chapter 5, 5.1–5.4; Chapter 8, 8.4)

Ter Haar, D. *Elements of Thermostatistics*. Holt, Rinehart and Winston, New York, 1966. (Chapter 2, 2.2, 2.3; Chapter 5, 5.2; Chapter 6, 6.1–6.3; Chapter 8, 8.1–8.3)

Wang, F. F. Y. *Introduction to Solid State Electronics*, 2nd ed. North-Holland, Amsterdam, 1989. (Chapter 5, 5.2–5.4)

Whalen, J. W. *Molecular Theormodynamics: A Statistical Approach*. Wiley, New York, 1991. (Chapter 3, 3.1, 3.2; Chapter 4, 4.1–4.4)

Wilson, A. H. *Thermodynamics and Statistical Mechanics*. Cambridge University Press, Cambridge, 1960. (Chapter 1, 1.2; Chapter 2, 2.3–2.4; Chapter 3, 3.1–3.3, 3.6; Chapter 5, 5.1–5.3, 5.5; Chapter 6, 6.1–6.5; Chapter 7, 7.1–7.3)

Wolf, H. F. *Semiconductors*. Wiley–Interscience, New York, 1971. (Chapter 5, 5.2–5.4; Chapter 6, 6.4; Chapter 8, 8.4)

INDEX